Lecture Notes in Computer Science 15273

Founding Editors

Gerhard Goos
Juris Hartmanis

Kareem A. Wahid · Cem Dede ·
Mohamed A. Naser · Clifton D. Fuller
Editors

Head and Neck Tumor Segmentation for MR-Guided Applications

First MICCAI Challenge, HNTS-MRG 2024
Held in Conjunction with MICCAI 2024
Marrakesh, Morocco, October 17, 2024
Proceedings

 Springer

Editors
Kareem A. Wahid
MD Anderson Cancer Center
Houston, TX, USA

Cem Dede
MD Anderson Cancer Center
Houston, TX, USA

Mohamed A. Naser
MD Anderson Cancer Center
Houston, TX, USA

Clifton D. Fuller
MD Anderson Cancer Center
Houston, TX, USA

ISSN 0302-9743   ISSN 1611-3349 (electronic)
Lecture Notes in Computer Science
ISBN 978-3-031-83273-4   ISBN 978-3-031-83274-1 (eBook)
https://doi.org/10.1007/978-3-031-83274-1

This work was supported by National Institutes of Health (NIH) and The University of Texas MD Anderson Cancer Center.

This Springer imprint is published by the registered company Springer Nature Switzerland AG
The registered company address is: Gewerbestrasse 11, 6330 Cham, Switzerland

If disposing of this product, please recycle the paper.

Preface

The Head and Neck Tumor Segmentation for Magnetic Resonance Guided Applications (HNTS-MRG) 2024 Challenge built upon the success of previous segmentation initiatives to address critical and evolving issues in radiation oncology. With its superior soft tissue contrast and longitudinal imaging capabilities, magnetic resonance imaging (MRI)-guided radiation therapy is transforming head and neck cancer treatment, improving tumor targeting and reducing treatment-related toxicity. The HNTS-MRG 2024 Challenge focused on advancing clinical workflows by leveraging artificial intelligence (AI) for automated segmentation of tumor regions in multi-timepoint MRI scans.

As a satellite event of the 27th International Conference on Medical Image Computing and Computer-Assisted Intervention (MICCAI), HNTS-MRG 2024 introduced two key segmentation tasks: pre-radiotherapy MRI tumor segmentation (Task 1) and mid-radiotherapy MRI tumor segmentation (Task 2). These tasks challenged participants to delineate gross tumor volumes for primary tumors and regional nodal metastases on MRI scans, utilizing 150 cases for training AI algorithms and 50 cases for final algorithm evaluation. This dataset, annotated by a team of expert clinicians, serves as a valuable resource for advancing methodologies in MRI-based radiotherapy segmentation.

The challenge attracted global participation, with 19 teams across North America, South America, Europe, and Asia submitting solutions to one or both tasks. Top-performing algorithms in both tasks exceeded benchmarks of clinician interobserver variability, showcasing the transformative potential of AI-driven solutions in clinical practice. The results and winners for each task were presented ast the half-day satellite MICCAI event on October 17, 2024.

This volume in the Lecture Notes in Computer Science series compiles the methods and results presented by the HNTS-MRG 2024 participants. Contributions from diverse teams highlight state-of-the-art innovations in AI-driven segmentation. A total of 20 short papers were submitted and accepted for publication after single-blind peer review, with each paper evaluated by three expert reviewers. The participant papers are complemented by an overview paper prepared by the challenge organizers.

We extend our gratitude to all participants and reviewers, and to the organizing committee for their hard work and dedication. Special thanks go to our financial sponsors, the National Institutes of Health (R01DE028290-05S1, R01CA257814-03S2), the MICCAI 2024 organizers, and the grand-challenge.org platform for their invaluable support.

We hope this collection inspires further advancements in the field, fostering collaboration and innovation in AI applications for head and neck cancer treatment.

December 2024

Kareem A. Wahid
Cem Dede
Mohamed A. Naser
Clifton D. Fuller

Organization

General Chairs

Kareem A. Wahid   MD Anderson Cancer Center, USA
Cem Dede   MD Anderson Cancer Center, USA
Mohamed A. Naser   MD Anderson Cancer Center, USA
Clifton D. Fuller   MD Anderson Cancer Center, USA

Program Committee

Moamen R. A. Abdelaal   MD Anderson Cancer Center, USA
Sara Ahmed   MD Anderson Cancer Center, USA
Qusai Alakayleh   MD Anderson Cancer Center, USA
Enoch Chang   MD Anderson Cancer Center, USA
John P. Christodouleas   Elekta, USA
Kelsey L. Corrigan   MD Anderson Cancer Center, USA
Stephanie O. Dudzinski   MD Anderson Cancer Center, USA
Dina M. El-Habashy   Corewell Health William Beaumont University
                               Hospital, USA
Renjie He   MD Anderson Cancer Center, USA
Serageldin Kamel   MD Anderson Cancer Center, USA
Yomna Khamis   University of Maryland School of Medicine, USA
Stephen Y. Lai   MD Anderson Cancer Center, USA
Lucas McCullum   MD Anderson Cancer Center, USA
Brigid A. McDonald   MD Anderson Cancer Center, USA
Abdallah S. R. Mohamed   Baylor College of Medicine, USA
Samuel L. Mulder   MD Anderson Cancer Center, USA
Michael K. Rooney   MD Anderson Cancer Center, USA
Andrew J. Schaefer   Rice University, USA
Carlos Sjogreen   MD Anderson Cancer Center, USA
Travis C. Salzillo   MD Anderson Cancer Center, USA

Additional Reviewers

Zaphanlene Kaffey
Saleh Remenzani
Natasha Topolski
Laia Humbert-Vidan
Natalie West

Contents

Overview of the Head and Neck Tumor Segmentation for Magnetic Resonance Guided Applications (HNTS-MRG) 2024 Challenge

Kareem A. Wahid[1,2] , Cem Dede[1] , Dina M. El-Habashy[1,3] ,
Serageldin Kamel[1] , Michael K. Rooney[1] , Yomna Khamis[1,4,5] ,
Moamen R. A. Abdelaal[1] , Sara Ahmed[1] , Kelsey L. Corrigan[1] ,
Enoch Chang[1] , Stephanie O. Dudzinski[1] , Travis C. Salzillo[1] ,
Brigid A. McDonald[1] , Samuel L. Mulder[1] , Lucas McCullum[1,6] ,
Qusai Alakayleh[1] , Carlos Sjogreen[1] , Renjie He[1] ,
Abdallah S. R. Mohamed[1,7] , Stephen Y. Lai[8] , John P. Christodouleas[9] ,
Andrew J. Schaefer[10] , Mohamed A. Naser[1(✉)] , and Clifton D. Fuller[1(✉)]

[1] Department of Radiation Oncology, The University of Texas MD Anderson Cancer, Houston, TX, USA
{manaser,cdfuller}@mdanderson.org
[2] Department of Imaging Physics, The University of Texas MD Anderson Cancer, Houston, TX, USA
[3] Transitional Year Program, Corewell Health Wiliam Beaumont, Royal Oak, MI, USA
[4] Department of Radiation Oncology, University of Maryland School of Medicine, Baltimore, MD, USA
[5] Department of Clinical Oncology and Nuclear Medicine, Faculty of Medicine, Alexandria University, Alexandria, Egypt
[6] UT MD Anderson Cancer Center UTHealth Houston Graduate School of Biomedical Sciences, Houston, USA
[7] Department of Radiation Oncology, Baylor College of Medicine, Houston, TX, USA
[8] Department of Head and Neck Surgery, The University of Texas MD Anderson Cancer, Houston, TX, USA
[9] Elekta, Atlanta, GA, USA
[10] Department of Computational Applied Mathematics and Operations Research, Rice University, Houston, TX, USA

Abstract. Magnetic resonance (MR)-guided radiation therapy (RT) is enhancing head and neck cancer (HNC) treatment through superior soft tissue contrast and longitudinal imaging capabilities. However, manual tumor segmentation remains a significant challenge, spurring interest in artificial intelligence (AI)-driven automation. To accelerate innovation in this field, we present the Head and Neck Tumor Segmentation for MR-Guided Applications (HNTS-MRG) 2024 Challenge, a satellite event of the 27th International Conference on Medical Image Computing and Computer Assisted Intervention. This challenge addresses the scarcity of large, publicly available AI-ready adaptive RT datasets in HNC and

---

K. A. Wahid and C. Dede—Co-first authors.

K. A. Wahid et al. (Eds.): HNTS-MRG 2024, LNCS 15273, pp. 1–35, 2025.
https://doi.org/10.1007/978-3-031-83274-1_1

explores the potential of incorporating multi-timepoint data to enhance RT auto-segmentation performance. Participants tackled two HNC segmentation tasks: automatic delineation of primary gross tumor volume (GTVp) and gross metastatic regional lymph nodes (GTVn) on pre-RT (Task 1) and mid-RT (Task 2) T2-weighted scans. The challenge provided 150 HNC cases for training and 50 for final testing hosted on grand-challenge.org using a Docker submission framework. In total, 19 independent teams from across the world qualified by submitting both their algorithms and corresponding papers, resulting in 18 submissions for Task 1 and 15 submissions for Task 2. Evaluation using the mean aggregated Dice Similarity Coefficient showed top-performing AI methods achieved scores of 0.825 in Task 1 and 0.733 in Task 2. These results surpassed clinician interobserver variability benchmarks, marking significant strides in automated tumor segmentation for MR-guided RT applications in HNC.

Keywords: Head and neck cancer · magnetic resonance imaging · data challenge · segmentation · contouring · radiotherapy · image-guided · artificial intelligence · deep learning

1 Introduction: Research Context

Radiation therapy (RT) is a cornerstone of cancer treatment for a wide variety of malignancies. Chief among the beneficiaries of RT as a treatment modality is head and neck cancer (HNC). Recent years have seen an increasing interest in MRI-guided RT planning. As opposed to more traditional computed tomography (CT)-based RT planning, MRI-guided approaches afford superior soft tissue contrast, allow for functional imaging through special multiparametric sequences (e.g., diffusion-weighted imaging), and permit daily adaptive RT through intra-therapy imaging using MRI-Linac devices [1]. Subsequently, improved treatment planning through MRI-guided adaptive RT approaches would help further maximize tumor destruction while minimizing side effects in HNC [2, 3]. Given the great potential for MRI-guided adaptive RT planning, it is anticipated that these technologies will transform clinical practice paradigms for HNC [4].

The extensive data volume for MRI-guided HNC RT planning, particularly in adaptive settings, makes manual tumor segmentation (also referred to as contouring) by physicians—the current clinical standard—often impractical due to time constraints [5]. This is compounded by the fact that HNC tumors are among the most challenging structures for clinicians to segment [6]. Artificial intelligence (AI) approaches that leverage RT data to improve patient treatment have been an exceptional area of interest for the research community in recent years. The use of deep learning (DL) in particular has made significant strides in HNC tumor auto-segmentation [7]. These innovations have largely been driven by public data science challenges such as the HECKTOR Challenge [8] and the SegRap Challenge [9]. However, to-date, there exist no large publicly available AI-ready adaptive RT HNC datasets for public distribution. It stands to reason that community-driven AI innovations would be a remarkable asset to developing technologies for the clinical translation of MRI-guided RT.

In this public data science challenge—The Head and Neck Tumor Segmentation for MR-Guided Applications 2024 Challenge (HNTS-MRG 2024, pronounced "hunts"-"merge")—we focus on the segmentation of HNC tumors for MRI-guided adaptive RT applications. The challenge is composed of two tasks focused on automated segmentation of tumor volumes on 1) pre-RT MRI images and 2) mid-RT MRI images. An overview of HNTS-MRG 2024 is shown in Fig. 1.

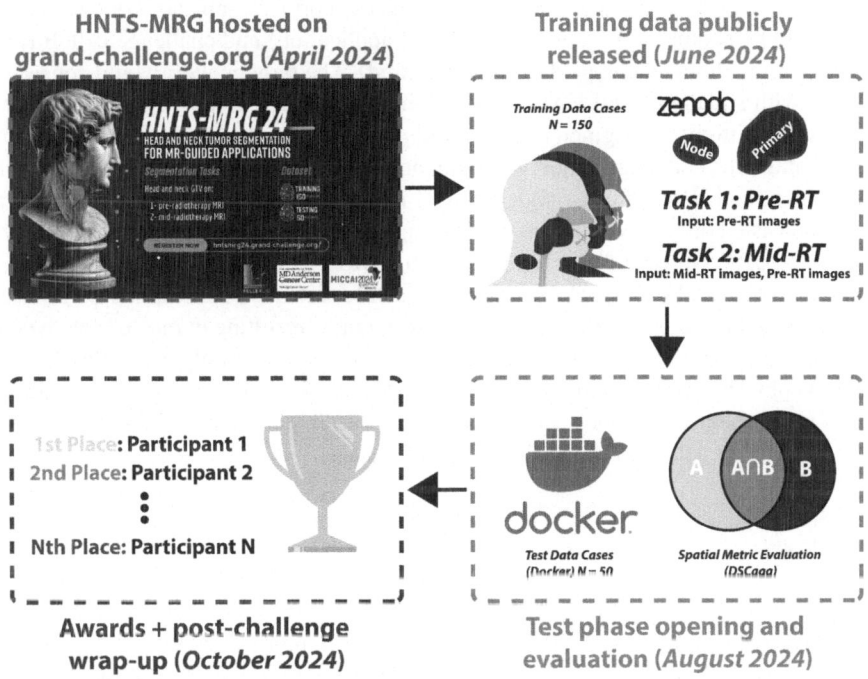

Fig. 1. General overview of the HNTS-MRG 2024 data science challenge. Two tasks focusing on pre-RT (Task 1) and mid-RT (Task 2) tumor segmentation using MRI scans were evaluated. Training data from 150 patients were publicly released, followed by an internal evaluation of algorithms on 50 final test patients. Subsequently, a post-challenge virtual wrap up session was held where winners were publicly announced.

2 Dataset and Challenge Details

2.1 Mission of the Challenge

Biomedical Application

This data science challenge followed the Biomedical Image Analysis Challenges (BIAS) statement reporting guidelines by Maier-Hein et al. [10] and was accepted as a satellite event for the 27th International Conference on Medical Image Computing and Computer Assisted Intervention (MICCAI). The algorithms submitted by participating teams were primarily designed for three main target applications: diagnosis, treatment planning, and medical research. These algorithms focused on two core tasks within medical imaging: segmentation and detection. Specifically, they were designed to analyze MRI images and classify/annotate individual voxels into three distinct categorical labels: primary gross tumor volume (GTVp), metastatic lymph node gross tumor volume (GTVn), or background tissue.

Cohorts

Following the BIAS guidelines, we define a target cohort (i.e. subjects from whom the data would be acquired in the final biomedical application) and challenge cohort (i.e., subjects from whom challenge data were acquired). The target cohort would consist of patients with squamous cell HNC who are referred to RT planning clinics. For these patients, the automated segmentation algorithms could potentially be used directly in RT planning. The challenge cohort were patients with a confirmed histological diagnosis of squamous cell HNC who had undergone RT at our institution. This patient cohort primarily consisted of individuals with oropharyngeal cancer (OPC) or cancer of unknown primary (CUP). We included CUP patients for two key reasons. Firstly, these cases are often undetected OPC [11]. Secondly, our dataset included patients whose primary tumors had achieved complete response by mid-therapy, resulting in images where only residual mid-RT lymph nodes remained visible. This scenario closely resembles the presentation of CUP cases, making their inclusion valuable for training algorithms to detect and segment metastatic nodes across a spectrum of clinical presentations. Section 2.3 describes further details of the challenge cohort dataset.

Target Entity

Data Origin
Extant images used in this study were all acquired from the head and neck region of HNC patients. However, the exact area captured in each scan (i.e., field of view) varied somewhat between images. While some scans extended inferiorly to include parts of the lungs or superiorly to include the top of the skull, all scans consistently captured at least the area from the clavicles up to the oropharyngeal region. To make the target regions more uniform across all images, we applied a cropping technique. The specific details of this cropping method are explained in Sect. 2.3.

Algorithm Target
The structures of interest for this study were GTVp and GTVn structures which are conventionally segmented by physicians for HNC RT planning. They represent the position and extent of gross tumor at the primary site and metastatic lymph nodes visible on medical imaging [12, 13].

Task Definition

The challenge consisted of two tasks, pre-RT segmentation (Task 1) and mid-RT segmentation (Task 2).

Task 1 required participants to predict GTVp and GTVn tumor segmentations on unseen pre-RT scans without annotations. This is a task analogous to previous conventional tumor segmentation challenges, such as Task 1 of the 2022 HECKTOR Challenge [14] and Task 2 of the 2023 SegRap Challenge [9]. Participants were free to use mid-RT data for training their pre-RT auto-segmentation algorithms if desired.

Task 2 simulated a real-world adaptive RT scenario, providing an unseen mid-RT image alongside a pre-RT image with corresponding pre-RT segmentation. Registered and original versions of the pre-RT data would be provided during model inference (more details in Sect. 2.3). The goal was to predict GTVp and GTVn segmentations on the new mid-RT images. This task is somewhat analogous to previous challenges that utilize multiple image inputs such as the 2023 SegRap Challenge (non-contrast CT + contrast CT) [9] and the 2023 HaN-Seg Challenge (CT + MRI) [15]. To our knowledge, no previous challenges utilized patient-specific multi-timepoint MRI for segmentation purposes, making this aspect of our challenge particularly unique. Participants were free to use any combination of input images/masks to develop their mid-RT auto-segmentation algorithms.

To foster innovation while maintaining fairness, we allowed participants to leverage pre-trained model weights, foundation models, and additional external data to augment their training. However, we stipulated that all such resources must be publicly accessible and properly cited in the participants' paper submissions. This approach encouraged the use of state-of-the-art techniques while ensuring transparency and reproducibility in the challenge.

2.2 Online Hosting of the Challenge

HNTS-MRG 2024 was hosted on grand-challenge.org, an open-source platform that has become a de facto standard for online biomedical image analysis competitions. This platform offers essential features for running online data challenges, including an application programming interface, user management, a discussion forum, support for multi-phase competitions with separate leaderboards, and an online image results viewer, among other functionalities. Moreover, the platform utilizes Docker frameworks [16] for containerized algorithm code submissions and automated algorithm evaluation. The online webpage for HNTS-MRG 2024 was launched in April 2024, offering participants a comprehensive environment to engage in this challenge [17].

2.3 Challenge Cohort Dataset

Institutional Review Board
Ethics approval was obtained from the University of Texas MD Anderson Cancer Center Institutional Review Board with protocol number RCR03-0800. This is a retrospective data collection protocol with a waiver of informed consent.

Data Source
All data for this study were collected from a single institution: The University of Texas MD Anderson Cancer Center. T2-weighted (T2w) anatomical MRI sequences were the focus of our challenge due to their ubiquity and importance in MRI-based HNC segmentation for RT [18]. Raw T2w images in Digital Imaging and Communications in Medicine (DICOM) format were automatically extracted from a centralized institutional imaging repository (Evercore). Notably, T2w images were a mix of fat-suppressed and non-fat-suppressed images. Images include pre-RT (0–3 weeks before the start of RT) and mid-RT (2–4 weeks intra-RT) scans. No exogenous contrast enhancement agents

were used for these scans. All patients were immobilized using a thermoplastic mask to aid in consistent anatomical positioning. Pre-RT and mid-RT image pairs for a given patient were consistently either fat-suppressed or non-fat-suppressed. In total, data from 202 squamous cell HNC patients were curated. T2w images of the head and neck region were acquired using a range of imaging devices and protocols. Images were acquired on the following devices: 1.5T Siemens Aera (n = 297), 1.5T Elekta Unity (n = 78), 3T Siemens Magnetom Vida (n = 19), 1.5T Siemens Magnetom Sola Fit (n = 10). A full list of imaging protocols are described in Table 1. Examples of T2w images for two patients are shown in Fig. 2.

Table 1. Magnetic resonance imaging acquisition parameters for this study. Median values with ranges shown. Values are calculated across the entire datasets for all timepoints (pre- and mid-radiotherapy).

Parameter	Median (range)
Repetition Time (ms)	4800 (1400–6250)
Echo Time (ms)	80 (74–375)
In-plane Resolution (mm)	0.5 (0.4–0.98)
Slice Thickness (mm)	2.0 (1.0–2.5)
Slice Gap (mm)	2.0 (1.0–2.5)
Number Of Axial Slices	120 (80–300)
Field Of View (mm)	256x256 (256–520)
Number of Averages	1 (1–2)

Annotation Characteristics

Each MRI scan was annotated for GTVp (maximum one per patient, potentially zero) and GTVn (variable number per patient, potentially zero). A team of 3 to 4 expert physicians each independently segmented these structures for all pre-RT and mid-RT cases. This approach aligns with recent findings from our group, suggesting that a minimum of 3 annotators is necessary to produce acceptable segmentations in these structures [19, 20] when combined using the simultaneous truth and performance level estimation (STAPLE) algorithm [21].

13 unique annotators independently contributed segmentation annotations to this study. All annotators were medical doctors with at least two years of experience in head and neck cancer segmentation. All annotators had access to patient medical histories and any previous relevant imaging (e.g., diagnostic positron emission tomography (PET)/CT imaging) via the patient's chart. Annotators were instructed to segment targets as they would normally in their clinical workflows. For mid-RT segmentations, annotators were permitted to use their registered pre-RT segmentations as a reference if desired. Segmentations were generated in Velocity AI (v.3.0.1; Varian Medical Systems; Palo Alto, CA, USA) and Raystation (v.11; RaySearch Laboratories, Stockholm, Sweden) using American Association of Physicists in Medicine Task Group 263 nomenclature [22]. A

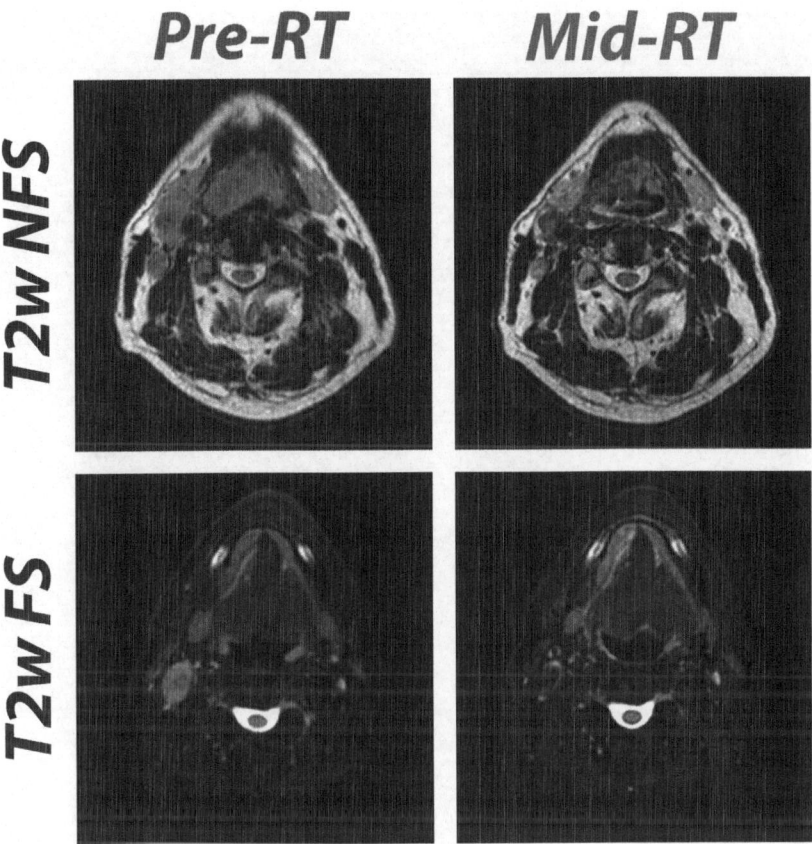

Fig. 2. Comparison of T2 weighted (T2w) MRI scans before radiotherapy (pre-RT) and during radiotherapy (mid-RT), showing images without fat suppression (T2w non fat suppressed [NFS], top row) and with fat suppression (T2w fat suppressed [FS], bottom row). Pre-RT scans are co-registered to the corresponding mid-RT scans.

senior radiation oncology faculty member with over 15 years of experience (C.D.F.) performed final quality verification of the segmentations, where annotators were instructed to modify certain segmentations if needed (e.g., in the case of missing a lymph node).

The STAPLE algorithm implementation in SimpleITK [23] was used to combine individual segmentations into a final ground truth (also referred to as reference standard) segmentation for each case (Fig. 3). In exceptional cases where significant discrepancies arose among annotators—such as disagreements over multiple nodal volumes or conflicting assessments of complete versus non-complete response—we deferred to the expert judgment of the senior faculty member (C.D.F.) for generating the final segmentation. The resulting ground truth label mask uses three values: 0 for background, 1 for GTVp, and 2 for GTVn (with multiple lymph nodes consolidated into a single label). An example of an image with the aforementioned labeling scheme is shown in Fig. 4.

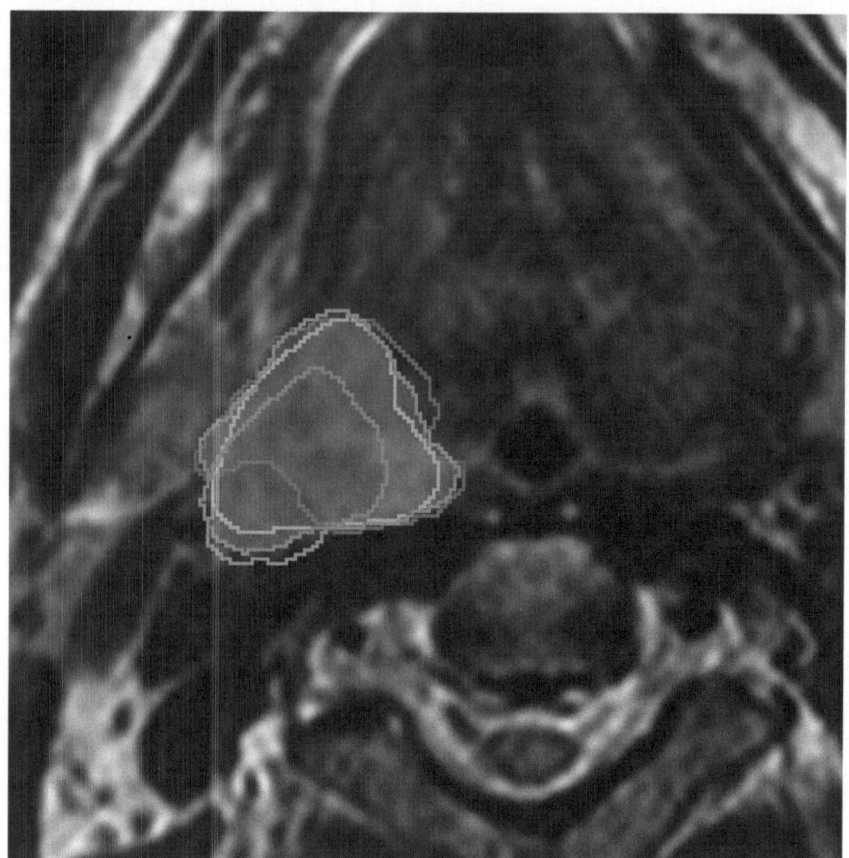

Observer 1, Observer 2, Observer 3, Observer 4, STAPLE Consensus

Fig. 3. Example of the Simultaneous Truth and Performance Level Estimation (STAPLE) algorithm consensus process combining multiple independent annotator segmentations (red, yellow, blue, purple outlined structures) into a single final consensus segmentation (green filled in structure) for a primary gross tumor volume. (Color figure online)

Data Preprocessing Methods

Anonymized DICOM files (MRI image and structure files) were converted to Neuroimaging Informatics Technology Initiative (NIfTI) format for ease of use by participants. Conversions were performed using DICOMRTTool v. 1.0 [24]. We chose the NIfTI format for our data due to its widespread adoption, standardized structure, and its compatibility with a broad range of analysis tools commonly used in medical imaging challenges [25]. All images were cropped from the top of the clavicles to the bottom of the nasal septum (oropharynx region to shoulders) by using manually selected inferior/superior axial slices. This allowed for more consistent image fields of view and

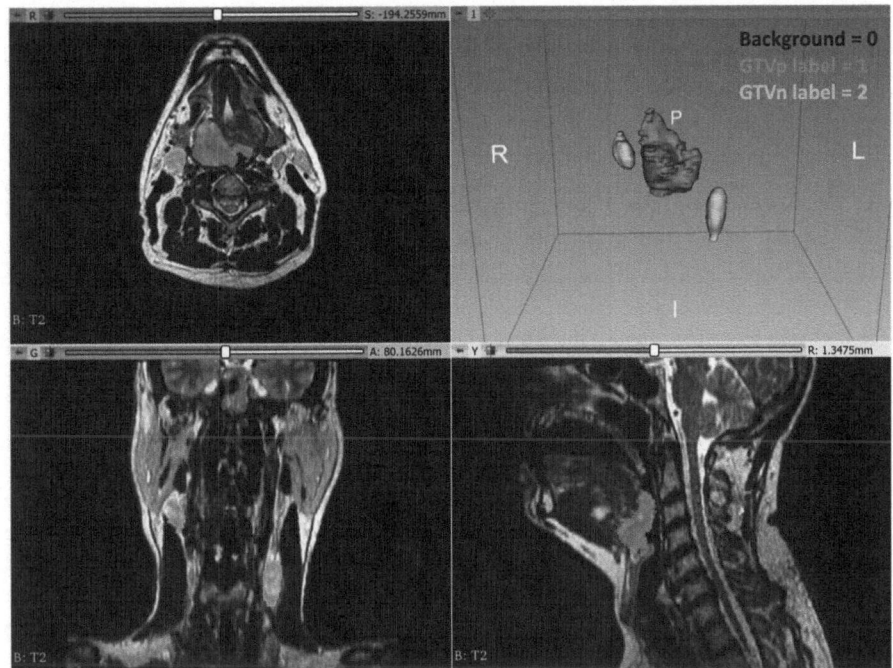

Fig. 4. A visual example of the mask labeling scheme for this challenge. Background = 0, primary gross tumor volume (GTVp) = 1 (green overlay), metastatic lymph node (GTVn) = 2 (yellow overlay). Masks shown are consensus segmentations from multiple independent annotators. Visualization performed in 3D Slicer. (Color figure online)

removal of identifiable facial structures (i.e., eyes, nose, ears); cropping did not impact any of the segmented volumes.

Registered data (i.e., for Task 2) were generated using SimpleITK [23], where the mid-RT image served as the fixed image and the pre-RT image served as the moving image. Specifically, we utilized the following steps: 1. Apply a centered transformation, 2. Apply a rigid transformation, 3. Apply a deformable transformation with Elastix using a preset parameter map (Parameter map 23 in the Elastix Model Zoo which utilizes a B-spline transformation [26]). This particular deformable transformation was selected as it is open-source and was benchmarked in a previous similar application [27]. In a small minority of cases where excessive warping occurred during deformable registration, we defaulted to using only the rigid transformation. To ensure transparency and reproducibility, we provided a detailed example of our registration process on our GitHub repository [28].

Sources of Errors - Interobserver Variability
The largest source of error naturally emerges from differences in annotator segmentations. We mitigated this by combining segmentations via the STAPLE algorithm which we have shown can yield acceptable segmentations given a minimal number of annotator inputs [19].

We evaluated the interobserver variability (IOV) for GTVp and GTVn structures in our dataset using traditional geometric measures. For each structure (i.e., GTVp and GTVn) and each timepoint (i.e., pre-RT and mid-RT), we calculated pairwise IOV. To calculate pairwise IOV, for each patient, metrics for all possible pairwise combinations between available annotator segmentations were calculated followed by computing the median value across all combinations. Naturally, the patient cases where only the senior faculty member observer contributed segmentations due to significant discrepancies among observers (see **Annotation Characteristics**) were excluded. Our analysis included four metrics: Dice Similarity coefficient (DSC), 95% Hausdorff distance (HD95), average surface distance (ASD), and surface DSC at a 2mm tolerance (SDSC). Metrics were calculated using the Surface Distances Python package [29] and in-house Python code. Pre-RT IOV DSC values showed median (interquartile range) of 0.747 (0.165) for GTVp and 0.845 (0.070) for GTVn. Mid-RT IOV DSC values were 0.558 (0.272) for GTVp and 0.808 (0.118) for GTVn. IOV based on all geometric measures are shown in Fig. 5.

To provide a more comprehensive assessment of IOV relevant to our challenge, we extended our analysis beyond traditional geometric measures to include aggregated Dice Similarity Coefficient [30] (DSCagg, see Sect. 2.4) IOV calculations. We first computed intermediate metrics in a pairwise fashion across all observers for each patient case independently. These metrics were then aggregated across all cases for each unique annotator pair, enabling the calculation of DSCagg-GTVp, DSCagg-GTVn, and subsequently DSCagg-mean (the mean of DSCagg-GTVp and DSCagg-GTVn) for each annotator pair. Finally, we derived the overall IOV DSCagg values by calculating a weighted average of these annotator pair DSCagg values. The weighting factor was based on the number of cases compared for each annotator pair, ensuring appropriate annotator representation in the final score. As before, cases where only the senior faculty member observer contributed final segmentations were excluded. Final weighted IOV DSCagg values for pre-RT segmentations were DSCagg-mean $= 0.806$, DSCagg-GTVp $= 0.757$, DSCagg-GTVn $= 0.854$. Final weighted IOV DSCagg values for mid-RT segmentations were DSCagg-mean $= 0.714$, DSCagg-GTVp $= 0.600$, DSCagg-GTVn $= 0.828$.

Training and Test Case Characteristics

Tasks 1 and 2 share a common training dataset, consisting of 150 patient cases, ensuring consistency across both challenges. Training data were publicly released on Zenodo [31] under a CC BY 4.0 license. For each patient case, we provided a comprehensive set of data in NIfTI format. This dataset included six files per patient: the original pre-treatment T2-weighted MRI volume with its corresponding segmentation mask, the original mid-treatment T2-weighted MRI volume with its corresponding segmentation mask, and a registered version of the pre-treatment T2-weighted MRI volume with its corresponding segmentation mask (more details on registration in **Data Preprocessing Methods**). Each of these six files was linked to a unique anonymized case identifier, ensuring that all data for a given patient could be easily accessed and correctly associated.

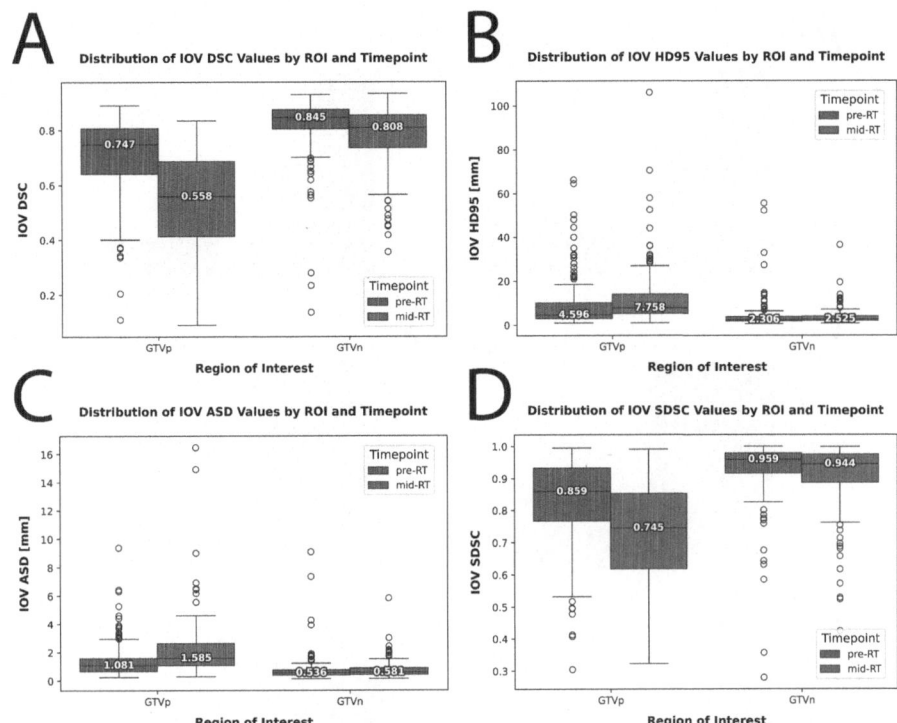

Fig. 5. Interobserver variability (IOV) data for primary gross tumor volume (GTVp) and nodal gross tumor volume (GTVn) regions of interest stratified by pre-radiotherapy (pre-RT) and mid-radiotherapy (mid-RT) timepoints. (A) Dice Similarity coefficient (DSC), (B) 95% Hausdorff distance (HD95), (C) average surface distance (ASD), (D) surface DSC at a 2 mm tolerance (SDSC). Each datapoint corresponds to the median metric value across all pairs of observers for a given patient image. Each box represents the interquartile range, with the horizontal line indicating the median score. Outliers are shown as individual points outside the whiskers. Higher values indicate greater agreement for DSC and SDSC, while lower values indicate greater agreement for HD95 and ASD.

The held-out private evaluation data comprised 52 additional cases, with two cases used for the challenge's preliminary debugging phase, leaving 50 cases for the final test phase (more information in Sect. 2.5). Only the challenge organizers had access to the ground truth segmentations for the test cases until final publication of the full dataset.

Training and held-out private evaluation data were partitioned to contain similar distributions based on dataset characteristics such as image fat-suppression status, tumor response, and staging. The distributions based on various parameters are shown in Fig. 6.

2.4 Assessment Method

Both tasks were evaluated in the same general manner using the aggregated Dice Similarity Coefficient (DSCagg). DSCagg was employed by Andrearczyk et al. for the segmentation task of the 2022 edition of the HECKTOR Challenge [14]. Specifically,

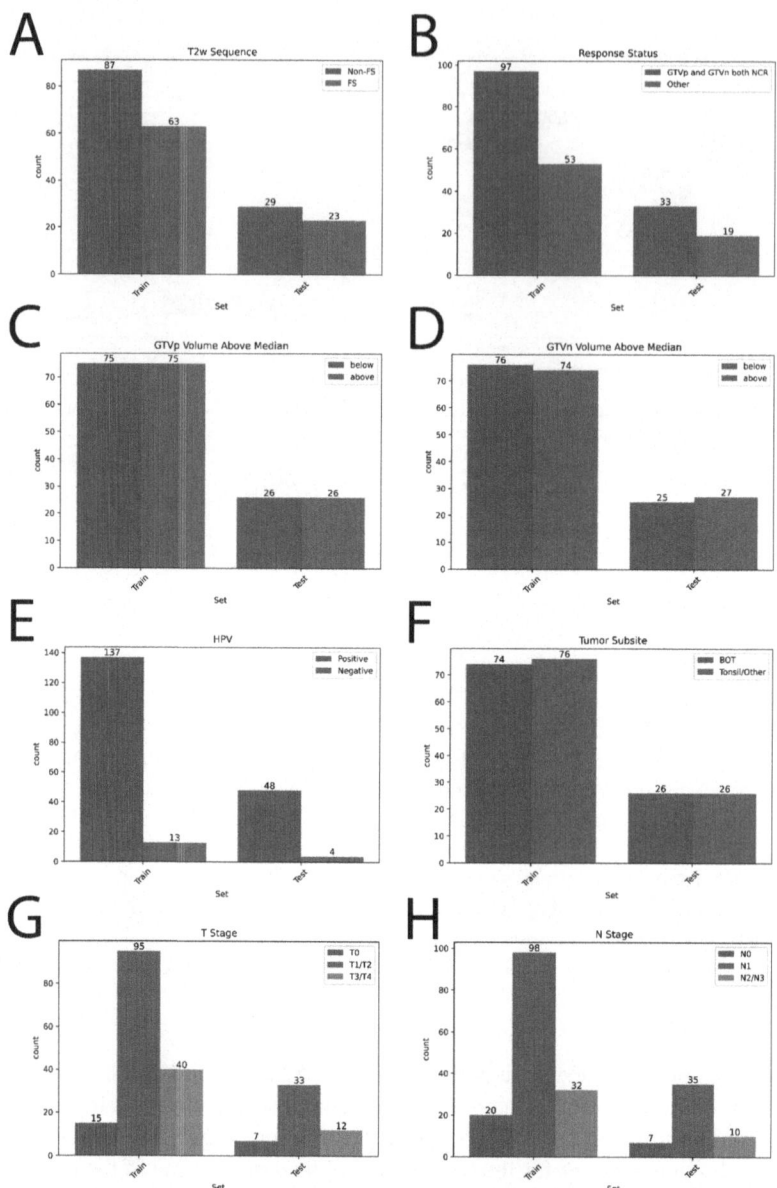

Fig. 6. Distribution of key parameters in training and held-out private evaluation sets (written as test). (A) T2-weighted MRI sequence type (Non-FS: non-fat suppressed, FS: fat suppressed). (B) Tumor response status at mid-therapy (NCR: non-complete response, Other: any combination of complete and non-response of primary and node). (C) Primary gross tumor volume (GTVp) and (D) nodal gross tumor volume (GTVn), both categorized as above or below the dataset median. (E) Human papillomavirus (HPV) status. (F) Tumor anatomic subsite (BOT: base of tongue). (G) T-stage and (H) N-stage as per the eighth edition American Joint Committee on Cancer staging system.

the DSCagg metric is defined as:

$$DSCagg = \frac{2\sum_i |Ai \cap Bi|}{\sum_i |Ai| + |Bi|} \tag{1}$$

where Ai and Bi are the ground truth and predicted segmentation for image i, where i spans the entire test set. Namely, DSCagg calculates intermediate metrics for each case individually and then aggregates the measurements across the test set (i.e., yields one value). DSCagg was initially described in detail by Andrearczyk et al. [30].

Conceptually, the 2022 edition of the HECKTOR Challenge had similar segmentation outputs (i.e., GTVp and GTVn for HNC patients) as our proposed challenge, so we deem DSCagg an appropriate metric. Since the presence of GTVp and GTVn were not consistent across all cases, the proposed DSCagg metric is well-suited for this task. Unlike conventional volumetric DSC, which can be overly sensitive to false positives when the ground truth mask is empty—resulting in a DSC of 0—the DSCagg metric is more robust. It effectively handles cases where certain structures may or may not be present, providing a more balanced evaluation across diverse scenarios encountered in our data. Notably, DSCagg was shown to be a stable metric with respect to final ranking from a secondary analysis of the HECKTOR 2021 results [32], further highlighting its appropriateness for the challenge.

The metric was computed individually for GTVp (DSCagg-GTVp) and GTVn (DSCagg-GTVn), and the mean average of the two (DSCagg-mean) was used for the final challenge ranking (similar to HECKTOR 2022). The metric was calculated for Task 1 (pre-RT segmentation) and Task 2 (mid-RT segmentation) separately. We provided an example of how the DSCagg was calculated for this challenge on our GitHub repository [28].

2.5 Docker Submission and Challenge Phases

Algorithm submissions for the challenge were managed through grand-challenge.org, with participants required to submit their solutions as Docker container images [16]. For algorithm submissions, Task 1 participants received only an unseen pre-RT image, while Task 2 participants were provided with an unseen mid-RT image, a pre-RT image with its corresponding segmentation, and a registered pre-RT image with its corresponding registered segmentation. Toy examples of algorithm inputs for both tasks are shown in Fig. 7. By utilizing a Docker framework, we maintained data integrity and challenge fairness, enabling us to use identical patient cases for both tasks without disclosing Task 1's ground truth segmentation masks to Task 2 participants. To ensure practical implementation and efficient evaluation, we established specific technical constraints, namely that algorithms were required to complete processing within 20 min per patient case through the Grand Challenge runtime environment (using an NVIDIA T4 graphics processing unit). To assist participants, we provided detailed examples of Docker image containerization on our GitHub repository [28]. We launched two distinct phases for each task on August 15th, 2024: a preliminary development phase and a final test phase.

The preliminary development phase served as a "practice" round, allowing participants to debug their algorithms and familiarize themselves with the Docker submission framework. During this optional but highly recommended phase, teams could make up to five valid submissions. We used data from two patients not included in the training set, selecting straightforward cases with easily identifiable segmentation targets to facilitate the debugging process. The two patients selected for the preliminary phase were both human papillomavirus (HPV)-positive with large GTVp and GTVn targets, with one patient's images featuring fat suppression and the other's without, providing participants exposure to different MRI acquisition techniques commonly encountered in the dataset. Results from this phase were immediately displayed on the leaderboard but did not impact the final rankings.

The final test phase, which was composed of 50 cases, determined the official evaluation and ranking of participants' algorithms. In contrast to the development phase, each team was limited to a single valid submission. This restriction ensured a fair comparison of each team's best-performing algorithm. The test set for this phase was entirely separate from the development phase data, providing a true measure of the algorithms' performance on unseen cases.

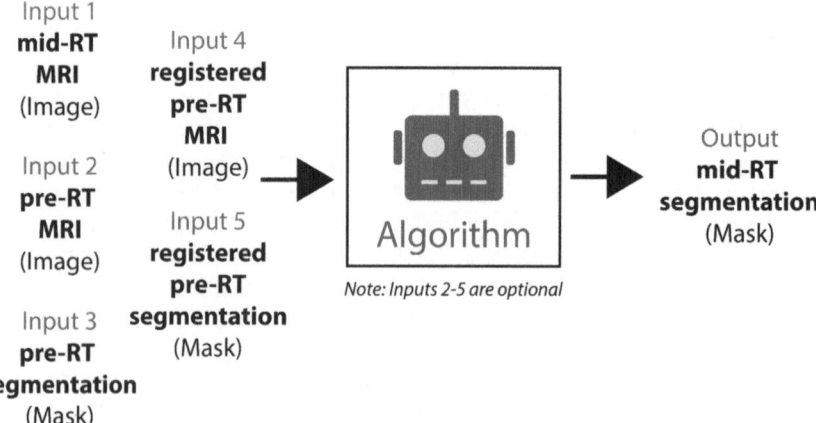

Fig. 7. Toy examples of model input and outputs for Task 1 (pre-radiotherapy segmentation, top) and Task 2 (mid-radiotherapy segmentation, bottom).

2.6 Baseline Models

To establish performance benchmarks for Tasks 1 and 2, we developed baseline algorithms using nnU-Net [33], widely regarded as the current DL gold standard for medical image segmentation [34]. For Task 1, we implemented an nnU-Net v2 model with default parameters (full 3D resolution, 1000 epochs, 5-fold cross-validation), using only the pre-RT training images (n = 150) as input. We applied an identical nnU-Net approach for Task 2, but utilized mid-RT images (n = 150) for training instead. No post-processing was applied to baseline models. Model training was performed on a Lamda workstation with 4 NVIDIA RTX A6000 graphics processing units. DL training took approximately 24 h per model. Additionally, we created a simple "null" algorithm for Task 2, which uses unmodified pre-treatment segmentations as mid-RT predictions. This approach mimics a typical starting point for segmentation adjustments in routine clinical workflows.

2.7 Post-challenge Publications and Conference

To be eligible for the final ranking and prizes, participants were required to submit a concise paper detailing their methods. Teams that participated in both tasks (pre-RT and mid-RT segmentation) had the option to submit either a single comprehensive paper or two separate papers describing their approaches. These submissions were subsequently published in our post-challenge proceedings, providing a valuable resource for the research community. Following the challenge's conclusion, we hosted a live virtual webinar event on the Zoom video conference platform, where top-performing teams were invited to present their innovative methods. This event culminated in the official announcement of the challenge winners.

3 Challenge Algorithm Results

3.1 Participation

As of September 18, 2024 (submission deadline), the number of registered teams for the challenge (regardless of the tasks) was 107. For each task, each team could submit up to five valid submissions for the preliminary development phase and one valid submission for the final test phase. By the submission deadline, we received a total of 164 valid entries across both tasks: 95 for Task 1 (75 in the preliminary development phase, 20 in the final test phase) and 69 for Task 2 (54 in the preliminary development phase, 15 in the final test phase). After accounting for eligibility, 19 unique teams were identified. The geographical distribution of initial registrants is shown in Fig. 8A, while the distribution of final eligible participants is shown in Fig. 8B. The geographical distribution of initial registrants and final participants followed similar patterns, except for Europe and Asia, where the relative proportions reversed between the initial registrants and the final eligible participants.

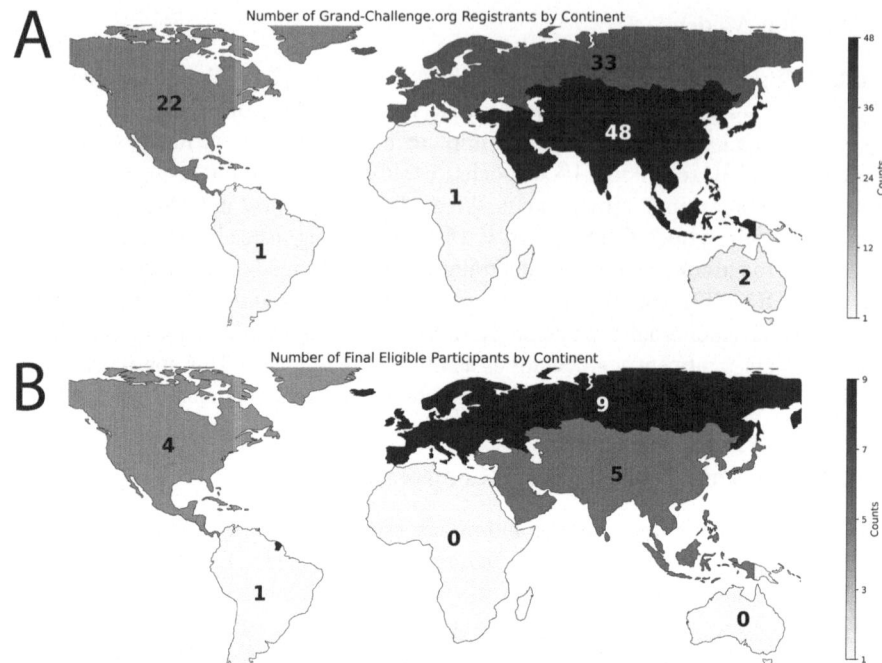

Fig. 8. Geographical distribution of initial registrants and final participants in the HNTS-MRG 2024 challenge by continent. (A) Number of initial registrants, showing the distribution of participants who signed up for the competition on the Grand Challenge website. (B) Number of final eligible participants, reflecting the participants who completed test phase submissions and submitted corresponding manuscripts. The color scale in both maps represents the count of individual Grand Challenge accounts per continent. For the final eligible participants, only the primary contact person from each team was considered.

3.2   Task 1 (Pre-RT Segmentation) Specific Results

Summary of Participants Methods
This section provides an overview of the methods proposed by each team for the automatic segmentation of the GTVp and GTVn in Task 1. The descriptions are presented in the order of the official rankings, beginning with the top-performing team. Each method is briefly outlined, focusing on the key distinguishing features of the method and corresponding submitted manuscript.

Team TUMOR [35] experimented with various nnU-Net methodologies alongside MedNeXt transformer-based models [36] of different kernel sizes. Their models were pre-trained on mid-RT and registered pre-RT images, then fine-tuned on the original pre-RT images. They also explored various ensembling strategies but found that averaging nnU-Net and MedNeXt solutions resulted in worse performance than using either model individually. Ultimately, their best approach was a "small" MedNeXt model with a kernel size of 3, which was used for the final test phase submission of Task 1. Interestingly, they also experimented with fine-tuning using a public meningioma RT dataset [37] but found it did not enhance performance, likely due to discrepancies in features learned during pre-training.

Team Hilab [38] explored a fully supervised learning approach enhanced with pre-trained weights and data augmentation techniques. Notably, they used the SegRap2023 challenge dataset [9] for fully supervised pre-training and applied histogram matching during preprocessing to align intensity differences between CT and MRI data, along with nonlinear transformations to image intensities. To mitigate the impact of negative samples and encourage the network to learn class distributions more effectively, they employed the MixUp technique [39], which augments the dataset by creating new training examples through linear interpolation between sample pairs. For their Task 1 submission, they ultimately used a combination of their base model and pre-training + MixUp model, incorporating ensembling from cross-validation folds.

Team Stockholm_Trio [40] explored a variety of architectures, including SegResNet, nnU-Net, ResEnc, MedNeXt, and U-Mamba. For Task 1, all models were trained exclusively on pre-RT data. The ResEnc and MedNeXt models consistently outperformed the others, particularly when trained on the preprocessed dataset (crop + intensity standardization). The best results were achieved by ensembling the ResEnc and MedNeXt models. However, the execution time of their Docker image on the evaluation platform exceeded the time limit, causing job submission failures. To meet the resource constraints, the final submitted models were significantly simplified by omitting certain preprocessing and ensembling steps.

Team CEMRG [41] introduced a two-stage self-supervised learning approach which leverages unlabeled data to develop robust pre-trained models. In the first stage, they utilized a Self-Supervised Student-Teacher Learning Framework, specifically DINOv2 [42] adapted for 3D data, to learn effective representations from a limited unlabeled dataset. In the second stage, they fine-tuned an xLSTM-based [43] UNet model designed to capture both spatial and sequential features. For Task 1, the team fine-tuned their model on pre-RT segmentation data, which was ultimately submitted for the final test phase.

Team mic-dkfz [44] utilized nnU-net with a residual encoder architecture [34] for their Task 1 solution. Importantly, they experimented with various training strategies, including extensive data augmentation (Aug++), pretraining, ensembling, post-processing, and test-time augmentation. The team leveraged transfer learning by pre-training their model on an unprecedentedly large set of public 3D medical imaging datasets and then fine-tuning on pre-RT data. For their submission to the final test phase, they used an ensemble of models that combined Aug++ and pretraining.

Team RUG_UMCG [45] initially experimented with a custom framework for Task 1, incorporating MONAI [46] with U-Net, 3D U-Net, Swin UNETR architectures, and MedSAM—a foundation model pretrained on a large medical dataset [47]. However, these approaches were less optimal in terms of training speed, validation results, and overall performance compared to a vanilla nnU-net. Ultimately, they transitioned to the nnU-Net framework, employing a 15-fold cross-validation ensemble. For Task 1, they enhanced their training data by incorporating mid-RT data as separate inputs. During inference, test time augmentation was disabled to meet the challenge runtime limit.

Team alpinists [48] conducted a comprehensive literature review to inform their approach, ultimately proposing a resource-efficient two-stage segmentation method using nnU-Net with residual encoders. In this two-stage approach, the segmentation results from the first training round guided the sampling process for a second refinement stage. For the pre-RT task, they achieved competitive results using only the first-stage nnU-Net, which was submitted as their final model. For the final test set submission, they retrained their selected pre-RT models on the full 150-patient dataset. Uniquely, the team used Code-Carbon [49] to monitor the computational efficiency of their approach.

Team SZTU_SingularMatrix [50] investigated the use of STU-Net, a model designed to improve scalability and transferability for medical segmentation [51]. Their approach involved large-scale pretraining on datasets such as TotalSegmentator [52] followed by fine-tuning on the challenge dataset. They explored various STU-Net variants with different parameter sizes and ultimately selected the STU-Net-B model (featuring 58.26 million parameters) for their final test phase submission.

Team SJTU & Ninth People's Hospital [53] explored the use of an nnU-Net model with a residual encoder, coupled with explicit selection of training data. Their initial experiments showed poor performance when the model encountered cases with high background ratios. To address this, for their Task 1 approach they retrained the model using a carefully selected subset of data consisting of cases with background ratios of 70–90% (i.e., the proportion of background voxels to tumor voxels). This approach aimed to improve segmentation performance, and they also incorporated registered data in the training process. While the model performed well on cases with lower background ratios, further optimization was needed for cases with higher background ratios. For the Task 1 final test phase, the authors submitted an ensemble model trained specifically on cases with high background ratios.

Team DCPT-Stine [54] integrated two promising segmentation frameworks, UMamba [55] and nnU-Net with a residual encoder, into a new approach called UMambaAdj. This method combines the feature extraction strengths of the residual encoder with the long-range dependency capabilities of Mamba blocks. The proposed approach demonstrated comparable segmentation accuracy to the original UMambaEnc, but with reduced training and inference times. Additionally, the team found that UMamba blocks significantly improved distance-based metrics, although these metrics were not considered in the final challenge rankings.

Team UW LAIR [56] implemented SegResNet [57] with deep supervision for Task 1. They trained the model using both pre-RT and mid-RT data, but only pre-RT data was used for model selection in the validation set. For each training/validation split, they used three random seeds, selecting the model with the highest DSCagg in the validation set for each of the five cross-validation folds. This process was repeated twice with different random seeds, resulting in a total of 10 models. The final Task 1 submission was an ensemble of these 10 models.

Team NeuralRad [58] developed an enhanced nnU-Net model augmented with an autoencoder architecture. During inference, they added an output channel to predict the original input images, allowing the model to generate both segmentation results and autoencoder predictions simultaneously. By introducing the original training images as additional input channels and incorporating mean squared error loss alongside dice loss, the model was able to learn additional image features, improving segmentation accuracy.

Team 1WM [59] benchmarked several state-of-the-art segmentation architectures to determine whether recent advances in deep encoder-decoder models are effective for low-data and low-contrast tasks. Interestingly, their results showed that traditional UNet-based methods outperform more modern architectures like UNETR, SwinUNETR, and SegMamba, suggesting that factors like data preparation, the underlying objective function, and preprocessing play a greater role than the network architecture itself. For Task 1, they focused on a single-channel pre-RT network, using the pre-RT volume as input. Ultimately, they submitted a ResUNet 5-fold cross-validation ensemble model for the final test phase of Task 1.

Team dlabella29 [60] explored SegResNet integrated into Auto3DSeg via MONAI. The models were pre-trained on both pre-RT and mid-RT image-mask pairs and then fine-tuned on pre-RT data without any preprocessing. Extensive exploratory analysis of the training data also played a key role in shaping their post-processing decisions which included removing smaller tumor and node predictions. For their final test phase submission, they used an ensemble of six SegResNet models, fusing predictions through weighted majority voting.

Team PocketNet [61] implemented a lightweight CNN architecture called PocketNet [62] using the medical image segmentation toolkit [63]. Unlike traditional networks that double the number of feature maps at lower resolutions, PocketNet maintains a constant number of feature maps across all resolution levels. This design results in significantly faster training time while reducing memory usage and requirements. The PocketNet model was trained on pre-RT images via 5-fold cross validation and the corresponding ensemble was submitted for the final test phase of Task 1.

Team andrei.iantsen [64] explored different variations of standard U-Net architectures in their solutions. They tested various processing configurations, including normalization, augmentation, and weighting techniques to find an optimal approach. For Task 1, they trained their networks on all available MRI images, including pre-RT, mid-RT, and registered pre-RT images. In the final test phase of Task 1, they submitted a 5-fold cross-validation ensemble that combined patch-wise normalization, scheduled augmentation, and Gaussian weighting.

Team FinoxyAI [65] proposed a dual-stage 3D UNet approach, called DualUnet, which uses a cascaded Unet framework for progressive segmentation refinement. In the first stage, the models produce an initial binary segmentation, which is then refined by an ensemble of second-stage models to achieve multiclass segmentation. Both pre-RT and mid-RT MRI scans were used as training inputs. This dual-stage approach consistently outperformed single-stage methods in segmentation performance in validation experiments. The approach was trained using 5-fold cross-validation and submitted to the final test phase as an ensemble of five coarse models and ten refinement models.

Team ECU [66] employed LinkNet [67] ensembles for their solution. They initially pre-trained a LinkNet model with weights from ImageNet [68] followed by fine tuning on the challenge dataset. From the training process, they selected eight high-performing model weights to create an ensemble. Each selected weight was used to generate a LinkNet architecture, resulting in eight networks whose predictions were averaged to produce the final segmentation. Their validation experiments demonstrated that the ensemble outperformed any individual model. Interestingly, they also found that increasing the number of networks beyond eight did not significantly enhance accuracy, suggesting a point of diminishing returns. They suggest their approach leverages the benefits of ensemble learning without the computational cost of training each network from scratch.

Challenge Ranking Results
The results for Task 1 are reported in Table 2. The DSCagg-mean results from the 18 participants ranged from 0.571 to 0.825 (overall mean = 0.783). Team TUMOR achieved the highest overall performance, with a DSCagg-mean of 0.825, including the top GTVn DSCagg score of 0.873. Team Stockholm_Trio secured the best GTVp DSCagg result, with a score of 0.795. Only the top 3 teams achieved DSCagg-mean results higher than the nnU-Net baseline (0.817). The top 9 teams (top 50%) achieved DSCagg-mean results higher than interobserver variability (0.806). Notably, two participants for Task 1 withdrew from the competition before submitting their methods manuscripts; their results are displayed for completeness in Table 2 but are not incorporated into any analysis.

3.3   Task 2 (Mid-RT Segmentation) Specific Results

Summary of Participants Methods
This section provides an overview of the methods proposed by each team for the automatic segmentation of the GTVp and GTVn in Task 2. The descriptions are presented in the order of the official rankings, beginning with the top-performing team. Each method is briefly outlined, focusing on the key distinguishing features of the method and/or corresponding manuscript.

Team UW LAIR [56] integrated novel mask-aware attention modules into a Seg-ResNet framework, allowing pre-RT masks to influence features learned from paired mid-RT data. The model took mid-RT MRI images along with pre-RT masks as inputs. During training, paired pre-RT data was also included, with prior masks set to zeros. They also applied mask propagation through deformable registration, which excluded predicted segmentations on mid-RT MRI scans that had no overlap with registered pre-RT segmentations. Ultimately, the attention-based approach outperformed the baseline method which concatenated mid-RT images with pre-RT masks. As in their Task 1 approach, they generated several models using cross validation splits and random seeds then ensembled them for the final submission.

Team mic-dkfz [44] integrated registered pre-RT images and their segmentations as additional inputs into the nnU-Net framework, through a method they referred to as LongiSeg [69]. Interestingly, though they initially experimented with the residual encoder architecture, it did not yield improved results. They investigated several LongiSeg variants and ultimately submitted an ensemble of their LongiSeg Pre-Seg-C model for the Task 2 final test phase. In this model, a one-hot encoding of the registered prior scan's segmentation mask was added to the network input (in addition to image inputs), following the order (current scan, prior scan, prior mask). The model was trained in chronological order, with the mid-RT scan as the current scan and the pre-RT scan as the prior.

Table 2. Task 1 (pre-radiotherapy segmentation) results for participating teams. Results are shown in descending order by mean aggregated DSC (DSCagg). DSCagg scores for primary gross tumor volume (GTVp) and metastatic lymph nodes (GTVn) are also shown. Highest scores for each category are bolded. The performance of the mean, baseline nnU-Net, interobserver variability (derived in Sect. 2.3), anonymized withdrawn teams are also shown at the bottom of the table. Values are rounded to the nearest 3rd decimal place.

Position	Team name	DSCagg mean	DSCagg GTVp	DSCagg GTVn
1	TUMOR	**0.825**	0.778	**0.873**
2	HiLab	0.824	0.785	0.862
3	Stockholm_Trio	0.822	**0.795**	0.849
4	CEMRG	0.815	0.769	0.860
5	mic-dkfz	0.812	0.756	0.869
6	RUG_UMCG	0.812	0.773	0.851
7	alpinists	0.810	0.767	0.852
8	SZTU-SingularMatrix	0.807	0.759	0.854
9	SJTU&NINTH PEOPLE'S HOSPITAL	0.806	0.767	0.845
10	DCPT-Stine's group	0.796	0.751	0.842
11	UW LAIR	0.794	0.745	0.844
12	NeuralRad	0.792	0.732	0.852
13	IWM	0.771	0.717	0.826
14	dlabella29	0.771	0.720	0.822
15	PocketNet	0.770	0.732	0.808
16	andrei.iantsen	0.752	0.709	0.794
17	FinoxyAI	0.737	0.697	0.777
18	ECU	0.571	0.495	0.646
NA	Mean	0.783	0.734	0.829
NA	nnU-Net baseline	0.817	0.769	0.865
NA	Interobserver variability	0.806	0.757	0.854
NA	Withdrawn team 1	0.774	0.726	0.822
NA	Withdrawn team 2	0.793	0.751	0.836

Team HiLab [38] introduced an innovative training strategy with a novel network architecture, termed Dual Flow UNet, which features separate encoders for mid-RT images and registered pre-RT images along with their labels. In this setup, the mid-RT encoder progressively integrates information from pre-RT images and labels during forward propagation. Their submission to the Task 2 final test phase was an intricate ensemble of methods, combining folds from the Dual Flow UNet with base + pre-RT and pre-training + MixUp variants.

Team andrei.iantsen [64] used the same standard Unet based processing approaches for Task 2 as in Task 1. Notably, for Task 2 they trained models using four simultaneous input channels: the mid-RT image, registered pre-RT image, and two binary masks for the GTVp and GTVn on the registered pre-RT image. As in Task 1, their submitted 5-fold cross validation ensemble model for the Task 2 final test phase utilized all three modifications (patch-wise normalization, scheduled augmentation, gaussian weighting).

Team Stockholm_Trio [40] used the same architectures for Task 2 as in Task 1, with the addition of ablation studies to test different combinations of image data and segmentation masks. Due to difficulties in optimizing the MedNeXt model for this task, only the nnU-Net ResEnc model was ultimately used for training. Notably, they applied a dilation to the pre-RT masks, then derived signed distance maps from the dilated masks, incorporating them as prior information to guide the network's attention (preDistance-prior). Their results showed that the preDistance-prior settings outperformed other models. As with Task 1, to meet resource constraints, the final submitted models were simplified by omitting preprocessing and ensembling steps.

Team IWM [59] implemented similar experiments for Task 2 as in Task 1, this time using a three-channel mid-RT network with the concatenated mid-RT volume, registered pre-RT volume, and the associated registered pre-RT ground truth segmentation. As with Task 1, they found simple Unet models superior and subsequently submitted a ResUNet 5-fold cross-validation ensemble model for the final test phase of Task 2.

Team RUG_UMCG [45] applied a similar approach to their Task 1 solution, utilizing the nnU-Net framework with a 15-fold cross-validation ensemble. For Task 2, they implemented a 3-channel input, where the mid-RT MRI volume served as the first channel, the registered pre-RT MRI volume as the second channel, and the corresponding segmentation mask as the third channel. As in Task 1, test-time augmentation was disabled during inference to comply with the challenge's runtime limit.

Team TUMOR [35] applied similar training approaches and architectures for Task 2 as in their Task 1 experiments, with the key difference being the use of concatenated multi-channel inputs to improve segmentation performance. Interestingly, they observed that using registered pre-RT images alone, without their segmentation masks, did not contribute useful information for segmenting mid-RT images. However, including both registered pre-RT images and their segmentation masks improved the DSCagg for mid-RT segmentation. Ultimately, an nnU-Net ensemble of the full-resolution and cascade models was selected as the final model for Task 2.

Team DCPT-Stine [70] employed a novel approach that computes gradient maps from pre-RT images and applies them to mid-RT images to enhance tumor boundary delineation. They applied connected component analysis to registered pre-RT tumor segmentations to initially create bounding boxes. These regions were then used to generate gradient maps on mid-RT T2w images, which served as additional input channels. Gradient maps from pre-RT images and their ground truth segmentation were also incorporated as extra training data. The method was built on nnU-Net with a residual encoder, and validation results showed that leveraging pre-RT information improved segmentation results. Experiments showed that using gradient maps led to more precise boundary localization than images alone.

Team alpinists [48] applied a similar two-stage approach as their Task 1 approach. However, to enhance segmentation performance, they incorporated prior knowledge from the registered pre-RT images and masks as an additional input for the second-stage refinement network. By leveraging the pre-RT data in the second stage, they were able to achieve more accurate mid-RT segmentations for their final submission. As with their Task 1 solution, they retrained the selected mid-RT model on the full 150-patient dataset.

Team NeuralRad [58] applied a similar approach for Task 2 as their Task 1 solution which coupled nnUnet to an autoencoder architecture. The main difference in their Task 2 submission was utilizing mid-RT data instead of pre-RT data.

Team dlabella29 [60] applied a similar methodology using SegResNet for Task 2 as they did in Task 1. Specific Task 2 preprocessing involved setting all voxels more than 1 cm from the registered pre-RT masks to background, followed by applying a bounding box to the image. The modified registered pre-RT and mid-RT MRI were used as input, and model training involved a single stage without any pre-training or fine tuning. Interestingly, they explored systematic radial reductions in the registered pre-RT masks and found that this simple technique performed surprisingly well. However, in keeping with the challenge's spirit, they avoided using simple mask reductions as their submission, as this method is not suitable for adaptive RT planning where patient-specific solutions are essential. Ultimately, they submitted an ensemble of five SegResNet models for the Task 2 final test phase submission.

Team CEMRG [41] employed the same two-stage self-supervised approach as in Task 1, but for Task 2, they fine-tuned their xLSTM-based UNet model on mid-RT segmentation data. This enabled the model to incorporate temporal dependencies specific to mid-treatment tumor response.

Team SJTU & Ninth People's Hospital [53] applied the same approach for Task 2 as they did for Task 1, using the nnU-Net residual encoder model coupled with selective training on specific data subsets. The key difference for Task 2 was the inclusion of mid-RT images instead of pre-RT images for model training. As in Task 1, they submitted an ensemble of folds for the final test phase.

Team TNL_skd [71] proposed an end-to-end coarse-to-fine cascade framework based on a 3D U-Net, inspired by future frame prediction in natural images and video [72]. The model has two interconnected components: a coarse segmentation network and a fine segmentation network, both sharing the same architecture. During coarse segmentation, a dilated pre-RT mask and mid-RT image are used to localize the region of interest and generate a preliminary prediction. During fine segmentation, resampling focuses on the region of interest, refining the prediction with the mid-RT image to produce the final mask. Notably, they also investigated training the networks separately but found the end-to-end combined model was superior.

Challenge Ranking Results

The results for Task 2 are reported in Table 3. The DSCagg-mean results from the 15 participants ranged from 0.562 to 0.733 (overall mean = 0.688). Team UW LAIR achieved the highest overall performance, with a DSCagg-mean of 0.733, including the top GTVp DSCagg score of 0.607. Team mic-dkfz secured the best GTVn DSCagg result, with a score of 0.875. All teams, with the exception of one, achieved DSCagg-mean results higher than the nnU-Net baseline (0.633) and the null algorithm (0.601). Only the top 4 teams (~top 25%) achieved DSCagg-mean results higher than interobserver variability (0.714).

Table 3. Task 2 (mid-radiotherapy segmentation) results for participating teams. Results are shown in descending order by mean aggregated DSC (DSCagg). DSCagg scores for primary gross tumor volume (GTVp) and metastatic lymph nodes (GTVn) are also shown. Highest scores for each category are bolded. The performance of the mean, baseline nnU-Net, null algorithm (simple structure propagation from registered images), and interobserver variability (derived in Sect. 2.3) are also shown at the bottom of the table. Values are rounded to the nearest 3rd decimal place.

Position	Team name	DSCagg mean	DSCagg GTVp	DSCagg GTVn
1	UW LAIR	**0.733**	**0.607**	0.859
2	mic-dkfz	0.727	0.579	**0.875**
3	HiLab	0.725	0.579	0.871
4	andrei.iantsen	0.718	0.592	0.845
5	Stockholm_Trio	0.710	0.554	0.866
6	lWM	0.707	0.579	0.836
7	RUG_UMCG	0.701	0.543	0.859
8	TUMOR	0.700	0.549	0.852
9	DCPT-Stine's group	0.700	0.534	0.867
10	alpinists	0.698	0.539	0.858
11	NeuralRad	0.685	0.526	0.843
12	dlabella29	0.655	0.499	0.811
13	CEMRG	0.654	0.534	0.773
14	SJTU&NINTH PEOPLE'S HOSPITAL	0.638	0.446	0.831
15	TNL_skd	0.562	0.500	0.625
NA	Mean	0.688	0.544	0.831
NA	nnU-Net baseline	0.633	0.422	0.844
NA	null algorithm	0.601	0.460	0.743
NA	Interobserver variability	0.714	0.600	0.828

3.4 General Results Summary

A boxplot summarizing both Task 1 and Task 2 performance is shown in Fig. 9.

A correlation analysis was conducted to assess the relationship between participant performance in Task 1 and Task 2, including only those participants who completed both tasks. Correlations were generally weak with no significant relationships identified.

Kendall's Tau correlation coefficients and corresponding p-values were: DSCagg-mean (-0.01, p = 1.00), DSCagg-GTVp (0.01, p = 1.00), and DSCagg-GTVn (0.18, p = 0.38).

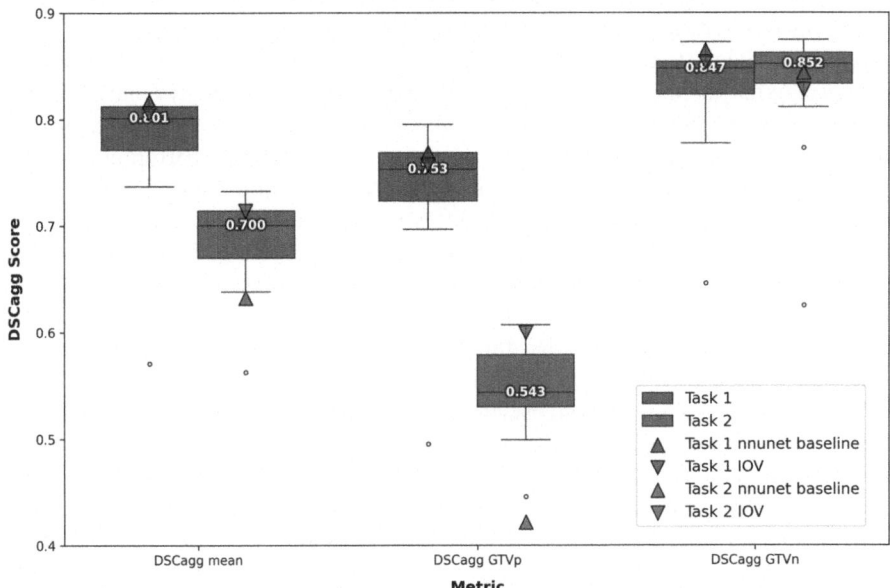

Fig. 9. Boxplot comparison of aggregated Dice Similarity Coefficient (DSCagg) scores across Task 1 (pre-radiotherapy) and Task 2 (mid-radiotherapy) for three metrics: DSCagg mean, DSCagg primary gross tumor volume (GTVp), and DSCagg metastatic lymph nodes (GTVn). Each box represents the interquartile range, with the horizontal line indicating the median score. Outliers are shown as individual points outside the whiskers. Task 1 is represented in red, and Task 2 in blue. Scatter symbols indicate the nnU-Net baseline (triangle), and interobserver variability (IOV, inverted triangle). (Color figure online)

4   Discussion: Putting the Results into Context

4.1   Outcome and Findings

Data challenges play a crucial role in advancing research and facilitating the clinical implementation of AI technologies [73]. Our challenge represents the first crowdsourced initiative for MR-based segmentation in HNC, with a unique focus on investigating whether incorporating prior timepoint data enhances auto-segmentation performance in RT applications. This approach addresses a critical gap in the field and provides valuable insights for adaptive RT workflows.

Task 1 (pre-RT segmentation) results demonstrated the high performance of auto-segmentation algorithms, with most solutions predominantly based on nnU-Net architectures. It was shown that the top 50% of submitted methods achieved DSCagg-mean scores comparable to or exceeding our measured IOV (DSCagg-mean ~ 0.80), indicating their potential for clinical application. Our results align closely with those of HECK-TOR 2022 [14], which also used DSCagg as a metric and saw top-performing algorithms achieve scores around 0.80, though their more heterogeneous test set potentially posed a more complex segmentation end-goal. Generally, GTVp structures were harder for algorithms to segment than GTVn structures. While teams experimented with various underlying training strategies and DL architectures, there didn't seem to be a clear optimal strategy for maximizing performance, though it is worth noting two out of the top three teams used MedNeXt—a transformer-driven architecture [36]—in their approach. Moreover, our baseline nnU-Net algorithm already achieved high-performing results, suggesting that current state-of-the-art methods provide a solid foundation for further improvements. The minimal quantitative differences observed between top-performing models echo findings from previous HNC challenges like HECKTOR [14], SegRap [9], and H&N-Seg [15]. This consistency across challenges underscores the robustness of current pre-RT segmentation algorithms, particularly given the strong baseline performance of nnU-Net. It suggests that for this specific task, we may be approaching a performance plateau with current DL architectures and available training data.

Task 2 (mid-RT segmentation) presented a more challenging problem, as clearly evidenced by the lower overall algorithmic performance compared to Task 1. This aligns with our expectations and the higher IOV observed in mid-RT annotations (DSCagg-mean ~ 0.71). As with Task 1, algorithms found GTVp structures more challenging to segment than GTVn structures. However, in Task 2, this difficulty gap was wider, aligning with the trends observed in our interobserver variability data. Notably, along with volumetric changes, tumor shrinkage is often accompanied by other radiation-induced biological effects [74], such as inflammation and necrosis. These changes can be visible on imaging and may complicate the accurate contouring of intra-treatment scans [75], particularly for GTVp structures. Subsequently, our baseline nnU-Net model for this task fell short of the DSCagg-mean IOV, likely due to the challenge of accurately capturing these complex, evolving tumor characteristics. Interestingly, a simple "null" model mimicking static contour propagation performed surprisingly similar to the baseline nnU-Net model (DSCagg-mean ~ 0.60). It's worth noting that our measured IOV may be slightly inflated due to the exclusion of particularly challenging cases (see Sect. 2.3, Sources of Errors - Interobserver Variability), potentially setting a higher benchmark than typically expected. Importantly, the vast majority of submitted algorithms thoroughly outperformed the baselines, with some even surpassing the IOV threshold. This achievement underscores the potential value of advanced auto-segmentation methods in adaptive RT workflows. However, the fact that only about 25% of teams were able to surpass IOV for Task 2, compared to 50% for Task 1, highlights the novel nature of this segmentation challenge and the need for innovative approaches. Moreover, GTVp IOV was only crossed by the winning algorithm, further illustrating the need to focus on GTVp auto-contouring improvements. Interestingly, the average GTVn segmentation performance was higher in Task 2 than in Task 1, likely because most OPC GTVn remain large and do not achieve a complete response by mid-RT [76], simplifying the segmentation process, especially if prior segmentation masks were incorporated. As expected, the most successful methods thoughtfully incorporated registered pre-RT data (i.e., images and masks) typically through novel DL architectural modifications, demonstrating the utility of leveraging prior timepoint information in adaptive RT auto-segmentation solutions. Although this challenge only utilized prior information from pre-RT to mid-RT scans, the same frameworks could potentially be extended to incorporate imaging data from additional intra-treatment timepoints.

4.2 Limitations of the Challenge

While we have striven for a comprehensive data challenge with adequate documentation, curation efforts, and execution, our study is not without extant limitations.

Firstly, a primary limitation of our data challenge was the relatively modest patient cohort size, derived from a single institutional data source. Our total cohort size (~ 200 cases) is on par with some previous HNC challenges like SegRap 2023 [9], but falls considerably short of larger-scale initiatives such as HECKTOR 2022 [14] (~ 900 cases). It is worth noting that, despite this limitation, DL auto-segmentation algorithms have demonstrated remarkable performance even with limited data [77, 78], as evidenced by the high performance achieved on our test sets. Nevertheless, expanding our dataset over time, following the example set by challenges like HECKTOR, would be beneficial for future iterations. On a related note, another limitation is our focus on oropharyngeal regional tumors, which restricts the diversity of HNC subsites represented in our study. While broadening the range of HNC regional subsites would be valuable, it's important to consider the potential advantages of a more focused approach. Recent recommendations in DL auto-segmentation suggest that decomposing tasks to reduce class imbalance (e.g., focusing on oropharyngeal region) may lead to more effective data utilization and superior models [79]. This data-centric approach could potentially yield better results than a single, all-encompassing model for diverse HNC subsites.

A second significant limitation was the high degree of IOV in our annotations, particularly evident in mid-RT GTVp structures. This variability, while expected, aligns with existing literature on human IOV in HNC tumor segmentation using MRI [80]. To address this issue in future challenges, it would be beneficial to implement strict annotation guidelines for clinician annotators. While such guidelines exist for clinical target volumes [81, 82], they are notably absent for gross tumor volumes, highlighting an area for improvement in future iterations. Furthermore, our study's reliance solely on MRI, while valuable for MR-centric adaptive approaches (e.g., MR-Linac), may have limited the accuracy of tumor delineation. Incorporating additional systematically co-registered imaging modalities such as PET and CT could enhance both the generation of ground truth segmentations by physicians and overall model performance, as supported by previous research [83, 84]. Although we initially planned to include multiple MRI sequences, particularly diffusion weighted sequences (i.e., apparent diffusion coefficient maps), data curation constraints prevented this inclusion without significantly reducing our sample size. Future challenges should explore the integration of additional MRI sequences (i.e., multiparametric MRI), as they could provide crucial information for more precise tumor segmentation [85].

Finally, while we aimed for a robust evaluation using DSCagg, a metric previously validated by Andrearczyk et al. [30, 32], our choice of evaluation metrics could be expanded in future iterations. Recent tumor segmentation challenges involving multiple objects, such as BraTS-METS 2024 [86], have employed more sophisticated measures like lesion-wise DSC. This approach uses ground truth label dilation to better understand lesion extent and rigorously penalizes false positives and negatives with a score of 0. Furthermore, related metrics developed for multiple sclerosis lesions, such as the object-normalized DSC proposed by Raina et al. [87], might offer advantages in handling volume discrepancies. This adaptation of DSC scales precision at a fixed recall rate, addressing bias related to the occurrence rate of the positive class in the ground truth. Notably, for RT-related tasks, incorporating surface distance measurements [88] or spatially accounting for healthy tissue proximity [89] could also provide a more comprehensive evaluation. Additionally, treating the nodal component of these tasks as an object detection problem in conjunction with segmentation could offer a more nuanced assessment of algorithm performance. In future iterations, adopting a broader range of evaluation metrics would likely provide a more holistic understanding of algorithm performance and better align with the specific intricacies of HNC segmentation for MRI-guided RT.

4.3 Future of the Challenge

While we initially released training data (i.e., MRI images and STAPLE consensus segmentations in NIfTI format) through Zenodo [31], we have plans for a more comprehensive data release. This expanded dataset will include raw DICOM data, individual observer segmentations, and relevant clinical metadata for both training and held-out evaluation sets. The inclusion of individual observer segmentations may be particularly valuable in ambiguity modeling experiments for deep learning uncertainty quantification [90, 91]. This extensive data release will be accompanied by a detailed data descriptor to facilitate its use by the research community. Furthermore, we intend to publish a post-challenge summary paper in a high-impact, field-specific journal. This paper will delve into meta-analytic approaches to comprehensively characterize algorithm results, including combined participant algorithms, inter-algorithm variability, additional subanalysis, and ranking stability, in a similar vein to previous post-challenge analyses [8, 32]. Eligible participants will be invited to co-author this manuscript, fostering collaborative insight into the challenge outcomes.

While there are currently no concrete plans for a second edition of HNTS-MRG, we remain open to the possibility of future iterations that could significantly enhance the challenge's scope and impact. Such future editions could potentially incorporate a wider array of imaging sequences and timepoints (i.e., greater number of intra-treatment images) [92], leveraging the full capabilities of MRI in adaptive RT for HNC. We also envision the inclusion of data from multiple institutions, which would not only increase the dataset size but also introduce valuable diversity in imaging protocols and patient populations for added generalization ability of algorithms. This approach would mirror the successful strategy employed by the HECKTOR series of challenges [14, 93, 94], which has seen progressive data enlargement and diversification over the years. By broadening our dataset in these ways, future iterations of HNTS-MRG could offer even more robust insights into the performance of MRI-guided RT segmentation algorithms.

5 Conclusions

This paper presented a comprehensive overview of the HNTS-MRG 2024 challenge, focusing on the automated analysis of MRI images in HNC patients. The challenge explored two critical tasks: fully-automated pre-RT segmentation (Task 1) and mid-RT segmentation (Task 2). Utilizing a robust dataset of 200 HNC cases (150 for training, 50 for final testing), this challenge garnered significant interest from leading research teams worldwide, resulting in 20 high-quality papers showcasing a diverse array of innovative methods. Task 1 algorithm performance was generally high and consistent with previous similar tumor segmentation challenges (e.g., HECKTOR 2022). Top-performing algorithms for Task 1 achieved DSCagg-mean scores comparable to or exceeding clinician IOV, with minimal differences between leading methods. Task 2 proved more challenging, as expected, with lower model performance compared to Task 1. Notably, the best-performing algorithms in Task 2 surpassed both our baseline models and clinician IOV, demonstrating the potential of advanced auto-segmentation methods in adaptive RT workflows. Across both tasks, algorithms consistently found GTVp structures more difficult to segment than GTVn structures, mirroring trends in clinician IOV. To further advance this field, future work should focus on harmonizing tumor segmentation guidelines for clinicians, investigating additional segmentation performance metrics, and expanding the patient cohort.

Acknowledgments. The organizing committee extends its sincere gratitude to all participating teams for their dedication, innovative approaches, and significant contributions to this challenge. This data challenge was directly supported by National Institutes of Health (NIH) Administrative Supplements to Support Collaborations to Improve the AI/ML-Readiness of NIH-Supported Data provided by the National Institute of Dental and Craniofacial Research (NIDCR) (R01DE028290-05S1) and National Cancer Institute (NCI) (R01CA257814-03S2) under parent grants R01DE028290 and R01CA257814, respectively. K.A.W. and B.A.M. are supported by the NCI through Image Guided Cancer Therapy (IGCT) T32 Training Program Fellowships (T32CA261856). L.M. is supported by a NIH Diversity Supplement (R01CA257814-02S2). M.A.N. receives funding from NIH NIDCR Grant (R03DE033550). C.D.F. receives related support from the NCI MD Anderson Cancer Center Core Support Grant Image-Driven Biologically-informed Therapy (IDBT) Program (P30CA016672-47). The authors also acknowledge support from the Tumor Measurement Initiative through the MD Anderson Strategic Research Initiative Development Program.

Disclosure of Interests. K.A.W. serves as an Editorial Board Member for Physics and Imaging in Radiation Oncology. C.D.F. has received travel, speaker honoraria, and/or registration fee waivers unrelated to this project from Siemens Healthineers/Varian, Elekta AB, Philips Medical Systems,The American Association for Physicists in Medicine, The American Society for Clinical Oncology, The Royal Australian and New Zealand College of Radiologists, Australian & New Zealand Head and Neck Society, The American Society for Radiation Oncology, The Radiological Society of North America, and The European Society for Radiation Oncology.

References

1. Pollard, J.M., Wen, Z., Sadagopan, R., Wang, J., Ibbott, G.S.: The future of image-guided radiotherapy will be MR guided. Br. J. Radiol. **90**, 20160667 (2017)

2. Mulder, S.L., et al.: MR-guided adaptive radiotherapy for OAR sparing in head and neck cancers. Cancers **14**, 1909 (2022)
3. Salzillo, T.C., et al.: Advances in imaging for HPV-related oropharyngeal cancer: applications to radiation oncology. Semin. Radiat. Oncol. **31**, 371–388 (2021)
4. Kiser, K.J., Smith, B.D., Wang, J., Fuller, C.D.: "Après Mois, Le Déluge": preparing for the coming data flood in the MRI-guided radiotherapy era. Front. Oncol. **9**, 983 (2019)
5. Thorwarth, D., Low, D.A.: Technical challenges of real-time adaptive mr-guided radiotherapy. Front. Oncol. **11**, 634507 (2021)
6. Segedin, B., Petric, P.: Uncertainties in target volume delineation in radiotherapy - are they relevant and what can we do about them? Radiol. Oncol. **50**, 254–262 (2016)
7. Hindocha, S., et al.: Artificial intelligence for radiotherapy auto-contouring: current use, perceptions of and barriers to implementation. Clin. Oncol. **35**, 219–226 (2023)
8. Oreiller, V., et al.: Head and neck tumor segmentation in PET/CT: the HECKTOR challenge. Med. Image Anal. **77**, 102336 (2022)
9. Luo, X., et al.: SegRap2023: A Benchmark of Organs-at-Risk and Gross Tumor Volume Segmentation for Radiotherapy Planning of Nasopharyngeal Carcinoma, http://arxiv.org/abs/2312.09576 (2023). https://doi.org/10.48550/ARXIV.2312.09576
10. Maier-Hein, L., et al.: BIAS: Transparent reporting of biomedical image analysis challenges. Med. Image Anal. **66**, 101796 (2020)
11. Pinkiewicz, M., Dorobisz, K., Zatoński, T.: A systematic review of cancer of unknown primary in the head and neck region. Cancer Manag. Res. **13**, 7235–7241 (2021)
12. Burnet, N.G., Thomas, S.J., Burton, K.E., Jefferies, S.J.: Defining the tumour and target volumes for radiotherapy. Cancer Imaging **4**, 153–161 (2004)
13. Jensen, K., et al.: Imaging for target delineation in head and neck cancer radiotherapy. Semin. Nucl. Med. **51**, 59–67 (2021)
14. Andrearczyk, V., et al.: Overview of the HECKTOR challenge at MICCAI 2022: automatic head and neck tumor segmentation and outcome prediction in PET/CT. In: Head and Neck Tumor Segmentation and Outcome Prediction, pp. 1–30. Springer, Cham (2023)
15. Podobnik, G., et al.: HaN-Seg: the head and neck organ-at-risk CT and MR segmentation challenge. Radiother. Oncol. **198**, 110410 (2024)
16. Boettiger, C.: An introduction to Docker for reproducible research. Oper. Syst. Rev. **49**, 71–79 (2015)
17. Head and neck tumor segmentation for MR-guided applications - grand challenge. https://hntsmrg24.grand-challenge.org/. Accessed 12 Aug 2024
18. McDonald, B.A., Dal Bello, R., Fuller, C.D., Balermpas, P.: The use of MR-guided radiation therapy for head and neck cancer and recommended reporting guidance. Semin. Radiat. Oncol. **34**, 69–83 (2024)
19. Lin, D., et al.: E pluribus unum: prospective acceptability benchmarking from the Contouring Collaborative for Consensus in Radiation Oncology crowdsourced initiative for multiobserver segmentation. J. Med. Imaging **10**, S11903 (2023)
20. Wahid, K.A., et al.: Large scale crowdsourced radiotherapy segmentations across a variety of cancer anatomic sites. Sci. Data **10**, 161 (2023)
21. Warfield, S.K., Zou, K.H., Wells, W.M.: Simultaneous truth and performance level estimation (STAPLE): an algorithm for the validation of image segmentation. IEEE Trans. Med. Imaging **23**, 903–921 (2004)
22. Mayo, C.S., et al.: American association of physicists in medicine task group 263: standardizing nomenclatures in radiation oncology. Int. J. Radiat. Oncol. Biol. Phys. **100**, 1057–1066 (2018)
23. Lowekamp, B.C., Chen, D.T., Ibáñez, L., Blezek, D.: The design of SimpleITK. Front. Neuroinform. **7**, 45 (2013)

24. Anderson, B.M., Wahid, K.A., Brock, K.K.: Simple python module for conversions between DICOM images and radiation therapy structures, masks, and prediction arrays. Pract. Radiat. Oncol. **11**, 226–229 (2021)
25. Wahid, K.A., et al.: Artificial intelligence for radiation oncology applications using public datasets. Semin. Radiat. Oncol. **32**, 400–414 (2022)
26. Leibfarth, S., et al.: A strategy for multimodal deformable image registration to integrate PET/MR into radiotherapy treatment planning. Acta Oncol. **52**, 1353–1359 (2013)
27. Naser, M.A., et al.: Quality assurance assessment of intra-acquisition diffusion-weighted and T2-weighted magnetic resonance imaging registration and contour propagation for head and neck cancer radiotherapy. Med. Phys. **50**, 2089–2099 (2023)
28. Wahid, K.: HNTSMRG_2024: Docker tutorial and example files related to HNTS-MRG 2024 Data Challenge. Github
29. Nikolov, S., et al.: Clinically applicable segmentation of head and neck anatomy for radiotherapy: deep learning algorithm development and validation study. J. Med. Internet Res. **23**, e26151 (2021)
30. Andrearczyk, V., Oreiller, V., Jreige, M., Castelli, J., Prior, J.O., Depeursinge, A.: Segmentation and classification of head and neck nodal metastases and primary tumors in PET/CT. In: Conference Proceedings IEEE Engineering in Medicine & Biology Society 2022, pp. 4731–4735 (2022)
31. Wahid, K., Dede, C., Naser, M., Fuller, C.: Training Dataset for HNTSMRG 2024 Challenge (2024). https://zenodo.org/doi/10.5281/zenodo.11199559. https://doi.org/10.5281/ ZENODO.11199559
32. Andrearczyk, V., et al.: Automatic head and neck tumor segmentation and outcome prediction relying on FDG-PET/CT images: findings from the second edition of the HECKTOR challenge. Med. Image Anal. **90**, 102972 (2023)
33. Isensee, F., Jaeger, P.F., Kohl, S.A.A., Petersen, J., Maier-Hein, K.H.: NnU-Net: a self-configuring method for deep learning-based biomedical image segmentation. Nat. Methods **18**, 203–211 (2021)
34. Isensee, F., et al.: nnU-Net Revisited: A Call for Rigorous Validation in 3D Medical Image Segmentation (2024). http://arxiv.org/abs/2404.09556
35. Moradi, N., et al.: Comparative analysis of nnUNet and MedNeXt for head and neck tumor segmentation in MRI-guided radiotherapy. In: Wahid, K.A., Dede, C., Naser, M.A., Fuller, C.D. (eds.) LNCS. Springer, Cham (2025)
36. Roy, S., et al.: MedNeXt: Transformer-driven scaling of ConvNets for medical image segmentation (2023). http://arxiv.org/abs/2303.09975
37. LaBella, D., et al.: A multi-institutional meningioma MRI dataset for automated multi-sequence image segmentation. Sci. Data. **11**, 496 (2024)
38. Wang, L., Liao, W., Zhang, S., Wang, G.: Head and neck tumor segmentation of MRI from pre- and mid-radiotherapy with pre-training, data augmentation and dual flow UNet. In: Wahid, K.A., Dede, C., Naser, M.A., Fuller, C.D. (eds.) LNCS. Springer, Cham (2025)
39. Zhang, H., Cisse, M., Dauphin, Y.N., Lopez-Paz, D.: mixup: Beyond Empirical Risk Minimization (2017). http://arxiv.org/abs/1710.09412
40. Astaraki, M., Toma-Dasu, I.: Enhancing Head and neck tumor segmentation in MRI: the impact of image preprocessing and model ensembling. In: Wahid, K.A., Dede, C., Naser, M.A., Fuller, C.D. (eds.) LNCS. Springer, Cham (2025)
41. Qayyum, A., Mazher, M., Niederer, S.A.: Assessing self-supervised xLSTM-UNet architectures for head and neck tumor segmentation in MR-guided applications. In: Wahid, K.A., Dede, C., Naser, M.A., Fuller, C.D. (eds.) LNCS. Springer, Cham (2025)
42. Oquab, M., et al.: DINOv2: Learning robust visual features without supervision (2023). http:// arxiv.org/abs/2304.07193

43. Alkin, B., Beck, M., Pöppel, K., Hochreiter, S., Brandstetter, J.: Vision-LSTM: xLSTM as Generic Vision Backbone (2024). http://arxiv.org/abs/2406.04303

44. Kächele, J., Zenk, M., Rokuss, M., Ulrich, C., Wald, T., Maier-Hein, K.H.: Enhanced nnU-Net architectures for automated MRI segmentation of head and neck tumors in adaptive radiation therapy. In: Wahid, K.A., Dede, C., Naser, M.A., Fuller, C.D. (eds.) LNCS. Springer, Cham (2025)

45. Mol, F.N., et al.: MRI-based head and neck tumor segmentation using nnU-Net with 15-fold cross-validation ensemble. In: Wahid, K.A., Dede, C., Naser, M.A., Fuller, C.D. (eds.) LNCS. Springer, Cham (2025)

46. Cardoso, M.J., et al.: MONAI: An open-source framework for deep learning in healthcare (2022). http://arxiv.org/abs/2211.02701

47. Ma, J., He, Y., Li, F., Han, L., You, C., Wang, B.: Segment anything in medical images. Nat. Commun. **15**, 654 (2024)

48. Tappeiner, E., Gapp, C., Welk, M., Schubert, R.: Head and neck tumor segmentation on MRIs with fast and resource-efficient staged nnU-nets. In: Wahid, K.A., Dede, C., Naser, M.A., Fuller, C.D. (eds.) LNCS. Springer, Cham (2025)

49. Courty, B., et al.: mlco2/codecarbon: v2.8.0. Zenodo (2024). https://doi.org/10.5281/ZEN ODO.14212766

50. Wang, Z., Lyu, M.: Head and neck tumor segmentation for MRI-guided radiation therapy using pre-trained STU-net models. In: Wahid, K.A., Dede, C., Naser, M.A., Fuller, C.D. (eds.) LNCS. Springer, Cham (2025)

51. Huang, Z., et al.: STU-Net: scalable and transferable medical image segmentation models empowered by large-scale supervised pre-training (2023). http://arxiv.org/abs/2304.06716

52. Wasserthal, J., et al.: TotalSegmentator: robust segmentation of 104 anatomic structures in CT images. Radiol. Artif. Intell. **5**, e230024 (2023)

53. Ji, K., Wu, Z., Han, J., Jia, J., Zhai, G., Liu, J.: Application of 3D nnU-Net with residual encoder in the 2024 MICCAI head and neck tumor segmentation challenge. In: Wahid, K.A., Dede, C., Naser, M.A., Fuller, C.D. (eds.) LNCS. Springer, Cham (2025)

54. Ren, J., Hochreuter, K., Kallehauge, J.F., Korreman, S.: UMambaAdj: advancing GTV segmentation for head and neck cancer in MRI-guided RT with UMamba and nnU-Net ResEnc planner. In: Wahid, K.A., Dede, C., Naser, M.A., Fuller, C.D. (eds.) LNCS. Springer, Cham (2025)

55. Ma, J., Li, F., Wang, B.: U-mamba: enhancing long-range dependency for biomedical image segmentation (2024). http://arxiv.org/abs/2401.04722

56. Tie, X., Chen, W., Huemann, Z., Schott, B., Liu, N., Bradshaw, T.J.: Deep learning for longitudinal gross tumor volume segmentation in MRI-guided adaptive radiotherapy for head and neck cancer. In: Wahid, K.A., Dede, C., Naser, M.A., Fuller, C.D. (eds.) LNCS. Springer, Cham (2025)

57. Myronenko, A.: 3D MRI brain tumor segmentation using autoencoder regularization (2018). http://arxiv.org/abs/1810.11654

58. An, Y., Wang, Z., Ma, E., Jiang, H., Lu, W.: Enhancing nnUNetv2 training with autoencoder architecture for improved medical image segmentation. In: Wahid, K.A., Dede, C., Naser, M.A., Fuller, C.D. (eds.) LNCS. Springer, Cham (2025)

59. Wodzinski, M.: Benchmark of deep encoder-decoder architectures for head and neck tumor segmentation in magnetic resonance images: contribution to the HNTSMRG challenge. In: Wahid, K.A., Dede, C., Naser, M.A., Fuller, C.D. (eds.) LNCS. Springer, Cham (2025)

60. LaBella, D.: Ensemble deep learning models for automated segmentation of tumor and lymph node volumes in head and neck cancer using pre- and mid- treatment MRI: application of Auto3DSeg and SegResNet. In: Wahid, K.A., Dede, C., Naser, M.A., Fuller, C.D. (eds.) LNCS. Springer, Cham (2025)

61. Twam, A., Celaya, A., Lim, E., Elsayes, K., Fuentes, D., Netherton, T.: Head and neck gross tumor volume automatic segmentation using PocketNet. In: Wahid, K.A., Dede, C., Naser, M.A., Fuller, C.D. (eds.) LNCS. Springer, Cham (2025)

62. Celaya, A., et al.: PocketNet: a smaller neural network for medical image analysis. IEEE Trans. Med. Imaging **42**, 1172–1184 (2023)

63. Celaya, A., et al.: MIST: A simple and scalable end-to-end 3D medical imaging segmentation framework (2024). http://arxiv.org/abs/2407.21343

64. Iantsen, A.: Improving the U-Net configuration for automated delineation of head and neck cancer on MRI. In: Wahid, K.A., Dede, C., Naser, M.A., Fuller, C.D. (eds.) LNCS. Springer, Cham (2025)

65. Saukkoriipi, M., Sahlsten, J., Jaskari, J., Al-Tahmeesschi, A., Ruotsalainen, L., Kaski, K.: Head and neck tumor segmentation using pre-RT MRI scans and cascaded DualUNet. In: Wahid, K.A., Dede, C., Naser, M.A., Fuller, C.D. (eds.) LNCS. Springer, Cham (2025)

66. Baldeon-Calisto, M.: Ensemble of LinkNet networks for head and neck tumor segmentation. In: Wahid, K.A., Dede, C., Naser, M.A., Fuller, C.D. (eds.) LNCS. Springer, Cham (2025)

67. Chaurasia, A., Culurciello, E.: LinkNet: exploiting encoder representations for efficient semantic segmentation. In: 2017 IEEE Visual Communications and Image Processing (VCIP), pp. 1–4. IEEE (2017)

68. Deng, J., Dong, W., Socher, R., Li, L.-J., Li, K., Fei-Fei, L.: ImageNet: a large-scale hierarchical image database. In: 2009 IEEE Conference on Computer Vision and Pattern Recognition, pp. 248–255. IEEE (2009)

69. Rokuss, M., et al.: Longitudinal segmentation of MS lesions via temporal Difference Weighting (2024). http://arxiv.org/abs/2409.13416

70. Ren, J., Hochreuter, K., Rasmussen, M.E., Kallehauge, J.F., Korreman, S.: Gradient map-assisted head and neck tumor segmentation: a pre-RT to mid-RT approach in MRI-guided radiotherapy. In: Wahid, K.A., Dede, C., Naser, M.A., Fuller, C.D. (eds.) LNCS. Springer, Cham (2025)

71. Ni, J., Yao, Q., Liu, Y., Qi, H.: A coarse-to-fine framework for mid radiotherapy head and neck cancer MRI segmentation. In: Wahid, K.A., Dede, C., Naser, M.A., Fuller, C.D. (eds.) LNCS. Springer, Cham (2025)

72. Gao, Z., Tan, C., Wu, L., Li, S.Z.: SimVP: Simpler yet Better Video Prediction (2022). http://arxiv.org/abs/2206.05099

73. Armato, S.G., 3rd., Drukker, K., Hadjiiski, L.: AI in medical imaging grand challenges: translation from competition to research benefit and patient care. Br. J. Radiol. **96**, 20221152 (2023)

74. Wang, J.-S., Wang, H.-J., Qian, H.-L.: Biological effects of radiation on cancer cells. Mil. Med. Res. **5**, 20 (2018)

75. Joint Head and Neck Radiation Therapy-MRI Development Cooperative, MR-Linac Consortium Head and Neck Tumor Site Group: Longitudinal diffusion and volumetric kinetics of head and neck cancer magnetic resonance on a 1.5 T MR-linear accelerator hybrid system: a prospective R-IDEAL stage 2a imaging biomarker characterization/pre-qualification study. Clin. Transl. Radiat. Oncol. **42**, 100666 (2023)

76. Ding, Y., et al.: Intravoxel incoherent motion imaging kinetics during chemoradiotherapy for human papillomavirus-associated squamous cell carcinoma of the oropharynx: preliminary results from a prospective pilot study: IVIM Kinetics during Chemoradiotherapy. NMR Biomed. **28**, 1645–1654 (2015)

77. Wahid, K.A., et al.: Evolving horizons in radiation therapy auto-contouring: distilling insights, embracing data-centric frameworks, and moving beyond geometric quantification. Adv. Radiat. Oncol. **9**, 101521 (2024)

78. Fang, Y., et al.: The impact of training sample size on deep learning-based organ auto-segmentation for head-and-neck patients. Phys. Med. Biol. **66**, (2021). https://doi.org/10.1088/1361-6560/ac2206
79. Rodríguez Outeiral, R., et al.: Strategies for tackling the class imbalance problem of oropharyngeal primary tumor segmentation on magnetic resonance imaging. Phys. Imaging Radiat. Oncol. **23**, 144–149 (2022)
80. Cardenas, C.E., et al.: Comprehensive quantitative evaluation of variability in magnetic resonance-guided delineation of oropharyngeal gross tumor volumes and high-risk clinical target volumes: an R-IDEAL stage 0 prospective study. Int. J. Radiat. Oncol. Biol. Phys. **113**, 426–436 (2022)
81. Grégoire, V., et al.: Delineation of the primary tumour Clinical Target Volumes (CTV-P) in laryngeal, hypopharyngeal, oropharyngeal and oral cavity squamous cell carcinoma: AIRO, CACA, DAHANCA, EORTC, GEORCC, GORTEC, HKNPCSG, HNCIG, IAG-KHT, LPRHHT, NCIC CTG, NCRI, NRG Oncology, PHNS, SBRT, SOMERA, SRO, SSHNO, TROG consensus guidelines. Radiother. Oncol. **126**, 3–24 (2018)
82. Grégoire, V., et al.: Delineation of the neck node levels for head and neck tumors: a 2013 update. DAHANCA, EORTC, HKNPCSG, NCIC CTG, NCRI, RTOG, TROG consensus guidelines. Radiother. Oncol. **110**, 172–181 (2014)
83. Ren, J., Eriksen, J.G., Nijkamp, J., Korreman, S.S.: Comparing different CT, PET and MRI multi-modality image combinations for deep learning-based head and neck tumor segmentation. Acta Oncol. **60**, 1399–1406 (2021)
84. Zhao, Y., et al.: Multi-modal segmentation with missing image data for automatic delineation of gross tumor volumes in head and neck cancers. Med. Phys. **51**, 7295–7307 (2024)
85. Wahid, K.A., et al.: Evaluation of deep learning-based multiparametric MRI oropharyngeal primary tumor auto-segmentation and investigation of input channel effects: Results from a prospective imaging registry. Clin Transl Radiat Oncol. **32**, 6–14 (2022)
86. Moawad, A.W., et al.: The Brain Tumor Segmentation - Metastases (BraTS-METS) Challenge 2023: Brain Metastasis Segmentation on Pre-treatment MRI. arXiv (2024)
87. Raina, V., et al.: Tackling Bias in the Dice Similarity Coefficient: Introducing nDSC for White Matter Lesion Segmentation (2023). http://arxiv.org/abs/2302.05432
88. Sherer, M.V., et al.: Metrics to evaluate the performance of auto-segmentation for radiation treatment planning: a critical review. Radiother. Oncol. **160**, 185–191 (2021)
89. McCullum, L., Wahid, K.A., Marquez, B., Fuller, C.D.: OAR-weighted dice score: a spatially aware, radiosensitivity aware metric for target structure contour quality assessment. Use Comput. Radiat. Ther. **2024**, 755–758 (2024)
90. Wahid, K.A., et al.: Harnessing uncertainty in radiotherapy auto-segmentation quality assurance. Phys. Imaging Radiat. Oncol. **29**, 100526 (2024)
91. Wahid, K.A., et al.: Artificial intelligence uncertainty quantification in radiotherapy applications − a scoping review. Radiother. Oncol. **201** (2024). https://doi.org/10.1016/j.radonc.2024.110542
92. El-Habashy, D.M., et al.: Dataset of weekly intra-treatment diffusion weighted imaging in head and neck cancer patients treated with MR-Linac. Sci. Data. **11**, 487 (2024)
93. Andrearczyk, V., et al.: Overview of the HECKTOR challenge at MICCAI 2020: automatic head and neck tumor segmentation in PET/CT. In: Head and Neck Tumor Segmentation, pp. 1–21. Springer (2021)
94. Andrearczyk, V., et al.: Overview of the HECKTOR challenge at MICCAI 2021: automatic head and neck tumor segmentation and outcome prediction in PET/CT images. In: Head and Neck Tumor Segmentation and Outcome Prediction, pp. 1–37. Springer (2022)

Gradient Map-Assisted Head and Neck Tumor Segmentation: A Pre-RT to Mid-RT Approach in MRI-Guided Radiotherapy

Jintao Ren[1,2], Kim Hochreuter[1,2], Mathis Ersted Rasmussen[1,2], Jesper Folsted Kallehauge[1,2], and Stine Sofia Korreman[1,2,3(✉)]

[1] Department of Clinical Medicine, Aarhus University, Nordre Palle Juul-Jensens Blvd. 11, 8200 Aarhus, Denmark
stine.korreman@clin.au.dk
[2] Aarhus University Hospital, Danish Centre for Particle Therapy, Palle Juul-Jensens Blvd. 25, 8200 Aarhus, Denmark
[3] Department of Oncology, Aarhus University, Palle Juul-Jensens Blvd. 35, 8200 Aarhus, Denmark

Abstract. Radiation therapy (RT) is a vital part of treatment for head and neck cancer, where accurate segmentation of gross tumor volume (GTV) is essential for effective treatment planning. This study investigates the use of pre-RT tumor regions and local gradient maps to enhance mid-RT tumor segmentation for head and neck cancer in MRI-guided adaptive radiotherapy. By leveraging pre-RT images and their segmentations as prior knowledge, we address the challenge of tumor localization in mid-RT segmentation. A gradient map of the tumor region from the pre-RT image is computed and applied to mid-RT images to improve tumor boundary delineation. Our approach demonstrated improved segmentation accuracy for both primary GTV (GTVp) and nodal GTV (GTVn), though performance was limited by data constraints. The final DSC_{agg} scores from the challenge's test set evaluation were 0.534 for GTVp, 0.867 for GTVn, and a mean score of 0.70. This method shows potential for enhancing segmentation and treatment planning in adaptive radiotherapy. Team: DCPT-Stine's group.

Keywords: Deep learning · Prior knowledge · Tumor Segmentation · Head and Neck Cancer · MRI

1 Introduction

Radiation therapy (RT) is a key treatment modality for head and neck cancer (HNC). However, tumor delineation for RT planning remains a significant challenge, involving the precise delineation of both the primary tumor volume (GTVp) and the involved nodal metastases (GTVn). Traditional RT planning

© The Author(s) 2025
K. A. Wahid et al. (Eds.): HNTS-MRG 2024, LNCS 15273, pp. 36–49, 2025.
https://doi.org/10.1007/978-3-031-83274-1_2

relies heavily on manual delineation of tumor volumes by clinicians, a process that is both time-consuming and prone to high inter-observer variability (IOV), particularly in HNC where complex anatomical structures and critical organs at risk lie in close proximity to the tumor [17,23,28]. MRI-guided RT has emerged as a promising approach offering superior soft tissue contrast and the advantage of avoiding additional ionizing radiation during imaging. Further, MRI-guided adaptive RT holds significant potential to improve clinical outcomes by maximizing tumor control while minimizing side effects [3,14,15]. In addition, the use of multimodality images is also recommended [9].

With the advancements in imaging techniques and artificial intelligence (AI), recent research has increasingly focused on automating the segmentation of HNC tumors and organs at risk using deep learning methods [13,19]. These AI-driven approaches aim to overcome the limitations of manual segmentation by providing faster and potentially more consistent delineations [5,6]. However, due to the lack of gold-standard imaging, these applications face a common challenge of uncertainty. Specifically, for HNC, tumors often have ambiguous borders, particularly in the GTVp, where the lack of clear distinction between healthy and malignant tissues further complicates tumor delineation for both human annotators and AI models. Strong IOV is particularly problematic for GTVp, where inconsistent manual annotations can significantly impact the training of deep learning models [4].

To further improve tumor segmentation performance for head and neck cancers, previous studies have explored methods that incorporate prior segmentations or prompts (e.g. bounding boxes, scribbles and clicks) to refine subsequent segmentation tasks. For example, Outeiral et al. applied bounding box cropping methods, which led to an increase in DSC for MRI images [18]. Ren et al. found that adding a bounding box as an additional channel improved nnUNet performance, raising GTVp/GTVn DSC from 0.68/0.63 to 0.88/0.89 on multimodal data (CT, PET, MRI) [21]. Wang et al. used a RetinaNet model to narrow segmentation fields on CT and PET scans, improving precision [33]. Wei et al. [34] demonstrated that incorporating minimal training steps after human interactions raised GTV accuracy from a DSC of 0.65 to 0.82. Interactions such as single or multiple clicks within the GTV have also demonstrated significant improvements in segmentation accuracy [20,25].

Building upon these advancements, this study addresses the challenges of MRI-guided adaptive RT for HNC, focusing on the second task of the HNTS-MRG 2024 challenge [29]. This task involves segmenting GTVp and GTVn using pre-RT and mid-RT T2-weighted (T2w) MRI images, with mid-RT images taken after RT treatment. The main challenge is accurately identifying all malignant tumor regions and determining the correct contours, complicated by similar soft tissue contrasts and the lack of definitive ground truth [24]. Tumor shrinkage or disappearance in mid-RT images further complicates boundary delineation.

To overcome these challenges, pre-RT images and their corresponding delineations can serve as valuable prior knowledge for assisting in the segmentation of tumors on mid-RT images. However, the optimal way to leverage this pre-RT information for mid-RT segmentation remains an open question [32].

Despite the temporal link, pre-RT and mid-RT images often differ in intensity, shape, and texture, presenting a challenge for consistent segmentation. These inter- and intra-patient variations may, however, enrich deep learning models by providing diverse training examples for better generalization.

In this study, we present a novel approach that utilizes pre-RT tumor delineations to improve mid-RT segmentation. We first use the deformably registered pre-RT tumor delineations to identify bounding boxes, defining Regions of Interest (ROIs) around the tumors. These ROIs are then employed to compute gradient maps on the mid-RT T2w images, which serve as additional input channels. Additionally, we generate gradient maps from the original pre-RT images and their ground truth (GT) delineations, incorporating them as extra training data. This approach aims to leverage both pre-RT and mid-RT information, thereby enhancing segmentation accuracy during the mid-RT phase.

2 Material and Methods

2.1 Data

The dataset used in this study was provided by the organizers of the HNTS-MRG 2024 challenge task2 which were 150 HNC patients, predominantly oropharyngeal cancer (OPC). Imaging provided for each patient consisted of T2-weighted (T2w) anatomical sequences of the head and neck region, including both fat-suppressed and non-fat-suppressed images, acquired at MD Anderson Cancer Center [29]. Images include pre-RT (1–3 weeks before the start of RT) and mid-RT (2–4 weeks intra-RT) scans. Multiple physician expert observers (n = 3 to 4) have independently segmented GTVp and GTVn structures for all cases (pre-RT and mid-RT) based on MRI images provided in the challenge. The ground truth was obtained via the Simultaneous Truth And Performance Level Estimation algorithm (STAPLE). Pre-RT images and delineations were deformably registered (DR) to the mid-RT images (DR pre-RT).

2.2 Incorporating Pre-RT Tumor Location with Gradient Maps for Mid-RT

We consider pre-RT images and their delineations as valuable prior knowledge for segmenting tumors on mid-RT images, especially for identifying tumor locations. To leverage this information, we performed connected component analysis on the GTV segmentation masks from DR pre-RT images to identify individual tumor instances (GTVp and GTVn). For each instance, 3D coordinates of the tumor boundaries were extracted, and a bounding box was created around the tumor, with random perturbations of 2–6 voxels in the x, y, and z directions to account for spatial variations. Next, we computed the gradient magnitude (to construct a gradient map) from mid-RT T2w MRI images within these bounding boxes to capture intensity changes around tumor boundaries. The preprocessing involved normalizing the T2w MRI based on its full intensity range and extracting the regions defined by the pre-RT mask. A gradient map was then generated using a Gaussian filter (sigma = 1) to highlight intensity changes. Gradient values were normalized to the range 0-1, with any values greater than 1 clipped to maintain consistency across all patients.

Although pre-RT and mid-RT images are temporally related, their higher-level features, such as tumor morphology and intensity patterns, can differ significantly, likely due to treatment effects or natural variability. To account for this, we treated the pre-RT and mid-RT images of the same patient as independent data points, effectively doubling the dataset size from 150 to 300 samples. Gradient maps were computed for both pre-RT and mid-RT images using a similar process. This approach allowed the model to learn from a broader range of spatial and intensity variations by expanding the training set.

At test time, the bounding boxes of tumors from DR pre-RT were used to construct the gradient maps of mid-RT T2w MR images. Therefore, both mid-RT T2w and the gradient maps were treated as two-channel inputs for the network. The overall process is outlined in detail in the flowchart, as illustrated in Fig. 1.

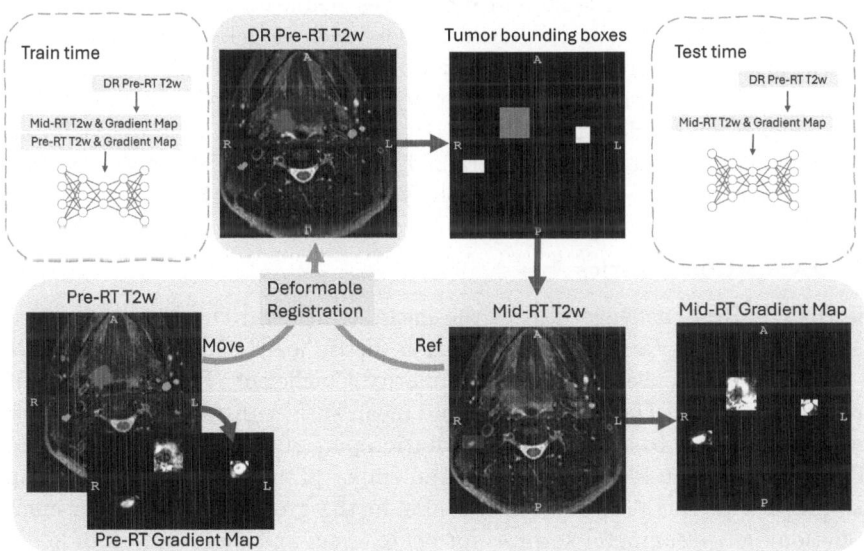

Fig. 1. Workflow for incorporating pre-RT tumor location and gradient maps in mid-RT segmentation.

2.3   Deep Learning Configurations

We utilized the nnUNet framework [7] to implement and train the model, employing the nnUNetResEncM planner to design the network architecture [8]. The model was based on a Residual Encoder U-Net with six stages. Each stage contained a specified number of convolutional blocks: Stage 1 had 1 block, Stage 2 had 3 blocks, Stage 3 had 4 blocks, and Stages 4 through 6 each had 6 blocks. The Conv3D layers employed a kernel size of $1 \times 3 \times 3$ for the first layer, followed by $3 \times 3 \times 3$ kernels for the subsequent layers. Strides were set to $1 \times 1 \times 1$ in the first stage and either $1 \times 2 \times 2$ or $2 \times 2 \times 2$ in the later stages to reduce dimensionality. InstanceNorm3D was used for normalization, and LeakyReLU served as the activation function.

The training process used a batch size of 8 with a patch size of $48 \times 192 \times 192$ voxels. Both the input T2w images and gradient maps were normalized using Z-Score normalization for intensity standardization. Cubic interpolation (order 3) was applied for data resampling, while first-order interpolation was used for segmentation masks. The median image size was $123 \times 512 \times 511$ voxels, with voxel spacing set at $1.2 \times 0.5 \times 0.5$ mm. Training was performed using the SGD optimizer with a PolyLR scheduler with exponent $= 0.9$, starting with a learning rate of 0.01. The loss function was a combination of cross-entropy and dice loss.

The 150 patients were randomly divided into 5 folds, with each fold containing 240 images/delineations (comprising pre-RT and mid-RT images for 120 patients) for training, while 30 mid-RT images/delineations for validation. The maximum number of epochs was set to 1000 for each model, and the final models from the last epoch were saved for prediction. An ensemble of all models trained across the five folds was used to generate predictions on the test set for the final challenge submission. In accordance with guidelines for reproducibility and verification [6], all source code and trained weights for gradient map generation and deep learning have been made publicly available on GitHub[1].

2.4   Evaluation Metrics

The HNTS-MRG challenge utilizes the mean aggregated Dice Similarity Coefficient (DSC_{agg}) as the primary evaluation metric for ranking [1]. In addition, we also used the conventional Dice Similarity Coefficient (DSC) and the 95th percentile Hausdorff Distance (HD_{95}) and mean surface distance (MSD) as supplementary metrics to assess the segmentation performance for both GTVp and GTVn. For HD_{95} and MSD, cases with an empty ground truth (no tumor) and false positive predictions, or with a tumor in the ground truth but an empty prediction, lack meaningful surfaces for comparison and were thus excluded.

2.5   System Environment

The experiments were performed on a system featuring dual AMD Ryzen Threadripper 3990X processors with 64 cores (128 threads) and 256GB of RAM. Training was conducted using an NVIDIA RTX A6000 GPU with 48GB of VRAM. The software setup consisted of Python 3.12.4, PyTorch 2.4.0, CUDA 12.6 and nnU-Net 2.5.1, while distance metrics were computed using MedPy 0.5.2.

---

[1] https://github.com/Aarhus-RadOnc-AI/GradientSegHNTS.

3 Results

3.1 5-Fold Cross-Validation Results

We evaluated our approach using 5-fold cross-validation on the training data ($n = 150$). For each fold, average DSC_{agg}, HD_{95}, and MSD scores were calculated for both GTVp and GTVn, as shown in Table 1 and Table 2. The results indicated variability in GTVp segmentation accuracy, with DSC_{agg} ranging from 0.469 to 0.697, while GTVn exhibited more consistent performance, with DSC_{agg} ranging from 0.786 to 0.871. The HD_{95} scores ranged from 9.3 to 15.6 mm for GTVp and 4.2 to 7.4 mm for GTVn, while MSD scores varied between 2.5 to 5.4 mm for GTVp and 1.0 to 1.8 mm for GTVn.

Table 1. GTVp performance on 5-Fold cross-validation

Metric	Fold 0	Fold 1	Fold 2	Fold 3	Fold 4	Average
DSC_{agg}	0.682	0.493	0.469	0.636	0.697	0.595
HD_{95} [mm]	9.9	15.6	14.6	9.3	12.3	12.3
MSD [mm]	3.6	5.4	4.2	2.5	3.9	3.9

Table 2. GTVn performance on 5-Fold cross-validation

Metric	Fold 0	Fold 1	Fold 2	Fold 3	Fold 4	Average
DSC_{agg}	0.871	0.786	0.868	0.859	0.827	0.842
HD_{95} [mm]	4.2	5.4	6.4	7.4	5.9	5.9
MSD [mm]	1.0	1.4	1.8	1.6	1.5	1.5

3.2 Comparison Between with and Without Gradient Map

We further compared the performance of using T2w images with (w/) and without (w/o) gradient maps on fold-0 of the validation set ($n = 30$). For GTVp, the mean DSC improved from 0.355 to 0.538 ($p < 0.005$), and for GTVn, it increased from 0.688 to 0.825 ($p < 0.001$), based on Wilcoxon signed-rank tests. In Fig. 2, the violin plots illustrate this comparison, showing a clear shift toward higher DSC values for both GTVp and GTVn when the gradient map is applied. The distributions highlight the overall improvement in segmentation accuracy when incorporating gradient maps.

Figure 3 presents two patient cases. In patient a, the use of the gradient map successfully segmented a previously missed GTVn on the T2w-only image, improving the DSC from 0.0 to 0.83. For patient b, both GTVp and GTVn segmentations improved, with DSC scores increasing from 0.14 to 0.80 and 0.73 to 0.87, respectively. However, part of the GTVp was missed as it extended beyond the bounding box range defined by the gradient map.

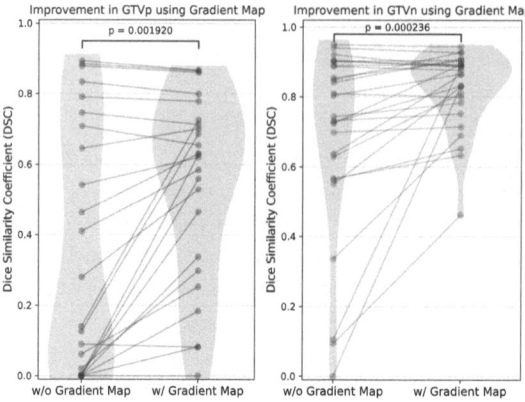

Fig. 2. Comparison of segmentation performance with (w/) and without (w/o) gradient maps on fold-0 of the validation set (n = 30).

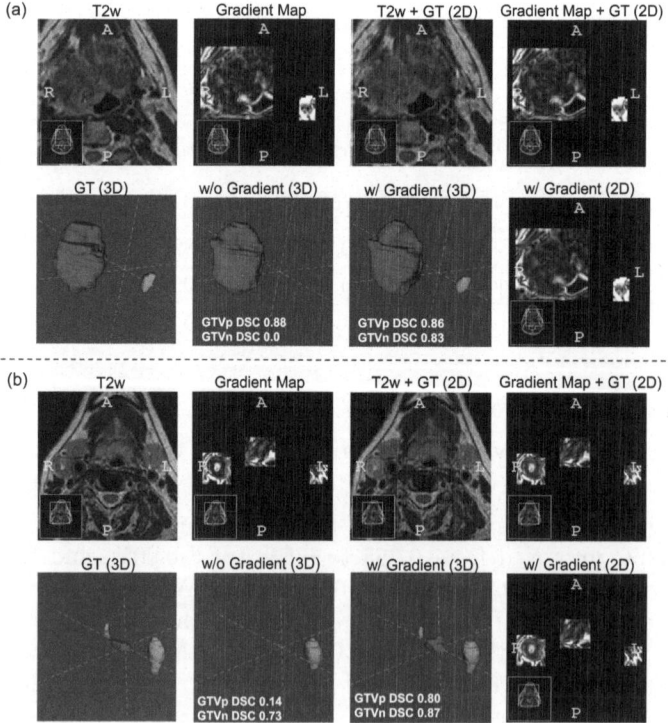

Fig. 3. Example cases demonstrating the impact of using gradient maps. *Patient a*: The gradient map enabled accurate segmentation of a previously undetected GTVn, improving the DSC from 0.0 to 0.83. *Patient b*: The use of the gradient map increased the DSC for both GTVp (from 0.14 to 0.80) and GTVn (from 0.73 to 0.87), although part of the GTVp was missed as it extended beyond the gradient map's bounding box.

3.3 Correlation Between Tumor Volume Change and DSC Scores

As shown in Fig. 4, cross-validation results (n = 150) revealed no strong correlation between tumor volume change and DSC scores for either GTVp or GTVn. Spearman's correlation coefficients were used to measure these relationships. We observed numerous instances where significant tumor shrinkage resulted in false predictions, particularly for GTVp, with DSC scores of 0.0. Moderate negative correlations were identified in specific volume ranges; for GTVp, a mid-RT volume bin between 0.8 and 3.0 cubic centimeters (cc) showed a Spearman's correlation coefficient of -0.41. Similarly, for GTVn, a volume bin ranging from 2.567 to 8.5 cc had a coefficient of -0.34.

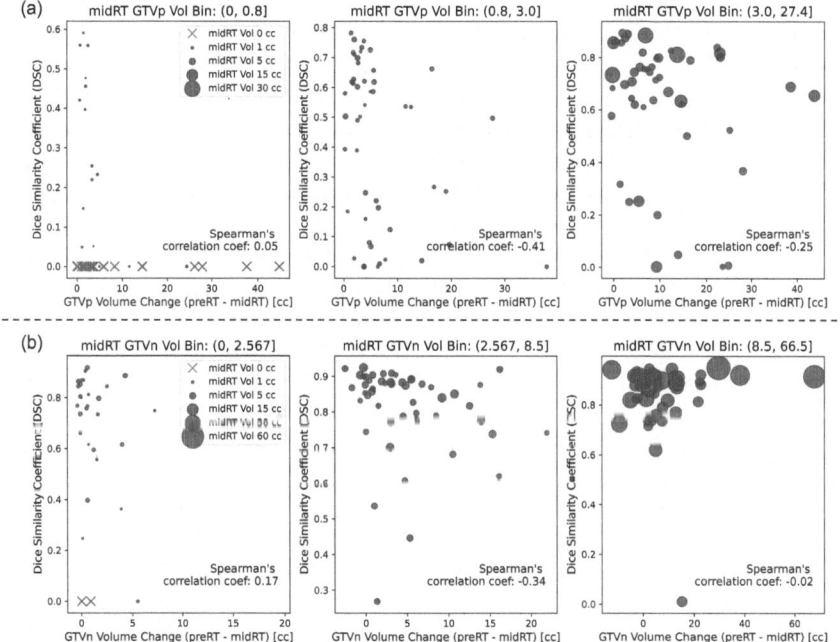

Fig. 4. Relationship between tumor volume change and DSC scores for GTVp and GTVn. (a) The first row shows three subplots representing different bins of mid-RT GTVp volumes. Each subplot's x-axis represents the change in volume (pre-RT volume minus mid-RT volume), and the y-axis represents the predicted DSC score. The size of the dots corresponds to the mid-RT GTV volume, while red X marks indicate instances where the current mid-RT volume is 0. (b) The second row displays similar subplots for GTVn, illustrating the relationship between volume change and segmentation accuracy. (Color figure online)

3.4   Final Test Score

After completing training and validation, we submitted our trained model's Docker container to the challenge platform. Predictions were generated using an ensemble of all five models trained across the 5-folds. The final DSC_{agg} scores on the test set, as evaluated by the challenge, were 0.534 for GTVp, 0.867 for GTVn, with a mean score of **0.70**.

4   Discussion

In this study, we integrated pre-RT tumor location and gradient maps to enhance tumor segmentation in mid-RT images for head and neck cancer. Results from the 5-fold cross-validation show that leveraging pre-RT information improved the delineation of both GTVp and GTVn. The average performance across folds, measured by DSC_{agg}, HD_{95}, and MSD, demonstrated consistent and reliable segmentation. The comparison of T2w without gradient map versus T2w with gradient map revealed that gradient maps provided more precise boundary localization. These findings highlight the potential of combining spatial and localized gradient intensity information to enhance segmentation accuracy in MRI-guided adaptive radiotherapy.

Deep learning-based AI approaches have garnered significant interest in enhancing RT patient treatment, with notable progress made in HNC tumor auto-segmentation [5]. Much of this advancement has been driven by MIC-CAI public data challenges, such as the HECKTOR challenges [1,2]. Numerous studies have participated and contributed diverse approaches to the challenge. Notably, Xie and Peng [35] achieved a DSC of 0.778 using a 3D SE UNet integrated with the nnUNet pipeline. Similarly, Naser et al. [16] employed a Resnet U-Net, achieving a mean Dice score of 0.69 and further validating the model on MRI, where the mean DSC reached 0.73 ± 0.12 for T2w+ T1w imaging [30].

Despite these advancements, there remains a scarcity of studies or publicly available datasets that leverage the temporal dependency between pre-RT and mid-RT images for head and neck tumor segmentation, especially using MRI data. The HNTS-MRG 2024 Challenge helps address this gap. A primary difficulty in this challenge is accurately identifying all malignant regions in the mid-RT MRI, where using pre-RT ground truth delineations as prior information can help overcome this difficulty. As demonstrated in our study, even a rough propagation of pre-RT information, such as incorporating a bounding box, led to significant improvements in segmentation accuracy. Specifically, the mean DSC for GTVp improved from 0.355 to 0.538 ($p < 0.005$), and for GTVn, it increased from 0.688 to 0.825 ($p < 0.001$). These results highlight the potential of utilizing pre-RT data; however, accurately tracking tumor evolution and precisely propagating contours from pre-RT to mid-RT images remains a challenging task, suggesting that more refined methods are needed to fully leverage this temporal information.

In addition to leveraging pre-RT information, our findings indicate that other factors may influence segmentation accuracy. Specifically, in cases where the tumor was absent after treatment (mid-RT GTVp volume = 0), the model failed to predict this absence, resulting in multiple cases of DSC = 0.0. This limitation may be exacerbated by the presence of gradient maps in the original tumor region, contributing to false-positive predictions. Furthermore, our results revealed moderate negative correlations in certain volume bins, indicating that as tumors shrink, particularly in the mid-RT moderate volume range GTVp (0.8–3 cc) and GTVn (2.567–8.5 cc), segmentation accuracy tends to decline. This decrease is likely due to the greater complexity in detecting and delineating smaller tumors during the mid-RT phase, where reduced volume leads to greater boundary ambiguity.

Deep learning-based tumor segmentation is inherently data-intensive, and the limited availability of annotated datasets presents a significant challenge. This issue is particularly pronounced in head and neck cancer for mid-RT, where tumor volumes diminish significantly during treatment. Although the dataset consists of 150 training cases, this may be insufficient for advanced techniques like deformation or image propagation networks, which require large datasets to optimize both image alignment and segmentation objectives [11]. Furthermore, variability in patient data distributions complicates training, reducing the model's ability to generalize across diverse cases.

Our approach aimed to mitigate these challenges by simplifying the problem through a data manipulation approach. Instead of propagating the full image evolution from pre-RT to mid-RT, which would require more complex modeling and a larger dataset, we focused on providing rough tumor localization through the use of bounding boxes derived from pre-RT images. By narrowing the region of interest, we reduced the complexity of the segmentation task. Moreover, we calculated the gradient of the mid-RT image within the bounding box to simplify the training process, allowing more detailed boundary information to be incorporated alongside the location data fed into the network.

A limitation of our approach is the use of a bounding box with an arbitrary expansion of 2–6 voxels around the deformable registered pre-RT image, which may not always align well with the mid-RT image. This misalignment becomes more critical if there are significant registration errors or substantial anatomical changes in the patient during the course of treatment. The arbitrary choice of expanding the bounding box by 2–6 voxels in all three dimensions carries inherent risks. If the expansion is too small, there is a risk of missing part of the actual tumor in the mid-RT image, as shown on Fig. 3(b). On the other hand, if the expansion is too large, it may reduce the precision of tumor localization, as the box would include more irrelevant tissue, thereby diluting the intended localization benefit.

In routine clinical practice, human involvement is often required to refine such critical tumor definition tasks. Clinicians may define tumor locations interactively, using tools such as scribbles, clicks, or bounding boxes, to provide more accurate localization cues. Studies have demonstrated that integrating such human-in-the-loop methods can significantly enhance detection and segmentation accuracy [21,25,34]. This approach is particularly useful when automated registration struggles with large anatomical changes or complex tumor morphologies, making precise localization difficult without manual input.

Our results demonstrate a significant difference in segmentation accuracy between GTVp and GTVn, with DSC_{agg} scores of 0.534 for GTVp and 0.867 for GTVn. The validation performance for GTVp ranged widely (0.47 to 0.69), whereas GTVn showed more consistency (0.79 to 0.87). This disparity is partly because GTVp often diminishes after treatment, making it barely visible on mid-RT images but still indicated by the gradient map. Additionally, the inherent ambiguity in defining GTVp tumor contours contributes to this difference. The challenge organizers used a STAPLE consensus from multiple annotators (3–4) as the ground truth, but due to low contrast at GTVp boundaries, there may exists considerable IOV, resulting in less agreement and higher uncertainty in GTVp segmentations [26].

To address this issue, the ground truth delineations should strictly adhere to established guidelines [10]. However, GTVp is often overestimated, leading to a high false positive rate compared to histology, highlighting flaws in delineations and the need for improved imaging techniques [27]. From a deep learning developer's view, an alternative approach could involve developing a probabilistic model based on multiple annotators' input for GTVp, rather than using a binary consensus map from STAPLE or averaging. Such probabilistic models can significantly improve uncertainty modeling and provide better confident calibration scores, leading to more reliable segmentation maps [12,22,31].

5   Conclusion

In conclusion, our novel approach effectively leverages pre-RT information to enhance mid-RT segmentation accuracy by incorporating gradient maps derived from pre-RT tumor delineations. Our results show that utilizing pre-RT delineations improves the model's ability to delineate both GTVp and GTVn for mid-RT, with significant gains in segmentation accuracy. However, the improvements for GTVp are less pronounced, likely due to reduced tumor volume, inherent ambiguity, and variability in tumor contours. We suggest that incorporating a semi-automatic, human-in-the-loop approach could help mitigate false predictions for GTVp, particularly when tumor boundaries are unclear in imaging. Future work could explore integrating probabilistic models into nnUNet frameworks to better address uncertainty and variability in segmentation. Overall, our approach demonstrates potential for enhancing segmentation accuracy in MRI-guided adaptive radiotherapy.

References

1. Andrearczyk, V., et al.: Overview of the hecktor challenge at MICCAI 2021: automatic head and neck tumor segmentation and outcome prediction in PET/CT images. In: 3D Head and Neck Tumor Segmentation in PET/CT Challenge, pp. 1–37. Springer (2021)
2. Andrearczyk, V., et al.: Overview of the hecktor challenge at MICCAI 2022: automatic head and neck tumor segmentation and outcome prediction in PET/CT. In: Andrearczyk, V., Oreiller, V., Hatt, M., Depeursinge, A. (eds.) Head and Neck Tumor Segmentation and Outcome Prediction. HECKTOR 2022, Lecture Notes in Computer Science, vol. 13626. Springer, Cham (2023). https://doi.org/10.1007/978-3-031-27420-6_1
3. Benitez, C.M., Chuong, M.D., Künzel, L.A., Thorwarth, D.: MRI-guided adaptive radiation therapy. In: Seminars in Radiation Oncology, vol. 34, pp. 84–91. Elsevier (2024)
4. Henderson, E.G., Osorio, E.M.V., van Herk, M., Brouwer, C.L., Steenbakkers, R.J., Green, A.F.: Accurate segmentation of head and neck radiotherapy CT scans with 3D CNNs: consistency is key. Phys. Med. Biol. **68**(8), 085003 (2023)
5. Hindocha, S., et al.: Artificial intelligence for radiotherapy auto-contouring: current use, perceptions of and barriers to implementation. Clin. Oncol. **35**(4), 219–226 (2023)
6. Hurkmans, C., et al.: A joint ESTRO and AAPM guideline for development, clinical validation and reporting of artificial intelligence models in radiation therapy. Radiother. Oncol. **197**, 110345 (2024)
7. Isensee, F., Jaeger, P.F., Kohl, S.A., Petersen, J., Maier-Hein, K.H.: nnU-Net: a self-configuring method for deep learning-based biomedical image segmentation. Nat. Methods **18**(2), 203–211 (2021)
8. Isensee, F., et al.: nnU-Net revisited: a call for rigorous validation in 3D medical image segmentation. arXiv preprint arXiv:2404.09556 (2024)
9. Jensen, K., et al.: Imaging for target delineation in head and neck cancer radiotherapy. In: Seminars in Nuclear Medicine, vol. 51, pp. 59–67. Elsevier (2021)
10. Jensen, K., et al.: The danish head and neck cancer group (DAHANCA) 2020 radiotherapy guidelines. Radiother. Oncol. **151**, 149–151 (2020)
11. Kawula, M., et al.: Prior knowledge based deep learning auto-segmentation in magnetic resonance imaging-guided radiotherapy of prostate cancer. Phys. Imaging Radiat. Oncol. **28**, 100498 (2023)
12. Kohl, S.A., et al.: A hierarchical probabilistic u-net for modeling multi-scale ambiguities. arXiv preprint arXiv:1905.13077 (2019)
13. McDonald, B.A., et al.: Investigation of autosegmentation techniques on T2-weighted MRI for off-line dose reconstruction in MR-Linac workflow for head and neck cancers. Med. Phys. **51**(1), 278–291 (2024)
14. McDonald, B.A., Dal Bello, R., Fuller, C.D., Balermpas, P.: The use of MR-guided radiation therapy for head and neck cancer and recommended reporting guidance. In: Seminars in Radiation Oncology, vol. 34, pp. 69–83. Elsevier (2024)
15. Mohamed, A.S., et al.: Prospective in silico study of the feasibility and dosimetric advantages of MRI-guided dose adaptation for human papillomavirus positive oropharyngeal cancer patients compared with standard IMRT. Clin. Transl. Radiat. Oncol. **11**, 11–18 (2018)

16. Naser, M.A., van Dijk, L.V., He, R., Wahid, K.A., Fuller, C.D.: Tumor segmentation in patients with head and neck cancers using deep learning based-on multi-modality PET/CT images. In: 3D Head and Neck Tumor Segmentation in PET/CT Challenge, pp. 85–98. Springer (2020)

17. Nielsen, C.P., et al.: Interobserver variation in organs at risk contouring in head and neck cancer according to the dahanca guidelines. Radiother. Oncol. **197**, 110337 (2024)

18. Outeiral, R.R., Bos, P., Al-Mamgani, A., Jasperse, B., Simões, R., van der Heide, U.A.: Oropharyngeal primary tumor segmentation for radiotherapy planning on magnetic resonance imaging using deep learning. Phys. Imaging Radiat. Oncol. **19**, 39–44 (2021)

19. Rasmussen, M.E., Nijkamp, J.A., Eriksen, J.G., Korreman, S.S.: A simple single-cycle interactive strategy to improve deep learning-based segmentation of organs-at-risk in head-and-neck cancer. Phys. Imaging Radiat. Oncol. **26**, 100426 (2023)

20. Ren, J., Nijkamp, J., Rasmussen, M., Eriksen, J., Korreman, S.: MO-0799 single-click user input reduces false detection in deep learning head and neck tumor segmentation. Radiother. Oncol. **182**, S669–S671 (2023)

21. Ren, J., Rasmussen, M., Nijkamp, J., Eriksen, J.G., Korreman, S.: Segment anything model for head and neck tumor segmentation with CT, PET and MRI multi-modality images. arXiv preprint arXiv:2402.17454 (2024)

22. Ren, J., et al.: Enhancing the reliability of deep learning-based head and neck tumour segmentation using uncertainty estimation with multi-modal images. Phys. Med. Biolo. **69**(16), 165018 (2024)

23. Riegel, A.C., et al.: Variability of gross tumor volume delineation in head-and-neck cancer using CT and PET/CT fusion. Int. J. Radiat. Oncolo. Biol. Phys. **65**(3), 726–732 (2006)

24. Rodriguez, J.D., Selleck, A.M., Razek, A.A.K.A., Huang, B.Y.: Update on MR imaging of soft tissue tumors of head and neck. Magn. Reson. Imaging Clin. **30**(1), 151–198 (2022)

25. Saukkoriipi, M., et al.: Interactive 3D segmentation for primary gross tumor volume in oropharyngeal cancer. arXiv preprint arXiv:2409.06605 (2024)

26. Segedin, B., Petric, P.: Uncertainties in target volume delineation in radiotherapy-are they relevant and what can we do about them? Radiol. Oncol. **50**(3), 254–262 (2016)

27. Smits, H.J., et al.: Improved delineation with diffusion weighted imaging for laryngeal and hypopharyngeal tumors validated with pathology. Radiother. Oncol. **194**, 110182 (2024)

28. van der Veen, J., Gulyban, A., Nuyts, S.: Interobserver variability in delineation of target volumes in head and neck cancer. Radiother. Oncol. **137**, 9–15 (2019)

29. Wahid, K., Dede, C., Naser, M., Fuller, C.: Training dataset for HNTSMRG 2024 challenge (2024). https://doi.org/10.5281/zenodo.11199559. [Data set]

30. Wahid, K.A., et al.: Evaluation of deep learning-based multiparametric MRI oropharyngeal primary tumor auto-segmentation and investigation of input channel effects: results from a prospective imaging registry. Clin. Transl. Radiat. Oncol. **32**, 6–14 (2022)

31. Wahid, K.A., et al.: Artificial intelligence uncertainty quantification in radiotherapy applications - a scoping review. Radiother. Oncol. **201**, 110542 (2024). https://doi.org/10.1016/j.radonc.2024.110542. https://www.sciencedirect.com/science/article/pii/S0167814024035205

32. Wang, X., Chang, Y., Pei, X., Xu, X.G.: A prior-information-based automatic segmentation method for the clinical target volume in adaptive radiotherapy of cervical cancer. J. Appl. Clin. Med. Phys. **25**(5), e14350 (2024)
33. Wang, Y., et al.: Comparison of deep learning networks for fully automated head and neck tumor delineation on multi-centric PET/CT images. Radiat. Oncol. **19**(1), 3 (2024)
34. Wei, Z., Ren, J., Korreman, S.S., Nijkamp, J.: Towards interactive deep-learning for tumour segmentation in head and neck cancer radiotherapy. Phys. Imaging Radiat. Oncol. **25**, 100408 (2023)
35. Xie, J., Peng, Y.: The head and neck tumor segmentation based on 3D U-net. In: 3D Head and Neck Tumor Segmentation in PET/CT Challenge, pp. 92–98. Springer (2021)

Enhanced nnU-Net Architectures for Automated MRI Segmentation of Head and Neck Tumors in Adaptive Radiation Therapy

Jessica Kächele[1,2,3]([✉]), Maximilian Zenk[1,2], Maximilian Rokuss[1,6], Constantin Ulrich[1,2,4], Tassilo Wald[1,5,6], and Klaus H. Maier-Hein[1,5,7]

[1] German Cancer Research Center (DKFZ), Heidelberg, Germany
jessica.kaechele@dkfz-heidelberg.de
[2] Medical Faculty Heidelberg, University of Heidelberg, Heidelberg, Germany
[3] German Cancer Consortium (DKTK), DKFZ, core center Heidelberg, Heidelberg, Germany
[4] National Center for Tumor Diseases (NCT), NCT Heidelberg, A partnership between DKFZ and University Medical Center Heidelberg, Heidelberg, Germany
[5] Helmholtz Imaging, DKFZ, Heidelberg, Germany
[6] Faculty of Mathematics and Computer Science, Heidelberg University, Heidelberg, Germany
[7] Pattern Analysis and Learning Group, Department of Radiation Oncology, Heidelberg University Hospital, Heidelberg, Germany

Abstract. The increasing utilization of MRI in radiation therapy planning for head and neck cancer (HNC) highlights the need for precise tumor segmentation to enhance treatment efficacy and reduce side effects. This work presents segmentation models developed for the HNTS-MRG 2024 challenge by the team mic-dkfz, focusing on automated segmentation of HNC tumors from MRI images at two radiotherapy (RT) stages: before (pre-RT) and 2–4 weeks into RT (mid-RT). For Task 1 (pre-RT segmentation), we built upon the nnU-Net framework, enhancing it with the larger Residual Encoder architecture. We incorporated extensive data augmentation and applied transfer learning by pretraining the model on a diverse set of public 3D medical imaging datasets. For Task 2 (mid-RT segmentation), we adopted a longitudinal approach by integrating registered pre-RT images and their segmentations as additional inputs into the nnU-Net framework. On the test set, our models achieved mean aggregated Dice Similarity Coefficient (aggDSC) scores of 81.2 for Task 1 and 72.7 for Task 2. Especially the primary tumor (GTVp) segmentation is challenging and presents potential for further optimization. These results demonstrate the effectiveness of combining advanced architectures, transfer learning, and longitudinal data integration for automated tumor segmentation in MRI-guided adaptive radiation therapy.

J. Kächele and M. Zenk—These authors contributed equally to this work.
T. Wald and K. H. Maier-Hein—These authors share last authorship.

K. A. Wahid et al. (Eds.): HNTS-MRG 2024, LNCS 15273, pp. 50–64, 2025.
https://doi.org/10.1007/978-3-031-83274-1_3

Keywords: Head and Neck Cancer · MRI-guided Radiation Therapy · Tumor Segmentation · nnU-Net · Transfer Learning · Longitudinal Data Integration

1   Introduction

Radiation therapy (RT) is fundamental in cancer treatment, with head and neck cancer (HNC) being one of the cancers that benefit most from this treatment. Recently, MRI-guided RT planning has gained traction due to MRI's superior soft tissue contrast compared to traditional CT-based approaches [1].

Despite the potential of MRI-guided adaptive RT, the extensive volume of MRI data presents significant challenges. Manual segmentation of tumors by physicians, the current clinical standard, is often impractical due to time constraints and the complex nature of HNC tumors [2,3]. This has spurred interest in artificial intelligence (AI) approaches to automate and improve the accuracy of tumor segmentation [4]. Recent advancements in AI-driven segmentation have been propelled by public data challenges like the HECKTOR [5] and SegRap [6] Challenges, which have fostered collaborative innovations in the field. However, there remains a lack of large, publicly available AI-ready datasets specifically tailored for adaptive RT in HNC, limiting the broader adoption and clinical translation of these technologies.

To address this gap, the annual Medical Image Computing and Computer Assisted Intervention Society (MICCAI) Head and Neck Tumor Segmentation for MR-Guided Applications (HNTS-MRG) 2024 challenge has been established, focusing on the segmentation of HNC tumors for MRI-guided adaptive RT applications. A unique aspect of HNTS-MRG 2024 is its emphasis on evaluating whether incorporating prior timepoint data into auto-segmentation algorithms enhances performance for RT applications.

This work presents the results of our segmentation models applied to the HNTS-MRG 2024 MRI training and testing datasets. For pre-radiotherapy (pre-RT) images, we utilized ResEnc models within the nnU-Net framework [7], enhanced through transfer learning and ensembling. Additionally, we developed a model for mid-radiotherapy (mid-RT) images that leverages the pre-RT images and their segmentations.

2   Methods

2.1   Imaging Data

The dataset employed in this study, made available for the HNTS-MRG 2024 Challenge, includes pre-RT and mid-RT MRI scans from 200 patients with HNC. This collection is divided into 150 cases for training and 50 cases for testing, as designated by the challenge organizers. Each training case is accompanied by consensus ground truth segmentations of the GTVp and GTVn regions, which were generated by multiple clinical experts. Additionally, the pre-RT MRI images and their corresponding segmentations are provided in both their original form and after being registered to the mid-RT images.

For the segmentation of pre-RT images, only the pre-RT MRI scans were utilized during the training process. In contrast, the segmentation of mid-RT images incorporated both pre-RT and mid-RT MRI scans along with the pre-RT segmentations, thereby enhancing model performance.

2.2 Task 1

Our approach for Task 1 builds upon the well-established nnU-Net framework [8], leveraging the enhanced capabilities of the larger ResEnc architecture preset as described in [7].

nnU-Net Configuration. For our experiments, we employed the 3d_fullres configuration of the nnU-Net framework. The input images were resampled to the median spacing of [1.12, 0.5, 0.5] mm, and intensities were normalized using Z-score normalization. The network was trained over 1000 epochs with a batch size of 2, utilizing an input patch size of 64x320x256. The default nnU-Net settings were applied for the learning rate, optimizer, and loss function.

Training Variants. Upon the baseline default nnU-Net and ResEnc models, we trained the following variants:

1. **extensive augmentations (Aug++) model:** We further enhanced the data augmentation strategy by incorporating additional and stronger augmentations beyond those provided by nnU-Net. These augmentations were designed to expose the model to a wider range of data variations, thereby improving its robustness and generalization capabilities. Notably, local intensity transformations such as patch blacking, sharpening, Gaussian intensity gradients, and local gamma adjustments significantly expanded the diversity of patches. Due to the spatial constraints in this manuscript, we refer to [9] for the full details and augmentation scheme. Moreover, an implementation of said scheme is included in the nnU-Net v2 repository as nnUNetTrainerDA5 [8]. The model was implemented using the ResEnc M U-Net architecture [7].
2. **pretrained model:** To incorporate prior knowledge and enhance our model's capabilities, we pre-trained it on an extensive and diverse collection of 3D medical images, combining various public datasets in a manner inspired by MultiTalent [10]. During this pretraining phase, we utilized separate segmentation heads for each dataset to accommodate their specific characteristics. The model was trained for 4000 epochs with a patch size of 192x192x192 and a batch size of 24 distributed to 8xA100 GPUs. A batch size of 24 was selected as it represents the maximum capacity supported by the available GPUs. Due to the dataset's substantial size, 4000 epochs were necessary to ensure the model's convergence, in line with findings from [11]. All images were resampled to an isotropic resolution of 1x1x1 mm^3 per voxel and normalized using Z-score normalization.

To ensure balanced training across datasets of different sizes, we sampled datasets inversely proportional to the square root of the number of images in each dataset. This approach prevented larger datasets from dominating the training process. The model was implemented using the nnU-Net framework [8] and the ResEnc L U-Net architecture [7].

The pretrained model weights are made publicly available[1]. A comprehensive overview of the datasets used for pretraining can be found in Table 4. The respective model was not explicitly trained and optimized for this challenge. After completing the pretraining phase, we fine-tuned the model on the HNTSMRG 2024 dataset. The initial preprocessing configuration was maintained, featuring a patch size of 64x320x256 and a spacing of [1.12, 0.5, 0.5] mm. We introduced an additional 50 epochs to warm up the network. During this warm-up period, the learning rate was linearly increased until it reached a learning rate of 1×10^{-3}, allowing for a smoother transition and better convergence during fine-tuning. After the warm-up, the default polynomial learning rate schedule with an initial learning rate of 1×10^{-3} was used.

Ensembling. We employed an ensembling strategy of multiple models to enhance the performance and robustness of our segmentation model. By averaging the predicted probabilities for each voxel across the best-performing models, we combined their strengths. This probabilistic averaging mitigates individual model biases and uncertainties, leading to more accurate and reliable segmentation results.

Postprocessing. Postprocessing steps were applied to refine the segmentation results. We utilized connected component analysis to remove low-volume predictions. Specifically, any connected components corresponding to GTVn with a volume less than $675 \, mm^3$ and GTVp with a volume less than $750 \, mm^3$ were removed from the final segmentation.

Test-Time Augmentation. Due to increased inference time associated with ensembling models, we considered reducing test-time augmentation (TTA). To assess whether this reduction would lead to worse performance compared to using a single model with TTA, we turned off TTA when evaluating the best-performing ensemble on the validation folds. This allowed us to determine if the ensembling without TTA could still perform competitively within the time constraints.

Alternative Approaches Explored. We experimented with several alternative approaches that did not yield improved results. Training the model on both pre-RT and mid-RT images resulted in a decline in performance, possibly

---

[1] https://zenodo.org/records/13753413.

because the mid-RT images introduced information that confused or misled the model during training.

We also attempted to align the preprocessing steps with those used during pretraining to ensure consistency between the datasets. This involved using a patch size of 192x192x192, and resampling the images to a spacing of [1, 1, 1] mm. While these modifications led to a higher DSC, the aggDSC worsened. This suggests that the model's performance improved on images with low tumor volumes, which do not have a substantial influence on the aggDSC.

Additionally, we experimented with batch dice loss, calculating the aggDSC for each batch to closely mimic the evaluation metric. However, this approach did not enhance performance.

Final Test Set Submission. Based on the results from 5-fold cross-validation, we chose to ensemble the Aug++ and the pretrained model for testing on the test set. During inference, we combined the models obtained from each of the five training folds and both models by averaging their softmax probabilities for each voxel. To comply with the 20-minute time constraint per case, we reduced TTA to mirroring along only one axis instead of all three axes typically used by nnU-Net.

2.3   Task 2

Method Overview. During the development phase, we evaluated a range of segmentation models. The baseline model is nnU-Net, which operates on a single input image. For training, we explored two approaches: using either only mid-RT scans or sampling minibatches from the combined set of mid-RT and pre-RT scans, while using the corresponding mask (mid-RT or pre-RT) as the target. In the latter case, the training and validation sets were split to ensure that scans from the same patient were not divided across sets. We refer to these models as **nnU-Net (only mid)** and **nnU-Net (pre & mid)** in the task-2 results.

Our main approach also builds on the nnU-Net framework but integrates longitudinal information by using both registered pre- and mid-RT scans as input, following the methodology of LongiSeg [12]. In short, this method predicts the segmentation mask of a *current* scan by using additional imaging data from a *prior* scan, which is acquired at an earlier time and provides context for improving the prediction of the current time point. We implemented the following "LongiSeg" variants:

1. **Base model:** Pre-RT and mid-RT scans are concatenated along the channel dimension and input into the network. The assignment of which scan serves as the current scan and which as the prior is randomized and the concatenation order is (current, prior). The network then outputs the segmentation of the current scan. This approach enhances the diversity of the training data but does not preserve the chronological order. The loss is computed with the segmentation of the current scan.

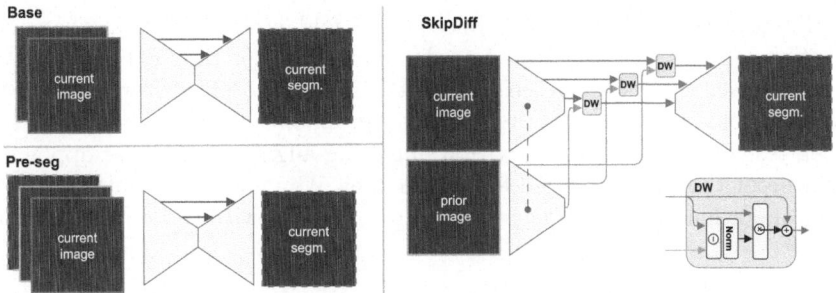

Fig. 1. Overview of the three longitudinal segmentation models developed for task 2. Red boxes indicate the data for the current time point, and green boxes for a prior time. Images have solid outlines and segmentations have dashed lines (the third spatial dimension is not shown for clarity). The *Base* model stacks the current image and prior images in the channel dimension, and the *Pre-seg* model also stacks the prior segmentation. *SkipDiff* uses a shared encoder for prior and current images; the features are merged with a difference weighting block (DW). Adapted from [12]. (Color figure online)

2. **SkipDiff:** This variant also encodes prior and current images separately using encoders with shared weights. Feature maps at each U-Net stage are merged using a difference weighting block and fed via skip connections into a single decoder, which generates the segmentation of the current image. Difference weighting multiply the current feature maps with the normalized difference between current and prior features, to capture the changes between the pre- and mid RT scans, which brought performance gains in [12].
3. **Pre-seg:** In addition to the image inputs, a one-hot encoding of the registered prior scan segmentation mask is added to the network input, resulting in the order (current, prior, prior mask). Two variations of this method were tested: In the first variant (**pre-seg**), current and prior scans were assigned randomly to pre-RT and mid-RT, as for the base model. In the second variant (**pre-seg-c**), training followed the chronological order, so that the mid-RT scan is always used as current scan and pre-RT as prior scan.

nnU-Net Configuration. All models were implemented in the nnU-Net framework and used the 3d_fullres configuration. The input images were resampled to a spacing of [1.12, 0.5, 0.5] mm, and image intensities were normalized using Z-score normalization. The network was trained with a batch size of 2 for 1000 epochs (250k minibatch updates), using an input patch size of 48x224x192. The loss function was a sum of batch DSC and cross-entropy loss. The default nnU-Net settings were applied for the data augmentations, learning rate, and optimizer. The U-Net architecture was adapted dynamically to the dataset by nnU-Net and had six stages. We also tried the residual encoder U-Net in preliminary experiments, but it did not improve results.

Final Test Set Submission. Based on 5-fold cross-validation, the model LongiSeg: **pre-seg-c** was selected as the submission for the final test set evaluation. During inference, we ensembled the models obtained from each of the 5 training folds by averaging their softmax probabilities. Additionally, for each model, test-time augmentation with 3 mirroring augmentations was applied, so that 4 predictions are averaged for each model. We reduced the number of mirroring operations from the default of 7 to 3, to stay within the inference time limits.

3   Results

The results presented originate from the evaluation on the validation set of the 5-fold cross-validation. For Task 2, only the mid-RT scans were taken into account, as they are the targets during testing as well. As metrics, the mean DSC and aggDSC were used for choosing the final model and are reported. For both the average per class was calculated and then aggregated across the classes. 90% confidence intervals for aggDSC were estimated with the bootstrapping method, using 1000 bootstrap samples.

3.1   Task 1

As shown in Table 1, the ResEnc architectures outperformed the baseline nnU-Net, with the ResEnc M architecture slightly surpassing the ResEnc L architecture. Through more data augmentation, the DSC improved marginally (from 0.621 to 0.624), although the aggregated aggDSC decreased slightly (from 0.821 to 0.82).

Pretraining led to a boost in performance, pushing the aggDSC from 0.814 to 0.823. Postprocessing steps contributed to a modest enhancement in performance, increasing the aggDSC to 0.824. The best results, with an aggDSC of 0.828, were achieved by ensembling the Aug++ model in conjunction with the pretrained model. Figure 1 presents representative samples from the validation folds, illustrating the overlay of tumor mask outlines on MRI images. It was observed that ensembling without TTA still performed better than a single model with TTA. This finding justifies the use of an ensembled model despite the computational constraints.

The results of the additional experiments discussed in Sect. 2.2 are presented in Table 5 in the appendix.

Test Set Results. On the official test set, our model achieved a mean aggDSC of 0.812, with individual aggDSC values of 0.869 for GTVn and 0.756 for GTVp. Compared to our cross-validation results, the GTVn score on the test set shows a slight improvement, whereas the GTVp score is considerably lower.

3.2 Task 2

Initial Model Exploration. We initially focused on fast experimentation, testing a variety of models on a single training fold. The results from these preliminary experiments are presented in Table 2. The LongiSeg pre-seg model, which leveraged both the registered pre-RT scan and its corresponding mask, outperformed all others in this phase. While the LongiSeg base model achieved slightly better performance than the best nnU-Net model, the difference is likely not statistically significant. Notably, there was a substantial performance gap between the nnU-Net trained solely on mid-RT scans and the one trained on both pre-RT and mid-RT scans, highlighting the advantage of a larger training dataset. Based on these findings, we decided to concentrate on the LongiSeg models, particularly the pre-seg variant and its potential optimizations.

Cross-Validation Results. The results from the 5-fold cross-validation are shown in Table 3. In the appendix, Fig. 2 reports DSC values for individual cases. We also trained the pre-seg-c variant, which maintained the chronological order of the scans during training. This variant improved the aggDSC for the GTVp region compared to the pre-seg model, although the performance on the GTVn region saw a slight decline. Both models significantly outperformed the

Table 1. Task 1 results for 5-fold cross-validation (150 validation cases) for different nnU-Net models trained on pre-RT scans, computed using the official evaluation code. aggDSC is used in the challenge ranking and a confidence interval is estimated with 1000 bootstrap samples. DSC is given as an auxiliary metric, computed on each case individually including cases with empty ground truth mask, too.

Setting	aggDSC			DSC
	GTVp	GTVn	mean	mean
Baseline				
nnU-Net	78.0 [75.6,80.8]	84.9 [83.0,87.0]	81.4 [79.7,83.3]	61.5
Model Architecture: ResEnc M				
nnU-Net ResEnc M	78.6 [76.2,81.4]	85.6 [83.9,87.5]	82.1 [80.5,84.0]	62.1
+ extensive augmentations (Aug++)	78.1 [75.8,80.9]	85.9 [84.4,87.7]	82.0 [80.5,83.8]	62.4
Model Architecture: ResEnc L				
nnU-Net ResEnc L	77.3 [74.9,80.2]	85.4 [83.6,87.5]	81.4 [79.7,83.3]	61.4
+ Pretraining	78.7 [76.4,81.5]	85.9 [84.3,87.8]	82.3 [80.8,84.0]	62.6
+ Postprocessing	78.7 [76.3,81.3]	86.0 [84.3,87.8]	82.4 [80.8,84.0]	62.3
Ensembling				
Aug++ model + pretrained model	**79.0** [76.8,81.4]	**86.5** [85.0,88.4]	**82.8** [81.3,84.3]	**62.7**
Without TTA				
Aug++ model + pretrained model	78.8 [76.6,81.8]	86.1 [85.5,88.0]	82.5 [80.9,84.3]	**62.7**

Table 2. Metric results on a single training fold (fold 0, 30 validation cases) for a range of compared segmentation models, computed using the official evaluation code. aggDSC is used in the challenge ranking and a confidence interval is estimated with 1000 bootstrap samples. DSC is given as an auxiliary metric, computed on each case individually including cases with empty ground truth mask, too.

Model	aggDSC			DSC
	GTVp	GTVn	mean	mean
nnU-Net: only mid	51.5 [36.0, 67.8]	81.1 [75.7, 84.0]	66.3 [58.4, 75.5]	42.49
nnU-Net pre & mid	57.0 [43.0, 71.0]	82.8 [77.8, 85.8]	69.9 [62.2, 76.5]	47.10
LongiSeg: base	58.1 [45.1, 70.9]	82.8 [77.6, 86.2]	70.4 [63.4, 77.1]	47.46
LongiSeg: skipdiff	53.3 [44.8, 61.3]	83.4 [78.8, 86.9]	68.4 [63.5, 73.3]	48.56
LongiSeg: pre-seg	**61.5** [53.6, 71.6]	**88.0** [85.8, 89.5]	**74.8** [70.6, 79.9]	**56.25**

LongiSeg base model. While we considered an ensemble of the two best models for our final submission, its aggDSC scores were lower in cross-validation. Therefore, we selected the LongiSeg pre-seg-c model as our final choice.

Table 3. Metric results for 5-fold cross-validation (150 validation cases) for different LongiSeg models, computed using the official evaluation code. aggDSC is used in the challenge ranking and a confidence interval is estimated with 1000 bootstrap samples. DSC is given as an auxiliary metric, computed on each case individually including cases with empty ground truth mask, too.

Model	aggDSC			DSC
	GTVp	GTVn	mean	mean
LongiSeg: base	54.9 [49.0, 60.6]	82.5 [80.3, 84.7]	68.7 [65.7, 71.9]	43.87
LongiSeg: pre-seg	61.1 [56.8, 65.8]	87.5 [86.2, 88.8]	74.3 [72.3, 76.9]	54.65
LongiSeg: pre-seg-c	**64.1** [59.8, 68.2]	87.3 [85.9, 88.5]	**75.7** [73.5, 78.1]	**55.92**
Ensemble pre-seg; -c	63.3 [59.1, 67.9]	**87.7** [86.4, 89.0]	75.5 [73.5, 78.0]	–

Test Set Results. On the official test set for task 2, our model achieved a mean aggDSC score of 0.7268, with individual aggDSC values of 0.8747 for GTVn and 0.5789 for GTVp. Compared to our cross-validation results, the GTVn score is as expected, while the GTVp score is considerably lower.

4   Conclusion

In this work, we developed and evaluated segmentation models for the segmentation of HNC tumors in MRI-guided adaptive radiation therapy applications, as part of the HNTS-MRG 2024 challenge. Our approaches, grounded in the nnU-Net framework and enhanced through techniques such as transfer learning, ensembling, and longitudinal data integration, demonstrated significant improvements over baseline models.

Our models achieved mean aggDSC scores of 0.812 for Task 1 and 0.727 for Task 2 on the test set, performing slightly worse than on the validation folds. While the performance for the GTVn region remained robust, the GTVp segmentation exhibited a notable drop compared to cross-validation results. Possibly there are more difficult cases in the test set or outliers that affect the test set scores. In general, the GTVp is significantly harder to segment, as shown by the higher inter-rater variability [13], which may also contribute to statistical noise in test set results. A closer analysis of failure cases on the validation and test sets could inspire application-specific refinements to enhance GTVp segmentation.

For future work, leveraging separate segmentations from individual raters would be beneficial to account for inter-rater variability, which is particularly high for GTVp. Furthermore, employing a probabilistic U-Net [14] could help manage segmentation uncertainties, enabling the model to capture a broader range of anatomical variations that appear to be present in GTVp.

Moreover, our exploration into reducing inference time indicates that optimizing preprocessing pipelines-for instance, by implementing data-specific preprocessing-can facilitate the deployment of larger models and enable more extensive ensembling without compromising computational efficiency.

A Appendix

Table 4. Overview over all datasets used for the supervised pretraining

Name	Images	Modality	Target	Link
Decathlon Task 3 [15,16]	131	CT	Liver, L. Tumor	http://medicaldecathlon.com/
Decathlon Task 6 [15,16]	63	CT	Lung Lesion	http://medicaldecathlon.com/
Decathlon Task 7 [15,16]	281	CT	Pancreas, P. Tumor	http://medicaldecathlon.com/
Decathlon Task 8 [15,16]	303	CT	Hepatic Vessel, H. Tumor	http://medicaldecathlon.com/
Decathlon Task 9 [15,16]	41	CT	Spleen	http://medicaldecathlon.com/
Decathlon Task 10 [15,16]	126	CT	Colon Tumor	http://medicaldecathlon.com/
BTCV [17]	30	CT	13 abdominal organs	https://www.synapse.org/Synapse:syn3193805/wiki/89480
LIDC [18]	1010	CT	Lung lesion	https://www.cancerimagingarchive.net/collection/lidc-idri/
BTCV 2 [19]	63	CT	9 abdominal organs	https://zenodo.org/records/1169361#.YiDLFnXMJFE
StructSeg Task1 [20]	50	CT	22 OAR Head & neck	https://structseg2019.grand-challenge.org
StructSeg Task2 [20]	50	CT	Nasopharynx cancer	https://structseg2019.grand-challenge.org/Home/
StructSeg Task3 [20]	50	CT	6 OAR Lung	https://structseg2019.grand-challenge.org/Home/
StructSeg Task4 [20]	50	CT	Lung Cancer	https://structseg2019.grand-challenge.org/Home/
SegTHOR [21]	40	CT	Heart, aorta, trachea, esophagus	https://competitions.codalab.org/competitions/21145
NIH-Pan [22–24]	82	CT	Pancreas	https://wiki.cancerimagingarchive.net/display/Public/Pancreas-CT
VerSe2020 [25–27]	113	CT	28 Vertebrae	https://github.com/anjany/verse

(continued)

Table 4. (*continued*)

Name	Images	Modality	Target	Link
RibSeg [28]	370	CT	Rips	https://github.com/M3DV/RibSeg?tab=readme-ov-file
KiTs23 [29]	489	CT	Kidneys, k. Tumors, Cysts	https://kits-challenge.org/kits23/
AbdomenAtlas1.0 [30,31]	5195	CT	8 abdominal organs	https://github.com/MrGiovanni/AbdomenAtlas?tab=readme-ov-file
TotalSegmentatorV2 [11]	1180	CT	117 classes of whole body	https://github.com/wasserth/TotalSegmentator
FLARE [32]	50	CT	13 abdominal organs	https://flare22.grand-challenge.org/
SegRap [33]	120	CT	45 OARs (Head&Neck)	https://segrap2023.grand-challenge.org/
SegA [34–36]	56	CT	Aorta	https://multicenteraorta.grand-challenge.org/data/
WORD [37,38]	120	CT	16 abdominal organs	https://github.com/HiLab-git/WORD
AbdomenCT1K [39]	996	CT	Liver, Kidney, Spleen, pancreas	https://github.com/JunMa11/AbdomenCT-1K
DAP-ATLAS [40]	533	CT	142 classes of whole body	https://github.com/alexanderjaus/AtlasDataset
CTORG [41]	140	CT	Lung, brain, bones, liver, kidneys and bladder	https://www.cancerimagingarchive.net/collection/ct-org/
HanSeg [42]	42	CT	OAR (Head&Neck)	https://han-seg2023.grand-challenge.org/
Decathlon Task 2 [15,16]	20	MRI	Heart	http://medicaldecathlon.com/
Decathlon Task 4 [15,16]	208	MRI	Hippocampus	http://medicaldecathlon.com/
Decathlon Task 5 [15,16]	32	MRI	Prostate	http://medicaldecathlon.com/
ISLES2015 [43]	28	MRI	Stroke Lesion	http://www.isles-challenge.org/ISLES2015/
Promise12 [44]	50	MRI	Prostate	https://zenodo.org/records/8026660
ACDC [45]	200	MRI	RV cavity, myocardium, LV cavity	https://www.creatis.insa-lyon.fr/Challenge/acdc/databases.html
ISBILesion2015 [46]	42	MRI	MS Lesion	https://iacl.ece.jhu.edu/index.php/MSChallenge
CHAOS [47]	60	MRI	Liver, Kidney (L&R), Spleen	https://zenodo.org/records/3431873
M&Ms [48,49]	300	MRI	L. ventricle, R. ventricle, L. ventri. myocardium	https://www.ub.edu/mnms/
ProstateX [50]	140	MRI	Prostate lesion	https://www.aapm.org/GrandChallenge/PROSTATEx-2/
MSLesion [51]	48	MRI	MS Lesion	https://data.mendeley.com/datasets/8bctsm8jz7/1
BrainMetShare [52]	84	MRI	Brain Metastases	https://aimi.stanford.edu/brainmetshare
CrossModa22 [53]	168	MRI	Vestibular schwannoma, cochlea	https://crossmoda2022.grand-challenge.org/
Atlas22 [54]	524	MRI	Stroke lesion	https://atlas.grand-challenge.org/
BraTs23 [55–58]	1251	MRI	Glioblastoma	https://www.synapse.org/Synapse:syn51156910/wiki/621282
AutoPet2 [59]	1014	PET, CT	Lesions	https://autopet-ii.grand-challenge.org/
Hecktor2022 [60]	524	PET, CT	Nodal Gross Tumor Volumes (Head&Neck)	https://hecktor.grand-challenge.org/
AMOS [61]	360	CT, MRI	15 abdominal organs	https://amos22.grand-challenge.org/
TopCow [62]	200	CT, MRI	Vessel components of CoW	https://topcow23.grand-challenge.org/

Table 5. Task 1 5-fold cross-validation results for different nnU-Net configurations, that did not yield improved results. Settings include training on pre & mid-RT images, applying pretraining preprocessing, and incorporating batch dice.

Setting	aggDSC			DSC
	GTVp	GTVn	mean	mean
pre & mid-RT	75.6 [73.0,78.5]	84.3 [82.4,86.4]	79.9 [78.9,81.8]	61.0
pretraining preprocessing	78.6 [76.5,81.1]	85.6 [83.8,87.9]	82.1 [80.5,83.9]	63.6
batch dice	78.7 [76.3,81.5]	85.5 [83.7,87.6]	82.1 [80.4,84.0]	61.1

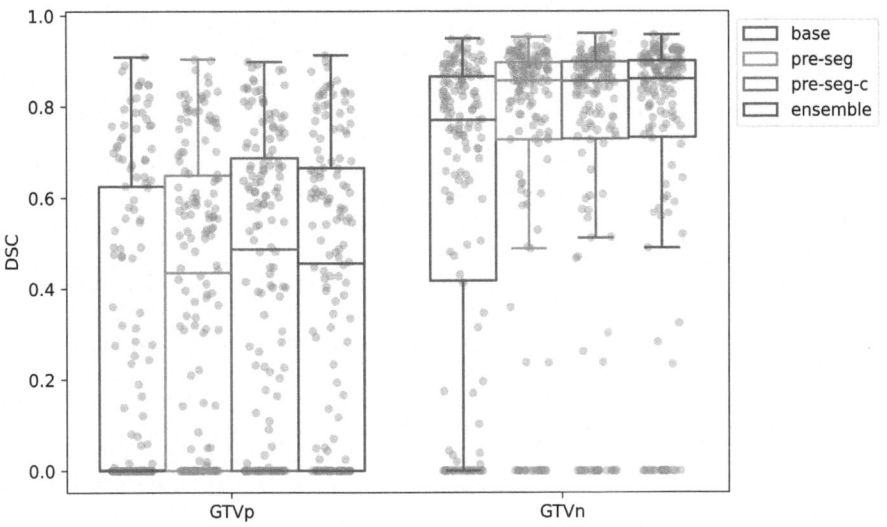

Fig. 2. Individual DSC scores for each model in the cross-validation evaluation for task 2. A box plot summarizes the quartiles of each distribution. Cases with zero DSC occur especially if the ground truth is empty and the prediction is not, or vice versa.

References

1. Pollard, J.M., Wen, Z., Sadagopan, R., Wang, J., Ibbott, G.S.: The future of image-guided radiotherapy will be MR guided. Br. J. Radiol. **90**(1073), 20160667 (2017)
2. Thorwarth, D., Low, D.A.: Technical challenges of real-time adaptive MR-guided radiotherapy. Front. Oncol. **11**, 634507 (2021)
3. Segedin, B., Petric, P.: Uncertainties in target volume delineation in radiotherapy - are they relevant and what can we do about them? Radiol. Oncol. **50**(3), 254–262 (2016)
4. Hindocha, S., et al.: Artificial intelligence for radiotherapy auto-contouring: current use, perceptions of and barriers to implementation. Clin. Oncol. (R Coll. Radiol.) **35**(4), 219–226 (2023)
5. Oreiller, V., et al.: Head and neck tumor segmentation in PET/CT: the HECKTOR challenge. Med. Image Anal. **77**(102336), 102336 (2022)
6. Luo, X., et al.: Segrap2023: a benchmark of organs-at-risk and gross tumor volume segmentation for radiotherapy planning of nasopharyngeal carcinoma (2023)
7. Isensee, F., et al.: nnu-net revisited: a call for rigorous validation in 3D medical image segmentation (2024)
8. Isensee, F., Jaeger, P.F., Kohl, S.A., Petersen, J., Maier-Hein, K.H.: nnU-Net: a self-configuring method for deep learning-based biomedical image segmentation. Nat. Methods **18**(2), 203–211 (2021)
9. Pflüger, I., et al.: Automated detection and quantification of brain metastases on clinical MRI data using artificial neural networks. Neurooncol. Adv. **4**(1), vdac138 (2022)

10. Ulrich, C., Isensee, F., Wald, T., Zenk, M., Baumgartner, M., Maier-Hein, K.H.: Multitalent: a multi-dataset approach to medical image segmentation. In: Medical Image Computing and Computer Assisted Intervention - MICCAI 2023: 26th International Conference, Vancouver, BC, Canada, 8–12 October 2023, Proceedings, Part III, pp. 648–658. Springer, Heidelberg (2023)
11. Wasserthal, J., et al.: Totalsegmentator: robust segmentation of 104 anatomic structures in CT images. Radiol. Artif. Intell. (2023)
12. Rokuss, M., et al.: Longitudinal segmentation of MS lesions via temporal difference weighting (2024)
13. Lin, D., et al.: E pluribus unum: prospective acceptability benchmarking from the contouring collaborative for consensus in radiation oncology crowdsourced initiative for multiobserver segmentation. J. Med. Imaging 10(S1), S11903–S11903 (2023)
14. Kohl, S.A.A., et al.: A probabilistic u-net for segmentation of ambiguous images. In: NIPS 2018, pp. 6965–6975. Curran Associates Inc, Red Hook (2018)
15. Antonelli, M., et al.: The medical segmentation decathlon. arXiv:2106.05735 (2021)
16. Simpson, A.L., et al.: A large annotated medical image dataset for the development and evaluation of segmentation algorithms. arXiv:1902.09063 (2019)
17. Landman, B., Zhoubing, X., Igelsias, J.E., Styner, M., et al.: MICCAI multi-atlas labeling beyond the cranial vault workshop and challenge. In: Proceedings of MICCAI Multi-Atlas Labeling Beyond Cranial Vault-Workshop Challenge (2015)
18. Armato III, S.G., et al.: Data from LIDC-IDRI (2015)
19. Gibson, E., et al.: Automatic multi-organ segmentation on abdominal CT with dense V-networks. IEEE Trans. Med. Imaging (2018)
20. Li, H., Zhou, J., Deng, J., Chen, M.: Automatic structure segmentation for radiotherapy planning challenge (2019). https://structseg2019.grand-challenge.org/. Accessed 25 Feb 2022
21. Lambert, Z., Petitjean, C., Dubray, B., Kuan, S.: Segthor: segmentation of thoracic organs at risk in CT images. arXiv:1912.05950 (2019)
22. Roth, H.R., et al.: Deeporgan: multi-level deep convolutional networks for automated pancreas segmentation. arXiv:1506.06448 (2015)
23. Clark, K., et al.: The cancer imaging archive (TCIA): maintaining and operating a public information repository. J. Digit. Imaging (2013)
24. Roth, H.R., et al.: Deeporgan: multi-level deep convolutional networks for automated pancreas segmentation. In: Medical Image Computing and Computer-Assisted Intervention – MICCAI 2015 (2015)
25. Sekuboyina, A., et al.: Verse: a vertebrae labelling and segmentation benchmark for multi-detector CT images. Med. Image Anal. (2021)
26. Löffler, M.T., et al.: A vertebral segmentation dataset with fracture grading. Radiol. Artif. Intell. (2020)
27. Liebl, H., et al.: A computed tomography vertebral segmentation dataset with anatomical variations and multi-vendor scanner data (2021)
28. Yang, J., Gu, S., Wei, D., Pfister, H., Ni, B.: Ribseg dataset and strong point cloud baselines for rib segmentation from CT scans. In: de Bruijne, M., et al. (eds.) Medical Image Computing and Computer Assisted Intervention – MICCAI 2021 (2021)
29. Heller, N., et al.: The KITS21 challenge: automatic segmentation of kidneys, renal tumors, and renal cysts in corticomedullary-phase CT (2023)
30. Li, W., Yuille, A., Zhou, Z.: How well do supervised models transfer to 3D image segmentation? In: The Twelfth International Conference on Learning Representations (2024)

31. Qu, C., Zhang, T., Qiao, H., Tang, Y., Yuille, A.L., Zhou, Z.: Abdomenatlas-8k: annotating 8,000 CT volumes for multi-organ segmentation in three weeks. In: Advances in Neural Information Processing Systems (2023)
32. Ma, J., et al.: Unleashing the strengths of unlabeled data in pan-cancer abdominal organ quantification: the flare22 challenge. arXiv preprint arXiv:2308.05862 (2023)
33. Luo, X., et al.: SegRap2023: a benchmark of organs-at-risk and gross tumor volume segmentation for radiotherapy planning of NasoPharyngeal carcinoma. arXiv (2023)
34. Radl, L., et al.: AVT: multicenter aortic vessel tree CTA dataset collection with ground truth segmentation masks. Data Brief (2022)
35. Jin, Y., et al.: AI-based aortic vessel tree segmentation for cardiovascular diseases treatment: status quo. arXiv (2021)
36. Pepe, A., et al.: Detection, segmentation, simulation and visualization of aortic dissections: a review. Med. Image Anal. (2020)
37. Luo, X., et al.: Word: a large scale dataset, benchmark and clinical applicable study for abdominal organ segmentation from CT image. Med. Image Anal. (2022)
38. Liao, W., et al.: Comprehensive evaluation of a deep learning model for automatic organs-at-risk segmentation on heterogeneous computed tomography images for abdominal radiation therapy. Int. J. Radiat. Oncol. Biol. Phys. **117**(4), 994–1006 (2023)
39. Ma, J., et al.: Abdomenct-1k: is abdominal organ segmentation a solved problem? IEEE Trans. Pattern Anal. Mach. Intell. (2022)
40. Jaus, A., et al.: Towards unifying anatomy segmentation: automated generation of a full-body CT dataset via knowledge aggregation and anatomical guidelines. arXiv preprint arXiv:2307.13375 (2023)
41. Rister, B., Shivakumar, K., Nobashi, T., Rubin, D.L.: CT-ORG: a dataset of CT volumes with multiple organ segmentations (2019)
42. Podobnik, G., Strojan, P., Peterlin, P., Ibragimov, B., Vrtovec, T.: HaN-Seg: the head and neck organ-at-risk CT and MR segmentation dataset. Med. Phys. (2023)
43. Maier, O., et al.: Extra tree forests for sub-acute ischemic stroke lesion segmentation in MR sequences. J. Neurosci. Methods (2015)
44. Litjens, G., et al.: Evaluation of prostate segmentation algorithms for MRI: the promise12 challenge. Med. Image Anal. (2014)
45. Bernard, O., et al.: Deep learning techniques for automatic MRI cardiac multi-structures segmentation and diagnosis: is the problem solved? IEEE Trans. Med. Imaging (2018)
46. Carass, A., et al.: Longitudinal multiple sclerosis lesion segmentation: resource and challenge. NeuroImage (2017)
47. Kavur, A.E., Selver, M.A., Dicle, O., Barış, M., Gezer, N.S.: CHAOS -combined (CT-MR) healthy abdominal organ segmentation challenge data (2019)
48. Campello, V.M., et al.: Multi-centre, multi-vendor and multi-disease cardiac segmentation: the MMS challenge. IEEE Trans. Med. Imaging (2021)
49. Martín-Isla, C., et al.: Deep learning segmentation of the right ventricle in cardiac MRI: the MMS challenge. IEEE J. Biomed. Health Inform. (2023)
50. Litjens, G., Debats, O., Barentsz, J., Karssemeijer, N., Huisman, H.: Spie-aapm prostatex challenge data (2017)
51. Muslim, A.M.: Brain MRI dataset of multiple sclerosis with consensus manual lesion segmentation and patient meta information (2022)
52. Grøvik, E., Yi, D., Iv, M., Tong, E., Rubin, D., Zaharchuk, G.: Deep learning enables automatic detection and segmentation of brain metastases on multisequence MRI. J. Magn. Reson. Imaging (2020)

53. Shapey, J., et al.: Segmentation of vestibular schwannoma from magnetic resonance imaging: an open annotated dataset and baseline algorithm (vestibular-schwannoma-SEG) (2021)
54. Liew, S.-L., et al.: A large, open source dataset of stroke anatomical brain images and manual lesion segmentations. Sci. Data (2018)
55. Karargyris, A., et al.: Federated benchmarking of medical artificial intelligence with medperf. Nat. Mach. Intell. (2023)
56. Bakas, S., et al.: Advancing the cancer genome atlas glioma MRI collections with expert segmentation labels and radiomic features. Sci. Data (2017)
57. Menze, B.H., et al.: The multimodal brain tumor image segmentation benchmark (BRATS). IEEE Trans. Med. Imaging (2015)
58. Baid, U., et al.: The RSNA-ASNR-MICCAI BraTS 2021 benchmark on brain tumor segmentation and radiogenomic classification. arXiv (2021)
59. Gatidis, S., Kuestner, T.: A whole-body FDG-PET/CT dataset with manually annotated tumor lesions (FDG-PET-CT-Lesions) (2022)
60. Andrearczyk, V., et al.: Overview of the HECKTOR challenge at MICCAI 2022: automatic head and neck TumOR segmentation and outcome prediction in PET/CT. Head Neck Tumor Chall. (2022) (2023)
61. Ji, Y., et al.: Amos: a large-scale abdominal multi-organ benchmark for versatile medical image segmentation. In: Advances in Neural Information Processing Systems (2022)
62. Yang, K., et al.: Benchmarking the cow with the topcow challenge: topology-aware anatomical segmentation of the circle of willis for CTA and MRA (2024)

Head and Neck Tumor Segmentation for MRI-Guided Radiation Therapy Using Pre-trained STU-Net Models

Zihao Wang[1,2] and Mengye Lyu[1,2(✉)]

[1] College of Health Science and Environmental Engineering,
Shenzhen Technology University, Shenzhen, China
lvmengye@sztu.edu.cn
[2] College of Applied Sciences, Shenzhen University, Shenzhen, China

Abstract. Accurate segmentation of tumors in MRI-guided radiation therapy (RT) is crucial for effective treatment planning, particularly for complex malignancies such as head and neck cancer (HNC). This study presents a comparative analysis between two state-of-the-art deep learning models, nnU-Net v2 and STU-Net, for automatic tumor segmentation in pre-RT MRI images. While both models are designed for medical image segmentation, STU-Net introduces critical improvements in scalability and transferability, with parameter sizes ranging from 14 million to 1.4 billion. Leveraging large-scale pre-training on datasets such as TotalSegmentator, STU-Net captures complex and variable tumor structures more effectively. We modified the default nnU-Net v2 by adding additional convolutional layers to both the encoder and decoder, improving its performance for MRI data. Based on our experimental results, STU-Net demonstrated better performance than nnU-Net v2 in the head and neck tumor segmentation challenge. These findings suggest that integrating advanced models like STU-Net into clinical workflows could remarkably enhance the precision of RT planning, potentially improving patient outcomes. Ultimately, the performance of the fine-tuned STU-Net-B model submitted for the final evaluation phase of Task 1 in this challenge achieved a DSCagg-GTVp of 0.76, a DSCagg-GTVn of 0.85, and an overall DSCagg-mean score of 0.81, securing ninth place in the Task 1 rankings. The described solution is by team SZTU-SingularMatrix for Head and Neck Tumor Segmentation for MR-Guided Applications (HNTS-MRG) 2024 challenge. Link to the trained model weights: https://github.com/Duskwang/Weight/releases.

Keywords: Medical image segmentation · STU-Net · nnU-Net · tumor segmentation · MRI-guided radiation therapy

1 Introduction

Radiation therapy (RT) plays a pivotal role in the management of various malignancies, particularly head and neck cancer (HNC) [1,2]. The introduction of

K. A. Wahid et al. (Eds.): HNTS-MRG 2024, LNCS 15273, pp. 65–74, 2025.
https://doi.org/10.1007/978-3-031-83274-1_4

MRI-guided RT planning has gained substantial interest in recent years, largely due to the superior soft tissue contrast that MRI provides compared to traditional CT-based approaches [3]. MRI not only enhances anatomical visualization but also enables functional imaging through advanced multiparametric sequences, such as diffusion-weighted imaging. These capabilities are further leveraged by MRI-Linac systems, which allow for daily adaptive RT through real-time intra-therapy imaging and adjustments, optimizing tumor targeting while minimizing collateral damage [4–6]. These technological advancements have the potential to transform clinical practices for HNC treatment.

However, the integration of MRI into RT planning comes with challenges, particularly in the context of image segmentation. Accurate tumor segmentation is crucial for effective RT planning, yet the manual segmentation of MRI data is time-consuming and complex, especially given the intricate anatomy of HNC [7, 8]. The high resolution and detailed contrast of MRI, while advantageous, result in large datasets that are labor-intensive for clinicians to process. Moreover, the variability in MRI appearances and the need for consistent, reproducible segmentation further complicate the task.

To address these challenges, artificial intelligence (AI) methods, particularly deep learning, have shown great promise in automating the segmentation of HNC tumors [9]. A notable advancement in this field is the development of STU-Net, a scalable and transferable model designed for medical image segmentation. STU-Net builds upon the popular nnU-Net framework but introduces key improvements to enhance scalability and transferability. It can be scaled to sizes ranging from 14 million to 1.4 billion parameters, making it the largest medical image segmentation model to date [10]. By leveraging large-scale pre-training on datasets, STU-Net is capable of generalizing across various medical imaging tasks, including those involving CT, MRI, and PET scans. Its ability to scale in both depth and width allows it to achieve superior performance, especially in handling the complex and variable structures encountered in HNC segmentation.

In this paper, we apply the STU-Net framework to the segmentation of HNC tumors in the context of MRI-guided adaptive RT and compare it with nnU-Net based models. Overall, the results highlight the clear advantage of STU-Net for precise tumor segmentation in this context.

2   Methods

2.1   Model Architecture

In this study, we used two state-of-the-art deep learning frameworks for fully automatic tumor volume segmentation on pre-RT MRI: nnU-Net v2 and STU-Net. Both architectures are designed for medical image segmentation, but they offer distinct advantages in terms of flexibility, scalability, and performance.

1. nnU-Net V2: The nnU-Net v2 [11] framework is an advanced version of the nnU-Net, featuring improved self-configuring capabilities and enhanced adaptability for medical image segmentation tasks. It builds upon the original U-Net architecture with refined self-configuration that includes optimized network depth, input patch size, and training parameters based on the input data's characteristics [11]. nnU-Net v2 incorporates advanced training strategies, flexible data preprocessing, and improved upsampling methods, which collectively enhance its performance and scalability. For this study, we utilized the default 3D full-resolution nnU-Net v2 configuration, which has shown exceptional robustness in medical image segmentation tasks, including tumor delineation in MRI images [12].

The nnU-Net v2 framework was trained using the HNC dataset in a supervised manner. The model's advanced self-adaptation capabilities without requiring manual tuning make it a solid baseline for comparison. However, given the complexity and variability of HNC tumors in MRI images, we identified opportunities for further improvement with a more scalable and context-aware approach.

2. STU-Net: STU-Net [10] is developed from the nnU-Net framework to address the challenges of scalability and transferability in large-scale medical image segmentation models. It improves upon nnU-Net by refining convolutional blocks, replacing upsampling methods, and fixing hyperparameters to enhance model scalability and transferability [10]. Additionally, by pretraining on the large-scale dataset, STU-Net demonstrates exceptional performance across various medical image segmentation tasks. By scaling both the width and depth of the network, STU-Net can capture both fine-grained local details and long-range dependencies [10], which are crucial in segmenting complex tumor structures like those found in HNC.

STU-Net's design allows for larger models, with parameter sizes ranging from 14 million to 1.4 billion. In this study, we trained STU-Net with official pretrained weights of different sizes, followed by fine-tuning on the HNC MRI data. This transfer learning approach allowed STU-Net to leverage global structural information from the pre-training phase, while fine-tuning on the HNC-specific data further refined the model for better segmentation accuracy.

2.2 Training Strategies

We conducted a series of experiments to evaluate the impact of different training strategies on segmentation performance:

1. Standard nnU-Net V2: In this study, we utilized the publicly available nnU-Net v2 framework, which was installed directly from its official GitHub repository. The segmentation process was carried out using the default configuration provided by the nnU-Net v2 framework, which automatically determines optimal network depth, input patch size, and training parameters based on the characteristics of the input data. No manual tuning or customization of hyperparameters was applied, ensuring that the results reflect the baseline performance of nnU-Net v2 for the given dataset.

2. nnU-Net V2*: We modified the default nnU-Net v2 configuration by increasing the number of convolutional layers in each stage of the encoder from 2 to 3. This adjustment aims to enhance the feature extraction capabilities at each resolution level, potentially improving the model's ability to capture complex tumor structures in MRI images of the head and neck. The added convolutional layers may help the model better delineate tumor boundaries, potentially improving segmentation accuracy, particularly for more challenging anatomical regions.

3. nnU-Net V2**: Based on nnU-Net v2*, we increased the number of convolutional layers in each stage of the decoder from 2 to 3. This further adjustment enhances the model's feature extraction and reconstruction capabilities, enabling it to better capture and restore complex details in MRI images of head and neck cancer, ultimately improving segmentation accuracy.

4. STU-Net-S: By simultaneously scaling the encoder and decoder structures in STU-Net, and scaling the depth and width at each resolution stage by the same ratio, different scales of STU-Net can be obtained [10]. We first used a small-scale model referred to as STU-Net-S, which has 14.60M parameters [10]. This design aims to enhance performance at a relatively lower computational cost, particularly in capturing finer details in head and neck tumor images.

5. STU-Net-B: Expanding the width of STU-Net-S results in the STU-Net-B model, which has 58.26M parameters [10]. STU-Net family also includes two even larger scale models (STU-Net-L and STU-Net-H), which were not investigated in this study considering the high hardware requirement. The purpose of expanding the width is to improve the model's feature extraction capabilities, potentially enabling it to better capture complex tumor features.

6. STU-Net-B*: To determine whether the improved segmentation performance is due to the STU-Net architecture itself or the effect of pre-training, we conducted experiments without loading the pre-trained weights. This allowed us to assess the contribution of each factor. We denote the STU-Net-B model without pre-trained weights as STU-Net-B*.

2.3   Evaluation Metrics

To evaluate the segmentation results for challenge Task 1 (pre-RT segmentation), we used the aggregated Dice Similarity Coefficient (DSCagg) metric provided officially. DSCagg is calculated separately for GTVp (DSCagg-GTVp) and GTVn (DSCagg-GTVn), with the final segmentation performance based on the average of these two values (DSCagg-mean). Additionally, we also use the traditional Dice Similarity Coefficient (DSC) as an evaluation metric [13]. The expected labels for the predicted masks are as follows: 1 for GTVp, 2 for GTVn, and 0 for the background.

2.4   Datasets

The official head and neck tumor dataset consists of 150 cases. For this study, we selected Task 1 and exclusively used pre-RT data for training. The dataset was randomly split, with 30 cases designated as the internal validation set, while the remaining 120 cases were used for model training.

2.5   Implementation Details

The experiments were implemented using Python and the PyTorch deep learning library. Training was conducted on a GPU Server. The training protocols are presented in Table 1. The hardware configuration and development environments are presented in Table 2.

Table 1. Training protocols.

Network initialization	normal initialization
Batch size	2
Patch size	$48 \times 225 \times 225$
Total epochs	1000
Optimizer	SGD with nesterov momentum ($\mu = 0.99$)
Weight decay	3e-5
Loss function	dice and cross entrop loss
Initial learning rate (lr)	0.001

Table 2. Development environments and requirements.

System	Ubuntu 22.04
CPU	Intel(R) Xeon(R) CPU E5-2680 v4
RAM	128G
GPU (number and type)	P40
CUDA version	12.1
Programming language	Python 3.10.10
Deep learning framework	torch 2.1.2, torchvision 0.16.2

3   Results

The comparative performance analysis between nnU-Net v2 and STU-Net revealed clear advantages for STU-Net in the segmentation of head and neck cancer tumors. Considering the limitations of computational resources and the prolonged training time due to the use of a less powerful GPU, this study conducted a single training-validation split for model training and evaluation. While this approach may not fully capture the variability across different data partitions, it was adopted as a practical compromise to ensure the feasibility of the experiments within the available resource constraints. Table 3 presents the quantitative results of all evaluation metrics derived from the model training conducted using a single training-validation split approach.

nnU-Net v2 achieved an average DSCagg-mean of 0.77, with DSCagg-GTVp at 0.75 and DSCagg-GTVn at 0.78. nnU-Net v2* improved slightly with a DSCagg-mean of 0.74, reflecting better feature extraction capabilities in the encoder. nnU-Net v2** with both the encoder and decoder enhanced showed a modest improvement in GTVn segmentation with a DSCagg-mean of 0.75.

STU-Net-S outperformed the nnU-Net variants, achieving a DSCagg-mean of 0.82, with remarkable improvements in both GTVp (0.79) and GTVn (0.84). STU-Net-B, the base version of STU-Net, performed even better, with a DSCagg-mean of 0.83, showcasing its ability to capture more complex features in HNC MRI data.

When training without pre-trained weights, STU-Net-B* achieved a DSCagg-mean of 0.81. This indicates that although STU-Net (without pre-trained weights) still performed better than nnU-Net v2, it did not reach the performance of its pre-trained counterparts (STU-Net-S and STU-Net-B). This suggests that the pre-training phase substantially contributes to the model's superior performance, particularly in the complex segmentation task of head and neck cancer.

In Table 3, "Test" refers to the performance of the trained STU-Net-B model submitted for the final evaluation phase of Task 1 in this challenge, achieving a DSCagg-mean of 0.81.

Table 3. Segmentation metrics by different methods.

	Class 1 (GTVp)		Class 2 (GTVn)		DSCagg-mean
	DSC	DSCagg-GTVp	DSC	DSCagg-GTVn	
Standard nnU-Net v2	0.56 ± 0.34	0.75	0.60 ± 0.35	0.78	0.77
nnU-Net v2*	0.57 ± 0.32	0.73	0.57 ± 0.35	0.75	0.74
nnU-Net v2**	0.55 ± 0.34	0.73	0.66 ± 0.31	0.77	0.75
STU-Net-B*	0.63 ± 0.33	0.79	0.61 ± 0.37	0.83	0.81
STU-Net-S	0.64 ± 0.31	0.79	0.66 ± 0.34	0.84	0.82
STU-Net-B	$\mathbf{0.66 \pm 0.30}$	**0.81**	$\mathbf{0.72 \pm 0.27}$	**0.84**	**0.83**
Test	\	0.76	\	0.85	0.81

These results indicate that STU-Net demonstrates promising segmentation performance, particularly in delineating the intricate boundaries of HNC tumors on the validation set, though further validation on large-scale independent datasets is necessary. The scalability of STU-Net allowed it to effectively model both local and global structures, contributing to its higher Dice scores. Furthermore, the pre-training on the large-scale dataset provided a solid foundation for transfer learning, allowing the model to generalize better across different imaging tasks.

Additionally, the average training time per epoch varied across strategies: 378.72 s for the default nnU-Net v2, 530.9 s for nnU-Net v2**, 418 s for STU-Net-B, and 198 s for STU-Net-S. These differences reflect the variations in model complexity and parameter counts, which influence the computational cost and training duration.

In qualitative analyses (Fig. 1), STU-Net demonstrated a more precise segmentation of tumor boundaries compared to the nnU-Net variants, particularly in cases with high variability in tumor shape and size. This suggests that the architectural enhancements in STU-Net, combined with its ability to scale in depth and width, offer substantial advantages in handling the complexities of MRI-guided RT planning for HNC.

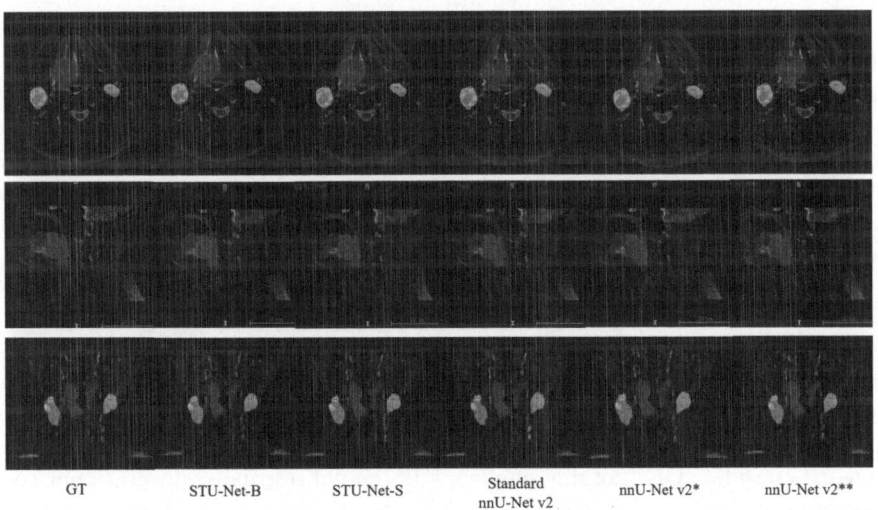

Fig. 1. Qualitative visualization of different models.

Overall, the results suggest that STU-Net, with its improved scalability and transferability, shows potential as a more accurate and reliable approach for tumor segmentation in MRI-guided RT compared to nnU-Net v2, but further verification is required.

4   Discussion

In this study, we explored the performance of two state-of-the-art deep learning models, nnU-Net v2 and STU-Net, for automatic tumor segmentation in pre-RT MRI of head and neck cancer (HNC). While both frameworks are designed to handle medical image segmentation, our findings indicate that STU-Net consistently outperformed nnU-Net v2 in this context, particularly when segmenting the complex and heterogeneous structures characteristic of HNC tumors.

The superior performance of STU-Net can be attributed to several key factors. First, the pre-training of STU-Net on the large-scale dataset provides substantial advantage over randomly initialized models. This pre-training allowed the model to leverage global structural information across multiple medical imaging modalities, enhancing its transferability and fine-tuning capacity on the HNC-specific MRI data. In comparison, the nnU-Net v2, while effective as a baseline model, lacked the same level of transfer learning benefits.

To investigate the contribution of pre-training, we trained a version of STU-Net without pre-trained weights, denoted as STU-Net-B*. While STU-Net-B* performed better than nnU-Net v2, its performance did not reach the levels of the pre-trained STU-Net models. This clearly indicates that pre-training plays a crucial role in enhancing STU-Net's ability to capture complex tumor features and produce more accurate segmentation results.

Additionally, the ability to scale both the depth and width of the network allowed STU-Net to effectively capture fine-grained details while also modeling long-range dependencies, which are crucial for delineating intricate tumor boundaries. In contrast, nnU-Net v2, though adaptable and self-configuring to certain degrees, was more constrained in its ability to handle such complexities due to its default architecture. This limitation became evident when dealing with the high variability in tumor morphology and tissue contrast present in MRI-guided RT data.

Another aspect potentially contributing to STU-Net's superior performance is the incorporation of transformer modules, which can improve global context awareness [14]. This suggests that for complex tumor segmentation tasks, models that incorporate advanced context-aware mechanisms like Transformers may be better suited than traditional fully convolutional architectures.

Despite the promising results achieved with STU-Net, there are still limitations to consider. One notable challenge is the substantial computational cost associated with training large-scale models like STU-Net, particularly when dealing with 3D high-resolution MRI data. While STU-Net offers scalability, training the largest model configurations may require costly hardware, which could limit its accessibility in clinical settings with limited resources. However, the scalability feature also allows for adjustments to be made based on available computational power, making STU-Net a versatile option for a range of applications.

In the context of Task 2, which involves predicting tumor segmentation on unseen mid-RT images, STU-Net's ability to model long-range dependencies and adapt to variable anatomical structures offers great potential. By leveraging its transfer learning capabilities and advanced architectural components, STU-Net could be well-suited for addressing the unique challenges posed by mid-RT segmentation. Future research should investigate the performance of STU-Net on this task, as well as its ability to generalize to other imaging modalities such as PET-CT or ultrasound, to validate its broader applicability.

In summary, the results of this study demonstrate that STU-Net provides a more effective solution for the segmentation of HNC tumors in MRI-guided RT planning compared to widely used nnU-Net v2 models. These findings suggest that integrating large-scale models like STU-Net into clinical workflows could improve the accuracy and efficiency of radiation therapy planning, potentially leading to better patient outcomes.

Acknowledgments. This work was in part supported by the National Natural Science Foundation of China (No. 62101348), Shenzhen Science and Technology Program (Shenzhen Higher Education Stable Support Program, No. 20220716111838002), and Natural Science Foundation of Top Talent of Shenzhen Technology University (No. 20200208).

Disclosure of Interests. The authors have no competing interests.

References

1. Anderson, G., Ebadi, M., Vo, K., Novak, J., Govindarajan, A., Amini, A.: An updated review on head and neck cancer treatment with radiation therapy. Cancers **13**(19), 4912 (2021)
2. Chin, S., et al.: Magnetic resonance-guided radiation therapy: a review. J. Med. Imaging Radiat. Oncol. **64**(1), 163–177 (2020)
3. Otazo, R., et al.: MRI-guided radiation therapy: an emerging paradigm in adaptive radiation oncology. Radiology **298**(2), 248–260 (2021)
4. Lagendijk, J.J., et al.: MRI/linac integration. Radiother. Oncol. **86**(1), 25–29 (2008)
5. Raaymakers, B.W., et al.: First patients treated with a 1.5 T MRI-linac: clinical proof of concept of a high-precision, high-field MRI guided radiotherapy treatment. Phys. Med. Biol. **62**(23), L41 (2017)
6. Ng, J., et al.: MRI-linac: a transformative technology in radiation oncology. Front. Oncol. **13**, 1117874 (2023)
7. Chen, L.-C., Papandreou, G., Kokkinos, I., Murphy, K., Yuille, A.L.: Deeplab: semantic image segmentation with deep convolutional nets, atrous convolution, and fully connected CRFs. IEEE Trans. Pattern Anal. Mach. Intell. **40**(4), 834–848 (2017)
8. Myronenko, A.: 3D MRI brain tumor segmentation using autoencoder regularization. In: Brainlesion: Glioma, Multiple Sclerosis, Stroke and Traumatic Brain Injuries: 4th International Workshop, BrainLes 2018, Held in Conjunction with MICCAI 2018, Granada, Spain, 16 September 2018, Revised Selected Papers, Part II 4, pp. 311–320. Springer (2019)
9. Zhong, Y., Yang, Y., Fang, Y., Wang, J., Hu, W.: A preliminary experience of implementing deep-learning based auto-segmentation in head and neck cancer: a study on real-world clinical cases. Front. Oncol. **11**, 638197 (2021)

10. Huang, Z., et al.: Stu-net: scalable and transferable medical image segmentation models empowered by large-scale supervised pre-training (2023)
11. Isensee, F., Jaeger, P.F., Kohl, S.A., Petersen, J., Maier-Hein, K.H.: nnU-Net: a self-configuring method for deep learning-based biomedical image segmentation. Nat. Methods **18**(2), 203–211 (2021)
12. Çiçek, Ö., Abdulkadir, A., Lienkamp, S.S., Brox, T., Ronneberger, O.: 3D U-net: learning dense volumetric segmentation from sparse annotation. In: Medical Image Computing and Computer-Assisted Intervention–MICCAI 2016: 19th International Conference, Athens, Greece, 17–21 October 2016, Proceedings, Part II 19, pp. 424–432. Springer (2016)
13. Menze, B.H., et al.: The multimodal brain tumor image segmentation benchmark (brats). IEEE Trans. Med. Imaging **34**(10), 1993–2024 (2014)
14. Alexey, D.: An image is worth 16x16 words: transformers for image recognition at scale. arXiv preprint arXiv:2010.11929 (2020)

Head and Neck Tumor Segmentation of MRI from Pre- and Mid-Radiotherapy with Pre-Training, Data Augmentation and Dual Flow UNet

Litingyu Wang[1], Wenjun Liao[1,3], Shichuan Zhang[1,3], and Guotai Wang[1,2]($\boxtimes$) (iD)

[1] University of Electronic Science and Technology of China, Chengdu, China
guotai.wang@uestc.edu.cn
[2] Shang AI Laboratory, Shanghai, China
[3] Department of Radiation Oncology, Sichuan Cancer Hospital & Institute, Sichuan Cancer Center, Chengdu, China

Abstract. Head and neck tumors and metastatic lymph nodes are crucial for treatment planning and prognostic analysis. Accurate segmentation and quantitative analysis of these structures require pixel-level annotation, making automated segmentation techniques essential for the diagnosis and treatment of head and neck cancer. In this study, we investigated the effects of multiple strategies on the segmentation of pre-radiotherapy (pre-RT) and mid-radiotherapy (mid-RT) images. For the segmentation of pre-RT images, we utilized: 1) a fully supervised learning approach, and 2) the same approach enhanced with pre-trained weights and the MixUp data augmentation technique. For mid-RT images, we introduced a novel computational-friendly network architecture that features separate encoders for mid-RT images and registered pre-RT images with their labels. The mid-RT encoder branch integrates information from pre-RT images and labels progressively during the forward propagation. We selected the highest-performing model from each fold and used their predictions to create an ensemble average for inference. In the final test, our models achieved a segmentation performance of 82.38% for pre-RT and 72.53% for mid-RT on aggregated Dice Similarity Coefficient (DSC) as HiLab. Our code is available at https://github.com/WltyBY/HNTS-MRG2024_train_code.

Keywords: HNTS-MRG2024 · Automatic Segmentation · Head and Neck Cancer

1 Introduction

Head and neck (H&N) cancers are among the most prevalent types of cancer. Imaging in H&N cancer serves multiple purposes, including quantitative assessment of tumors, evaluation of nodal disease, and differentiation between recurrent tumors and post-treatment changes [11]. In recent years, convolutional neural networks (CNNs) within the realm of deep learning have profoundly impacted

© The Author(s) 2025
K. A. Wahid et al. (Eds.): HNTS-MRG 2024, LNCS 15273, pp. 75–86, 2025.
https://doi.org/10.1007/978-3-031-83274-1_5

medical image analysis [2,4]. The detection and segmentation of H&N tumors and metastatic lymph nodes are crucial for diagnosis and treatment. In addition, deep learning aids in these processes by segmenting target areas to facilitate treatment and surgical planning. During radiation treatment planning, computed tomography (CT) is commonly utilized [12]. However, CT images often lack clear delineation between lymph nodes and surrounding tissues [3,5]. And, unfortunately, most existing datasets are based on CT, contrast-enhanced CT, or positron emission computed tomography (PET) [1,9]. In contrast, the HNTS-MRG2024 Challenge provides a magnetic resonance imaging (MRI) dataset for H&N images, representing an uncommon imaging modality in H&N cancer diagnosis and capturing a significant scenario of adaptive radiotherapy. This competition focuses on segmenting gross tumor volumes of primary tumors (GTVp) and lymph nodes (GTVn) on both pre-radiotherapy (pre-RT) and mid-radiotherapy (mid-RT) MRI images, designated as Task-1 and Task-2, respectively.

Our methods and innovations are summarized as follows. We designed a novel framework, incorporating model pre-training, multiple data augmentations, and network architectures. Firstly, under the competition rules allowing the use of external public datasets, we pre-trained our model on a CT dataset due to the absence of an appropriate H&N MRI dataset. Given the significant differences in image intensities between CT and MRI, we used histogram matching strategy to align the intensity distributions in the preprocessing stage and employed nonlinear intensity transformations in the model pre-training stage. Secondly, to address the scarcity of foreground voxels in our dataset, we employed the MixUp [18] data augmentation technique. This approach not only expanded the training dataset, but also enhanced the model's generalizability. Furthermore, different from the conventional single encoder-decoder network structure commonly used in segmentation tasks, we adopted a dual encoder-decoder architecture to leverage per-RT images to guide the segmentation in mid-RT images. Attention blocks were introduced to this innovative structure, allowing for the effective fusion of information from disparate encoder streams. After conducting a rigorous five-fold cross-validation, we identified the top-performing model in each fold. The selected models demonstrated impressive performance, with an average aggregated Dice Similarity Coefficient (DSC) of 80.65% for Task-1 and 74.68% for Task-2 across five-fold cross-validation. Furthermore, in the final test, they scored 82.38% for Task-1 and 72.53% for Task-2.

2   Methods

We proposed multiple training strategies for the two tasks, and selected the best-performing model in each of the five-fold cross-validation divisions. For the segmentation of pre-RT images in Task-1, we employed:

- Fully supervised learning utilizing an encoder-decoder architecture, as depicted in Fig. 1(a).
- The same fully supervised learning approach, enhanced with pre-trained weights and the MixUp data augmentation method.

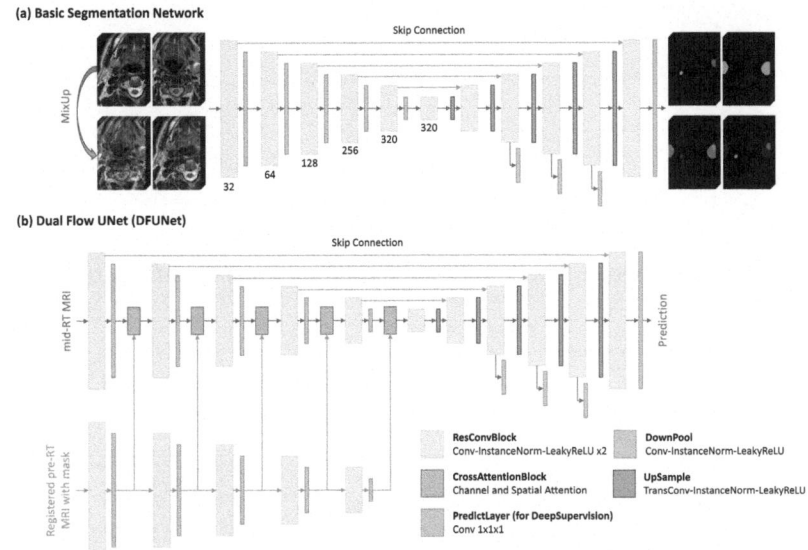

Fig. 1. Two architectures used for training. (a) An encoder-decoder network, named basic segmentation network, with MixUp diagram. (b) Two separate encoders with one decoder named Dual Flow UNet (DFUNet).

For Task-2, we introduced an additional training method:

- Fully supervised learning employing the Dual Flow UNet (DFUNet) architecture, illustrated in Fig. 1(b).

2.1 Networks

We designed two network architectures in this study. The first network is basic segmentation network, shown as Fig. 1(a)), which is an encoder-decoder architecture. For task-1, the input of basic segmentation network is the single-channel pre-RT images. For Task-2, the input of basic segmentation network is three-channel images, which are consisted of mid-RT image, registered pre-RT image and its mask. The second network is DFUNet, shown as Fig. 1(b), which is only used for Task-2. The DFUNet comprises two encoders, whose inputs are a single-channel mid-RT image and a two-channel input (registered pre-RT image and its mask), respectively. To facilitate information fusion, the information from the pre-RT image and its mask is integrated into the mid-RT stream at various encoder stages using a CNN-based cross attention block [13]. This CNN-based attention mechanism differs from the attention blocks described in [14] by employing both spatial and channel attention [16]. This approach enhances or fuses features while significantly reducing the computational demands of training and inference. The operational details of the cross attention block are illustrated in Fig. 2. Details of the network's configuration are elaborated in Sect. 3.3.

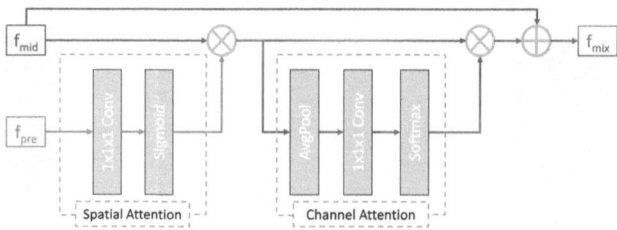

Fig. 2. CNN-based cross attention block to integrate secondary information f_{pre} into primary information f_{mid}.

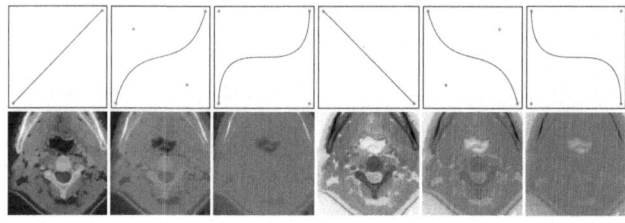

Fig. 3. Raw image (the first column) and corresponding augmented images using Bézier Curve [8].

2.2 Model Pre-Training

We utilized the SegRap2023 Challenge Dataset [9] and the basic segmentation network depicted in Fig. 1(a) for fully supervised pre-training. However, since SegRap2023 is based on CT images, we encountered challenges due to the inherent intensity differences between CT and MRI data. To minimize these differences, we employed histogram matching during preprocessing to align the intensity distributions of CT and MRI as closely as possible. Despite this effort, intensity variations across various anatomical structures remained a challenge. To further address these issues, we incorporated techniques from domain generalization [19], applying nonlinear transformations to the image intensities. This approach was designed to reduce the model's reliance on specific intensity values, thereby enhancing its generalization capabilities. The effectiveness of these nonlinear transformations is visualized in Fig. 3.

2.3 MixUp Augmentation Strategy

Since our training methodology is based on patches, patches in one mini-batch inevitably contain negative samples, which entirely belong to the background category. To mitigate the impact of these negative samples and to encourage the network to learn the distribution across different classes, we employ the MixUp [18] technique to augment our dataset. MixUp creates new training examples by linearly interpolating between pairs of samples, which helps the model to generalize better across class boundaries. The process is defined by the following equations:

$$\tilde{x} = \lambda x_i + (1 - \lambda)x_j$$
$$\tilde{y} = \lambda y_i + (1 - \lambda)y_j \tag{1}$$

where x_i and x_j are raw input patches from different images, y_i and y_j are the corresponding one-hot labels, and λ is a parameter in the range $[0, 1]$ that controls the interpolation. In our experiments, λ is sampled from the Beta distribution, which provides a flexible way to control the mixing ratio. During training, both the newly created examples from MixUp and the original samples are used in each batch, allowing the model to learn from a more diverse set of examples.

2.4 Loss

The inputs to our network can be categorized into two types: the raw input patches and the new cases generated by MixUp. For the raw input patches, we utilize a combination of CrossEntropy loss $\mathcal{L}_{CE}$ and Dice loss $\mathcal{L}_{Dice}$ for both pre-training and downstream training:

$$Loss_{raw} = \sum_{d=0}^{l} \frac{1}{2^d}(\mathcal{L}_{CE}(x_d, y_d) + L_{Dice}(x_d, y_d)) \tag{2}$$

where d represents the identifier for different resolutions, with smaller d values corresponding to higher resolutions of x_d and y_d. l represents the number of resolutions used in computing the loss. The low-resolution labels y_d are obtained by down-sampling the original labels. In contrast, for the cases generated through MixUp, we only compute $\mathcal{L}_{CE}$ and omit deep supervision:

$$Loss_{MixUp} = \mathcal{L}_{CE}(\tilde{x}, \tilde{y}) \tag{3}$$

3 Experiments

We conducted our experiments using the nnUNetv2 [6] framework, which streamlined the hyperparameter selection process. This toolkit automates many decisions, allowing us to focus on the specific architectural adjustments needed for our tasks. As depicted in Fig. 1, we selected a VNet-like [10] architecture for our network, which was trained using deep supervision.

3.1 Datasets

SegRap2023 Challenge Dataset. Segmentation of Organs-at-Risk and Gross Tumor Volume of NPC for Radiotherapy Planning Challenge Dataset [9] comprises 120 pairs of H&N CT and contrast-enhanced CT images. We ultimately used only the 120 enhanced CT scans because, during the training process, we found that the model's performance with enhanced CT was superior to that with CT or a combination of both.

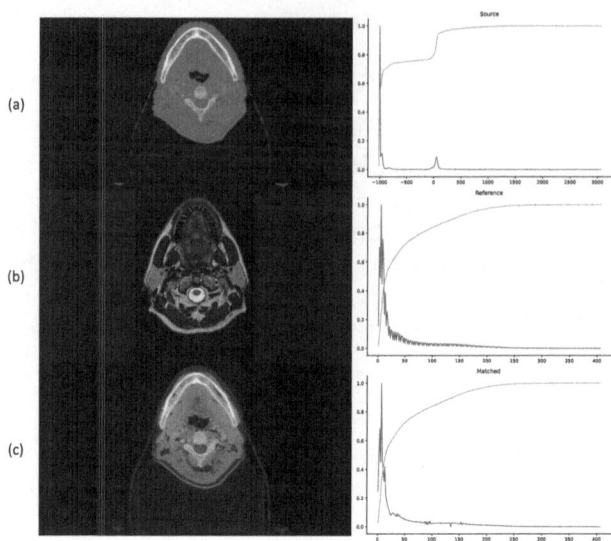

Fig. 4. Histogram matching on the SegRap2023 dataset. (a) shows the source image and its grayscale histogram from the SegRap2023 dataset. (b) displays the reference image from the HNTS-MRG2024 dataset, and (c) presents the matched image.

HNTS-MRG2024 Challenge Dataset. Head and Neck Tumor Segmentation for MR-Guided Applications 2024 Challenge Dataset includes three types of 3D T2-weighted (T2w) MRI scans: pre-RT, mid-RT, and registered pre-RT images of the head and neck. We randomly split the entire training set of 150 cases into 5 folds, training a separate model for each fold.

3.2   Data Preprocessing

SegRap2023. First, we performed morphological operations to crop the images to the region of interest within the human body based on intensity. Second, we randomly selected a pre-RT image from the HNTS-MRG2024 dataset and applied histogram matching to align the grayscale histograms, as illustrated in Fig. 4. Finally, we applied Z-score standardizing and resampled each volume into a resolution of $1.2\,\text{mm} \times 0.5\,\text{mm} \times 0.5\,\text{mm}$.

HNTS-MRG2024. We selected pixels with intensity values greater than 60 to identify the region of interest. We then located the largest connected component and applied morphological operations to create a mask of the human body. Next, we cropped the image along the x and y axes to match the dimensions of the head area. Finally, we normalized the image using Z-score standardization and resampled it to the same resolution used for the SegRap2023 dataset.

3.3   Models

As illustrated in Fig. 1, we proposed two distinct network architectures. Figure 1(a) displays a basic encoder-decoder model designed for semantic segmentation, which consists of six resolution stages. Each stage includes two convolutional layers with instance normalization, a LeakyReLU activation function, and a residual connection. This is followed by a DownPool block that contains an additional convolutional layer, instance normalization, and LeakyReLU activation. The structure of the decoder is symmetrical to that of the encoder, but it utilizes UpSample with transposed convolution. Features are pooled five times along the x and y axes, and three times along the z axis. In Fig. 1(b), the Dual Flow UNet (DFUNet) introduces an additional encoder and cross-attention blocks for merging two data streams. Both decoders share a design with four $1 \times 1 \times 1$ convolutions to output three channels for deep supervision.

3.4   Implementations

All the models were trained for 1000 epochs with a batch size of 2. When applying the MixUp [18] augmentation strategy, the batch size was increased to 4, as illustrated in Fig. 1(a). We used the Stochastic Gradient Descent (SGD) optimizer, with a momentum of 0.99, an initial learning rate of 0.01, and a weight decay of 3×10^{-5}. Training was conducted on a single NVIDIA RTX 4090 with 24GB of VRAM, using a patch size of $56 \times 224 \times 160$. Inference was performed with a sliding window strategy, enhanced by Test Time Augmentation (TTA) that included operations such as axes flips. Model performance was evaluated based on the aggregated Dice Similarity Coefficient (DSC).

4   Results

The results of the various training strategies based on a five-fold cross-validation of the training set, as outlined in the methods section, are detailed in Table 1 for Task-1 and Table 2 for Task-2. The bolding in Table 1 and Table 2 indicates the models selected for the corresponding data split, representing the models that yielded the best results in those particular folds. Overall, the average cross-validation performances for the selected models, as measured by the aggregated DSC, are 80.65% for Task-1 and 74.68% for Task-2.

For Task-1, as shown in Table 1, all strategies struggle to achieve significant improvements in GTVp segmentation, although they positively impact GTVn. The combination of pre-trained weights and MixUp marginally enhance GTVp segmentation by 0.47% in aggregated DSC, while GTVn sees a more substantial improvement of 1.30%. As illustrated in Fig. 5(a), pre-trained weights facilitate goal exploration to some extent, whereas MixUp significantly aids in the identification of foreground categories.

In Task-2, the situation is more complex. Although incorporating pre-RT images and their corresponding labels significantly enhances the segmentation

Table 1. The aggregated DSC (%) values from ablation study of our method, based on 5-fold cross-validation for **Task-1**. The terms "Base", "Pre-train", "MixUp", and "Pre-train+MixUp" correspond to fully supervised training, fully supervised training with pre-trained weights, fully supervised training with MixUp augmentation, and a combination of pre-trained weights with MixUp, respectively.

	Base		Pre-train		MixUp		Pre-train+MixUp	
	GTVp	GTVn	GTVp	GTVn	GTVp	GTVn	GTVp	GTVn
Fold1	78.63	83.89	78.60	83.44	81.64	84.81	**82.07**	**84.33**
Fold2	67.51	82.54	68.31	82.95	65.94	84.87	**67.75**	**84.88**
Fold3	70.68	84.86	71.10	86.55	70.21	85.79	**70.65**	**87.19**
Fold4	80.59	84.20	80.48	84.45	80.78	84.94	**80.88**	**84.72**
Fold5	**80.39**	**83.60**	77.66	84.70	79.80	84.02	78.82	84.48
Average	75.56	83.82	75.23	84.42	75.67	84.89	**76.03**	**85.12**

Table 2. The aggregated DSC (%) values of ablation study of our method, based on 5-fold cross-validation for **Task-2**. "Base" refers to fully supervised learning using mid-RT as inputs. "Base+pre-RT" signifies that the inputs have been expanded to include both mid-RT and registered pre-RT along with their labels. "DFUNet" indicates the substitution of the model with the DFUNet architecture compared to "Base+pre-RT". "Pre-train+MixUp" means the experiments were conducted on the basic segmentation network depicted in Fig.1 (a) with pre-trained weights and MixUp augmentation.

	Base		Base+pre-RT		DFUNet		Pre-train+MixUp	
	GTVp	GTVn	GTVp	GTVn	GTVp	GTVn	GTVp	GTVn
Fold1	40.46	74.10	59.80	86.62	**64.48**	**86.82**	62.08	87.63
Fold2	27.34	66.65	**59.01**	**85.00**	56.93	85.36	56.91	84.77
Fold3	38.21	75.47	**65.46**	**86.96**	58.74	86.65	64.98	87.45
Fold4	36.15	70.32	63.65	88.09	65.85	87.14	**64.89**	**88.32**
Fold5	45.85	79.72	58.66	86.59	55.87	86.92	**58.24**	**87.55**
Average	37.60	73.25	61.32	86.65	60.37	86.58	**61.42**	**87.14**

performance, leading to an 18.56% point increase in the average aggregated DSC, other methods show minimal improvement for GTVp segmentation, as demonstrated in Table 2. The segmentation results for the case shown in Fig. 6(a) follow the design of our method. Notably, DFUNet outperforms the basic segmentation network in fold 1 and 4 of Table 2, with average aggregated DSC improvements of 2.44% and 0.63%, respectively, which is not observed in other methods. In contrast, its performance is significantly worse in the other folds.

In the final testing phase, we selected the models that demonstrated the best compromise in performance and integrated their predictions, which are highlighted in Table 1 and Table 2. The final test allowed for a single submission, with our results for both tasks summarized in Table 3.

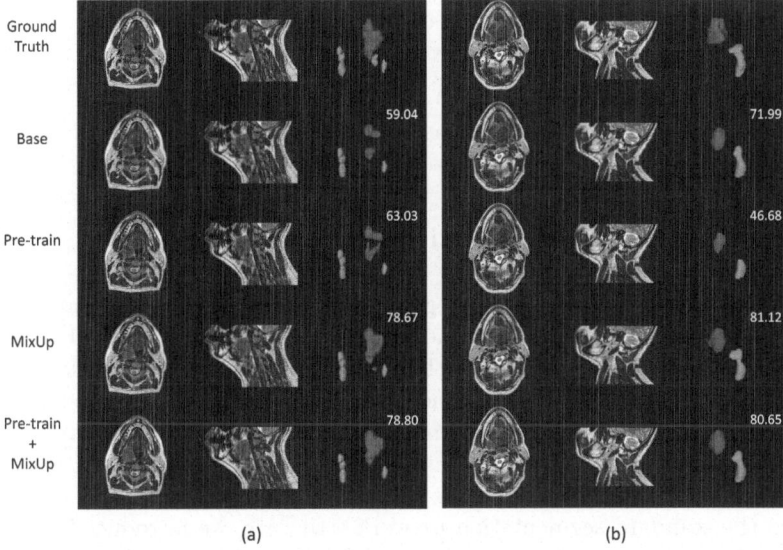

Fig. 5. Examples of **Task-1** illustrate the segmentation of GTVp (red) and GTVn (green) with mean DSC values (numbers in white). (a) represents a well-predicted case, while (b) shows a poorly predicted one. Classification of samples as well- or poor-predicted here refers to whether they meet the method's improvements.(Color figure online)

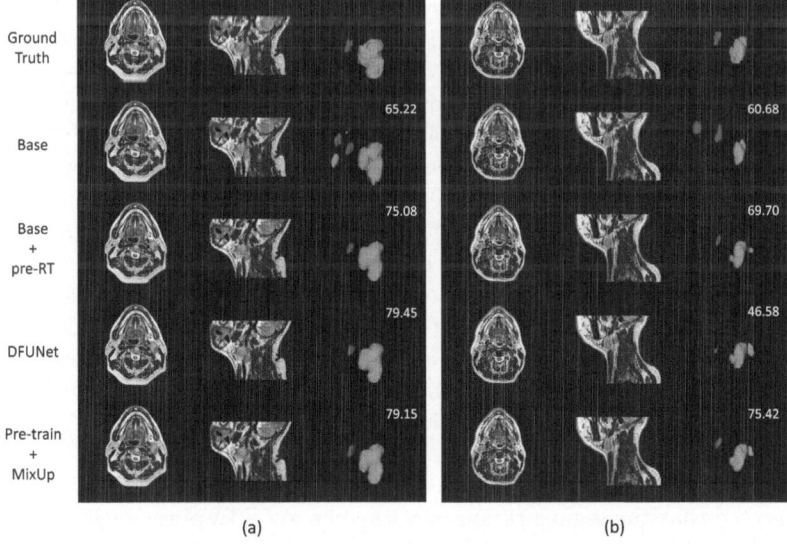

Fig. 6. Examples of **Task-2** illustrate the segmentation of GTVp (red) and GTVn (green) with mean DSC values (numbers in white). (a) represents a well-predicted case, while (b) is the bad one.

Table 3. Results of Final Test for the two tasks on aggregated DSC (%)

	GTVp	GTVn	Average
Task-1	78.51	86.24	82.38
Task-2	57.95	87.12	72.53

5   Discussion and Conclusion

In this study, we developed GTV segmentation models using two variations of the VNet-like [10] architecture on the MRI T2w H&N dataset from the HNTS-MRG2024 Challenge. Our investigation focused on expanding the dataset and refining network structures. To augment the dataset, we initially used an external public CT dataset [9] for pre-training and subsequently applied MixUp technology to generate more training data. For Task-2, we also utilized additional registered pre-RT images and masks to enhance the segmentation of mid-RT. Beyond the standard segmentation network structure, we introduced the DFU-Net, which includes two encoders, one decoder, and CNN-based cross attention blocks. This architecture enables the network to differentiate between the primary image for segmentation and supplementary information.

The 5-fold cross-validation results indicate that our proposed strategies have led to significant improvements in the segmentation of simpler categories, such as GTVn. However, the advancements in the segmentation of a more challenging category, like GTVp, have been marginal and could even introduce adverse effects, especially for Task-2, as shown in Fig. 6(b). We suspect that the suboptimal learning of GTVp is due to the severe class imbalance and the learning of full background samples. While nnUNetv2 [6] samples foreground categories during training, its methodology does not adequately address the imbalance among foreground categories. A potential solution might involve an expanded oversampling strategy that targets each foreground and background category individually. Although the DFUNet underperformed compared to the basic segmentation model in cross-validation, it demonstrated notable enhancements in GTVp segmentation in some folds. We are confident that with further refinements to the DFUNet architecture, it could outperform the basic model. Additionally, it is worth noting that pre-training is not universally effective, and in some cases, the use of pre-trained weights can significantly degrade segmentation performance. As illustrated in Fig. 5(b), the weaker segmentation performance of GTVn in CT scans may hinder the model's ability to segment MRI samples that are difficult to segment for GTVn.

Furthermore, during pre-training with the SegRap2023 Challenge dataset [9], we observed that the segmentation performance for GTVp was markedly higher on the CT dataset compared to GTVn. Interestingly, this trend was reversed for T2-weighted MRI in HNTS-MRG2024. Although the types of cancer in the two datasets are different, a model that can combine the advantages of the two modalities can improve the segmentation performance while greatly reducing the dependence on multi-modal data and minimizing the patient's exposure during image acquisition. Potential approaches to achieve this include cross-modal distillation [15], domain adaptation [7], or the use of generative models [17].

In conclusion, we investigated the enhancement of fully supervised learning through dataset expansion and domain generalization techniques. We observed a notable improvement in the performance of GTVn with pre-training and the application of the MixUp strategy. Additionally, DFUNet led to significant enhancements in GTVp segmentation in some folds. Our average aggregated DSC across the folds is 80.65% for Task-1 and 74.68% for Task-2. In the final test, we achieve scores of 82.38% for Task-1 and 72.53% for Task-2. Future work could concentrate on refining the DFUNet architecture, effectively initializing its weights, and leveraging pre-RT data to boost the segmentation performance of mid-RT.

Acknowledgments. This work was supported by the Radiation Oncology Key Laboratory of Sichuan Province Open Fund (2022ROKF04).

Disclosure of Interests. The authors have no competing interests to declare that are relevant to the content of this article.

References

1. Andrearczyk, V., et al.: Overview of the hecktor challenge at miccai 2021: automatic head and neck tumor segmentation and outcome prediction in pet/ct images. In: 3D Head and Neck Tumor Segmentation in PET/CT Challenge, pp. 1–37. Springer (2021)
2. Badrigilan, S., Nabavi, S., Abin, A.A., Rostampour, N., Abedi, I., Shirvani, A., Ebrahimi Moghaddam, M.: Deep learning approaches for automated classification and segmentation of head and neck cancers and brain tumors in magnetic resonance images: a meta-analysis study. Int. J. Comput. Assist. Radiol. Surg. **16**, 529–542 (2021)
3. Brændengen, M., Hansson, K., Radu, C., Siegbahn, A., Jacobsson, H., Glimelius, B.: Delineation of gross tumor volume (gtv) for radiation treatment planning of locally advanced rectal cancer using information from mri or fdg-pet/ct: a prospective study. Int. J. Radiation Oncol. Biol. Phys. **81**(4), e439–e445 (2011)
4. Chen, J., et al.: Transunet: transformers make strong encoders for medical image segmentation. arXiv preprint arXiv:2102.04306 (2021)
5. Dai, Y., King, A.: State of the art mri in head and neck cancer. Clin. Radiol. **73**(1), 45–59 (2018)
6. Isensee, F., Jaeger, P.F., Kohl, S.A., Petersen, J., Maier-Hein, K.H.: nnu-net: a self-configuring method for deep learning-based biomedical image segmentation. Nat. Methods **18**(2), 203–211 (2021)
7. Liu, J., et al.: Automated cardiac segmentation of cross-modal medical images using unsupervised multi-domain adaptation and spatial neural attention structure. Med. Image Anal. **72**, 102135 (2021)
8. Liu, X., Wu, J., Luo, X., Liao, W., Zhang, S., Zhang, S., Wang, G.: Rpl-sfda: reliable pseudo label-guided source-free cross-modality adaptation for npc gtv segmentation. In: 2024 IEEE International Symposium on Biomedical Imaging (ISBI), pp. 1–5. IEEE (2024)

9. Luo, X., et al.: Segrap2023: a benchmark of organs-at-risk and gross tumor volume segmentation for radiotherapy planning of nasopharyngeal carcinoma. arXiv preprint arXiv:2312.09576 (2023)
10. Milletari, F., Navab, N., Ahmadi, S.A.: V-net: fully convolutional neural networks for volumetric medical image segmentation. In: 2016 Fourth International Conference on 3D Vision (3DV), pp. 565–571. IEEE (2016)
11. Rumboldt, Z., Gordon, L., Bonsall, R., Ackermann, S.: Imaging in head and neck cancer. Curr. Treat. Options Oncol. **7**, 23–34 (2006)
12. Sager, O., Dincoglan, F., Demiral, S., Gamsiz, H., Uysal, B., Ozcan, F., Colak, O., Dirican, B., Beyzadeoglu, M.: Evaluation of the impact of magnetic resonance imaging (mri) on gross tumor volume (gtv) definition for radiation treatment planning (rtp) of inoperable high grade gliomas (hggs). Concepts in Magnetic Resonance Part A **2019**(1), 4282754 (2019)
13. Sun, J., Dai, Y., Zhang, X., Xu, J., Ai, R., Gu, W., Chen, X.: Efficient spatial-temporal information fusion for lidar-based 3d moving object segmentation. In: 2022 IEEE/RSJ International Conference on Intelligent Robots and Systems (IROS), pp. 11456–11463. IEEE (2022)
14. Vaswani, A.: Attention is all you need. Advances in Neural Information Processing Systems (2017)
15. Wang, H., et al.: Learnable cross-modal knowledge distillation for multi-modal learning with missing modality. In: International Conference on Medical Image Computing and Computer-Assisted Intervention, pp. 216–226. Springer (2023)
16. Woo, S., Park, J., Lee, J.Y., Kweon, I.S.: Cbam: convolutional block attention module. In: Proceedings of the European Conference on Computer Vision (ECCV), pp. 3–19 (2018)
17. Xia, Y., Feng, S., Zhao, J., Yuan, Z.: Robust cross-modal medical image translation via diffusion model and knowledge distillation. In: 2024 International Joint Conference on Neural Networks (IJCNN), pp. 1–8. IEEE (2024)
18. Zhang, H.: mixup: Beyond empirical risk minimization. arXiv preprint arXiv:1710.09412 (2017)
19. Zhou, Z., Qi, L., Yang, X., Ni, D., Shi, Y.: Generalizable cross-modality medical image segmentation via style augmentation and dual normalization. In: Proceedings of the IEEE/CVF Conference on Computer Vision and Pattern Recognition, pp. 20856–20865 (2022)

Head and Neck Tumor Segmentation on MRIs with Fast and Resource-Efficient Staged nnU-Nets

Elias Tappeiner$^{(\boxtimes)}$, Christian Gapp , Martin Welk ,
and Rainer Schubert

UMIT Tirol – Private University for Health Sciences and Health Technology,
Eduard-Wallnöfer-Zentrum 1, Hall in Tirol 6060, Austria
elias.tappeiner@umit-tirol.at

Abstract. MRI-guided radiotherapy (RT) planning offers key advantages over conventional CT-based methods, including superior soft tissue contrast and the potential for daily adaptive RT due to the reduction of the radiation burden. In the Head and Neck (HN) region labor-intensive and time-consuming tumor segmentation still limits full utilization of MRI-guided adaptive RT. The HN Tumor Segmentation for MR-Guided Applications 2024 challenge (HNTS-MRG) aims to improve automatic tumor segmentation on MRI images by providing a dataset with reference annotations for the tasks of pre-RT and mid-RT planning.

In this work, we present our approach for the HNTS-MRG challenge. Based on the insights of a thorough literature review we implemented a fast and resource-efficient two-stage segmentation method using the nnU-Net architecture with residual encoders as a backbone. In our two stage approach we use the segmentation results of a first training round to guide the sampling process for a second refinement stage. For the pre-RT task, we achieved competitive results using only the first-stage nnU-Net. For the mid-RT task, we could significantly increase the segmentation performance of the basic first stage nnU-Net by utilizing the prior knowledge of the pre-RT plan as an additional input for the second stage refinement network. As team alpinists we achieved an aggregated Dice Coefficient of 80.97 for the pre-RT and 69.84 for the mid-RT task on the online test set of the challenge. Our code and trained model weights for the two-stage nnU-Net approach with residual encoders are available at https://github.com/elitap/hntsmrg24.

Keywords: staged nnU-Net · MRI · Head and Neck Tumor · segmentation

1 Introduction

Tumors in the Head and Neck (HN) region are primarily treated with radiotherapy (RT) [1]. MRI-guided RT planning offers several advantages over traditional CT-based approaches, particularly in terms of superior soft tissue contrast and the potential for daily adaptive RT due to the reduction of radiation

K. A. Wahid et al. (Eds.): HNTS-MRG 2024, LNCS 15273, pp. 87–98, 2025.
https://doi.org/10.1007/978-3-031-83274-1_6

side effects [21]. Intra-therapy MRI imaging allows for more precise targeting of tumors while sparing surrounding healthy tissues, which is crucial given the complex anatomy and critical structures in the HN region [11]. Despite these advances, the process of manually segmenting HN tumors on MRI scans remains labor-intensive and time-consuming, limiting the feasibility of fully utilizing MRI-guided adaptive RT in clinical practice [14]. The HNTS for MR-Guided Applications 2024 challenge (HNTS-MRG) aims to address this gap by providing a comprehensive dataset and a competitive platform for the development and benchmarking of automated HN tumor segmentation algorithms.

Participants in the HNTS-MRG 2024 challenge are tasked with developing algorithms to segment primary gross tumor volumes (GTVp) and metastatic lymph nodes (GTVn) on MRI scans, based on a training set of 150 patients. The algorithms are evaluated over the Grand Challenge platform[1] on a hidden test set of 50 patients. The challenge is divided into two tasks. The first is focusing on pre-radiotherapy (pre-RT) planning and the second on mid-radiotherapy (mid-RT) planning. Notably, the second task encourages the exploration of data from prior timepoints to enhance segmentation performance.

Within this work, we present our approach to the HNTS-MRG 2024 challenge, which is based on the nnU-Net architecture with residual encoders of Isensee et al. [13]. Given the scarcity of high-end DL hardware since the rise of large language models in 2022 and the environmental impact of training these models [19], our primary focus is on the efficient use of computational resources while striving to achieve competitive results. Based on the hidden test data set of the challenge, we achieved an aggregated Dice Similarity Coefficient (aggDSC) of 80.97 for the pre-RT task using an nnU-Net with residual encoders and a target GPU training memory size of 24GB. For the mid-RT task we achieved an aggDSC of 69.84 by including the registered images of the pre-RT as additional input channels and applying a sampling strategy based on the available pre-RT segmentation mask. Being mindful and efficient in the use of computational resources our results show that convolutional neural networks (CNNs), especially the nnU-Net architecture, are still very well-suited for the given complex medical image segmentation using the given (relatively) small-sized datasets. For the mid-RT task, the incorporation of prior timepoint data significantly improved the training and inference times as well as the segmentation performance.

2   Dataset

The challenge dataset comprises 200 anonymized cases of patients with histologically confirmed HN cancer, primarily oropharyngeal cancer, treated with RT at the University of Texas MD Anderson Cancer Center. The dataset is split into 150 training cases, which are publicly available [28], and 50 hidden cases used for the online evaluation phase of the challenge. The dataset includes T2-weighted MRI scans captured pre-RT (1–3 weeks before treatment) for Task 1 of the challenge and mid-RT scans (2–4 weeks into treatment) for Task 2 of

---

[1] https://grand-challenge.org/.

the challenge. Along with the MRI images, both tasks include corresponding GTVp and GTVn segmentations. Task 2 additionally includes elastically registered versions of corresponding pre-RT MRI scans and their pre-RT tumor segmentations. The segmentation of GTVs was performed by multiple expert observers and combined using the STAPLE algorithm [30] to produce consensus ground truth segmentations, with final validation by experienced radiation oncology faculty. The dataset is designed to support the training of machine learning models, and both training and test sets are structured to reflect real-world clinical cases. Ethical approval for the use of this data was granted by the MD Anderson Cancer Center's Institutional Review Board. The hidden test set's ground truth segmentations are not available until the end of the challenge.

3   Related Work

Most similar works with interesting insights for our participation in the HNTS-MRG 2024 challenge are the review papers of similar challenges such as the HEad and neCK TumOR (HECKTOR) segmentation in PET/CT challenge 2020 [2], 2021 [3] and 2022 [4], the Segmentation of Organs-at-Risk and GTV of patients with nasopharyngeal carcinoma for Radiotherapy Planning (SegRap) challenge 2023 [16], and the Head and Neck Organ at Risk (OAR) MR and CT segmentation (HaN-Seg) challenge 2023 [20].

The HECKTOR 2020 challenge, was the first large HN tumor segmentation challenge. Participants were asked to segment the primary GTV in the oropharynx region based on 201 FDG-PET/CT images. The best-performing team submitted ensembles of 3D U-Nets [7], which were trained using a combination of the Dice loss [17] and the Focal loss [15]. The second placed team used a combination of a U-Net trained with Dice [17] and Top-K loss [31] functions and refined the results with an active contour approach. The third place was achieved using a classifier to identify slices with GTVs and a 2D U-Net [22] for the segmentation. The following HECKTOR 2021 challenge was expanded by two additional tasks and provided a training dataset of 224 PET and CT images. For Task 1, automatic HN GTV segmentation, the first, second, fourth and fifth ranked teams participated with ensembled nnU-Net [12] variants. The second-placed team implemented a staged approach that combined an initial rough segmentation with bounding box localization, followed by refined segmentations using a U-Net variant. In the most recent HECKTOR 2022 challenge, Task 1 was expanded to include the automatic segmentation of both primary GTV (GTVp) and metastatic lymph nodes (GTVn). The organizers provided an extensive training dataset of 524 segmented PET and CT images for the task. The first-place team utilized ensembled SegResNets [18] with automated parameter selection. The second-placed participants, similarly to the previous year, adopted a tumor localization and fine grained segmentation within the found bounding boxes. The final fine segmentation was achieved with ensembled nnU-Net and nnFormer [33] variants within this refined region. The third-placed team opted for an off-the-shelf nnU-Net [12] with basic pre- and post-processing rules.

The objective of the SegRap challenge [16] was the segmentation of 42 OARs (Task 1) in the HN region as well as GTVp and GTVn of nasopharyngeal carcinomas (Task 2) based on a training set of 140 non-contrast (ncCT) and contrast-enhanced CT (ceCT) images. For Task 2, the first-place team focused on efficient cropping and intensity clamping based on the HU values of ceCT and ncCT volumes, followed by training an nnU-Net-based segmentation network [12]. The second-place team used the UniSeg model [32], a pre-trained nnU-Net, with a bespoke pre-processing and ensemble strategy. The third-placed team also employed HU-based cropping before training different parameterized U-Net models [7], which are combined with additional test-time augmentation during inference.

The most recent challenge, the HaN-Seg 2023, differs from previous challenges by focusing solely on the segmentation of (OARs) in the HN region based on 40 MRI and CT images. We participated as team eli1 and achieved the first place using rigid registration of the MRI and CT images, followed by an nnU-Net-based [12] segmentation with an increased patch size. The second place was achieved using a YOLOv7 model [29] for the detection of the OARs, followed by OAR segmentation within the detected organ locations with a MONAI-based implementation of the nnU-Net named DynUNet. The third place was achieved using a pre-trained variant of the nnU-Net, trained with the class-adaptive Dice loss function [26] on rigidly registered MRI and CT images.

Amongst the studied challenges, the nnU-Net architecture has been a popular choice for many participants often leading to top results. The nnU-Net architecture introduced by Isensee et al. in 2021 [12] is a flexible and efficient deep learning framework for medical image segmentation based on the U-Net architecture [22]. Using a combination of dataset-dependent heuristics to define elements such as the spacing, patch-size, batch-size and the detailed network architecture as well as fixed hyperparameters such as the optimizer, learning rate, scheduler and loss functions, the nnU-Net architecture is able to achieve state-of-the-art performance without the need of extensive hyperparameter tuning. In a recent update, Isensee et al. [13] extended the base nnU-Net architecture with residual [10] encoders and reevaluated its performance on six different medical image segmentation tasks against other CNN-, transformer- [27] and mamba-based [9] architectures. The results show that the nnU-Net architecture with residual encoders is able to achieve high-quality performance while being more memory and runtime efficient than the other architectures. Although the newly presented nnU-Net clearly outperforms the observed transformer and mamba architectures, it is slightly outperformed by another CNN-based method, the MedNext [23]. According to Isensee et al., the MedNeXt's performance gains are mainly explained by a substantially increased training time and the target spacing selection. Concluding, the authors tie the dominance of CNNs to the evaluation setting of training methods from scratch on benchmarks with limited dataset sizes, a setting which we also face in the HNTS-MRG 2024 challenge.

Besides the nnU-Net as a common architecture choice, many challenge participants applied multi-staged approaches combining a localization and segmentation stage. In earlier work, we also investigated the usage of a two-staged training approach for the HN organ at risk segmentation [25]. To refine an initial rough U-Net-based segmentation mask we apply a second training stage that is guided by the initial rough segmentation mask. The second stage operates on a reduced search space and benefits from a reduced within-class imbalance [26], which lead to improvements of the final segmentation performance.

4   Methods

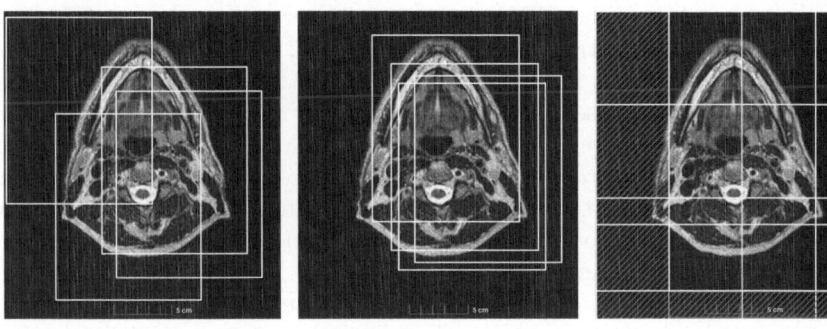

Fig. 1. Left: Training process of the first stage networks using the default nnU-Net sampling strategy. Middle: Training process of the second stage networks with mask-based oversampling, using masks from the first stage (pre-RT Task) or the pre-RT reference segmentations (mid-RT Task). Right: Sliding window inference with 50% overlap; marked areas are not inferred for the mid-RT Task.

Based on previous segmentation challenges and the reevaluation of the nnU-Net by Isensee et al. [13], which demonstrated its qualitative performance and computational efficiency, our method combines our previous work [25] using a two-stage training approach with the latest nnU-Net variant featuring residual encoders, the nnU-Net_ResEncM.

4.1   Pre-RT Task

For the pre-RT Task, we first train an nnU-Net_ResEncM with a GPU memory target of 24GB to utilize the full VRAM of the Nvidia Titan RTX available for training. Based on the dataset fingerprint of the challenge training dataset, the 3d_fullres variant of the nnU-Net_ResEncM is configured by the nnU-Net framework to have seven encoder/decoder stages, with a total of only 141 million trainable parameters. The network is trained with a dataset spacing of $0.5 \times 0.5 \times 1.2$ mm, a patch size of $320 \times 256 \times 64$, and a batch size of 2. By default, the optimizer used within the nnU-Net framework is Stochastic Gradient Descent with Nesterov momentum [6], trained with an initial learning rate of

0.01 and polynomial learning rate decay. The loss function of the nnU-Net is set as a combination of the Dice loss [17] and the Cross-Entropy loss, trained with deep supervision. During training the input data is augmented with the default nnU-Net augmentation techniques. Final segmentation results are obtained using a Gaussian-weighted sliding window approach with patch overlaps.

The second refinement stage is similar to our previous work [25], where a model is trained based on the segmentation output of the first stage. The first stage output is used as an additional input, guiding the patch sampling process during training by allowing only samples drawn from within the segmentation mask of the first stage. With the first stage providing candidate regions for the segmentation, the second stage can focus on refining the initial segmentation mask, as the patches are now likely to contain larger portions of the GTVp or GTVn than in the first stage. Compared to many staged methods applied in previous challenges, we do not use a separate localization network or bounding box detection to guide the segmentation process. This results in a straightforward approach that allows the reuse of the same network architecture in both stages. Notably, our method contrasts with the cascaded configuration of the nnU-Net, which uses the output of a first low-resolution training stage as an additional input channel for the second high-resolution training stage. Additionally, our patch sampling strategy differs from the default nnU-Net patching, which draws 33% of the training patches to include foreground label information based on the ground truth reference for training stability. Figure 1 (left and middle) illustrates the training process of the first and second stage networks.

4.2   Mid-RT Task

Unlike the pre-RT Task, where only a single MRI image is available, the mid-RT segmentation Task allows the utilization of the intra-therapy MRI, the registered pre-RT MRI, and the segmentation mask of the pre-RT GTVs. Assuming that the RT treatment is effective, reduced size GTVp and GTVn are expected to be in the same locations as found in the pre-RT MRI. Accordingly, for the mid-RT Task, we utilize the segmentation mask of the pre-RT phase as the additional input to guide the patch sampling process, requiring the training of only the second-stage nnU-Net_ResEncM. In contrast to the pre-RT Task, where the first stage output is not guaranteed to be accurate, we also use the pre-RT segmentation mask for the sliding window inference, by again only considering patches drawn from within the segmentation mask as shown in Fig. 1 (right).

For the training process, due to limited access to large amounts of high-end DL hardware and the narrow time frame between dataset release and challenge submission deadline, we randomly split the training data for the pre-RT and mid-RT tasks into a single training and validation set of 120 and 30 patients, respectively. Both the first and second stages of the nnU-Nets are trained as defined by the nnU-Net framework for 1000 epochs, with each epoch consisting of 250 training steps. The code of the nnU-Net with our implementation for the two-stage training as well as trained model weights are available at https:// github.com/elitap/hntsmrg24

5   Results

Since the HECKTOR 2022 challenge, the results of the automated GTVp and GTVn segmentation are evaluated based on the aggregated Dice Similarity Coefficient (aggDSC) [5]. The final ranking of the challenge is based on the average aggDSC of the GTVp and GTVn. Unlike the per-patient DSC, the aggDSC is robust against single false negative or positive results.

Table 1. aggDSC results for the epoch with the best validation metric (approximated DSC) and the fully trained final model evaluated on the validation data split, for the configurations of the pre-RT and mid-RT tasks. The best performing average results of each task and their configuration are marked in bold.

		best			final		
		GTVp	GTVn	avg (epoch)	GTVp	GTVn	avg
nnU-Net	pre-RT	79.56	87.45	83.51 (874)	79.72	86.95	83.33
2ndst_nnU-Net (c1)		78.25	86.36	82.31 (941)	76.69	76.81	76.75
2ndst_nnU-Net (c2)		79.73	87.45	83.59 (909)	79.97	87.61	**83.79**
nnU-Net (c1)	mid-RT	44.47	74.17	59.32 (104)	29.91	79.84	54.87
nnU-Net (c3)		62.90	78.21	70.55 (702)	52.19	82.90	67.55
2ndst_nnU-Net (c3)		59.24	85.95	**72.59** (283)	53.87	81.49	67.68

Based on our method, Table 1 presents the aggDSC results evaluated on our validation split of the dataset. To evaluate the impact of the second-stage training, Table 1 also contains the results of relevant ablation studies. For the pre-RT task, we also present the results of the first-stage nnU-Net, as well as two second-stage configurations. In the second stage nnU-Net (c1) configuration, we use the output of the first stage only for patch sampling, while for the second stage nnU-Net (c2) configuration, we use the output of the first stage as a full additional input channel for network training as well. For the mid-RT task, we train two first-stage configurations: one with the basic first-stage nnU-Net using the mid-RT MRI as the only input, and another with the registered pre-RT MRI and the pre-RT segmentation mask as additional input channels. Following the results from first-stage training, for the second stage the final configuration is trained using the mid-RT MRI, the registered pre-RT MRI and the pre-RT segmentation mask as input channels as well as the pre-RT mask for patch selection. Although the challenge is evaluated based on the aggDSC, individual per-patient measurements are of relevance for clinical applications. Accordingly, Fig. 2 also shows box plots of the average DSCs of each configuration for the pre-RT task (top) and the mid-RT task (bottom). Configurations marked with a star (*) significantly differ from the method with the best aggDSC marked in bold. The tests are based on a paired Wilcoxon signed-rank test with a significance level of 0.05.

Overall, we trained a total of six models for validation (including ablations) and an additional two models for the final submission. The training of a single nnU-Net in the given nnU-Net_ResEncM configuration took approximately 83h under full utilization of the used Nvidia Titan RTX. We also adopt Code-Carbon [24] to track the computational efficiency of our approach, measuring an energy consumption of around 25 kWh, which corresponds to 2.8 kg CO_2-equivalents, for the training of a single model. In summary, our entire challenge participation incurred marginal electricity costs of only 60 EUR for training in our region.

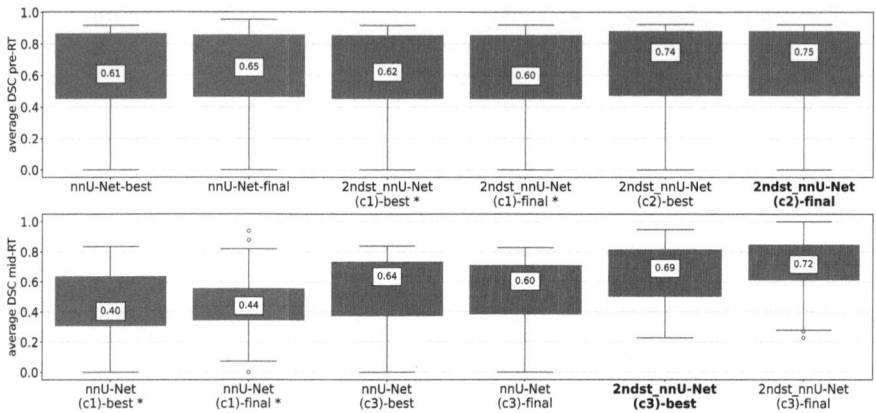

Fig. 2. Individual average per-patient DSC results for the pre-RT task (top) and the mid-RT task (bottom). Configurations marked with a star (*) are significantly different from the best performing configuration marked in bold.

6   Discussion and Conclusion

The model selections for the final submission to the online evaluation of the challenge are based on the validation results of Table 1 and Fig. 2, as well as resource requirements. To utilize the whole training dataset for the final submission, we retrained the selected pre-RT and mid-RT models on the full 150-patient dataset.

For the pre-RT task, the best validation scores are achieved towards the end of the training process. Models with checkpoints at the best validation score achieve similar validation results as the fully trained models, except for the two-staged nnU-Net (c1) configuration, where we observed the appearance of islands in the segmentation masks outside the regions of the first stage mask. Due to our introduced sampling process, the network is expected to always contain foreground regions in the patches, and accordingly introduces the islands during inference. The problem is solved in the two-staged nnU-Net (c2) configuration, where the first stage output is used as an additional input channel, allowing the network to learn where to find the GTVp and GTVn segments. Ultimately, the second stage of the nnU-Net (c2) outperforms the first-stage nnU-Net baseline in terms of the aggDSC. However, our statistical analysis on the individual per-patient DSCs shows no significant difference from the nnU-Net_ResEncM baseline configuration. Opting for resource efficiency, we submitted the basic first-stage nnU-Net configuration for the pre-RT task, achieving a final aggDSC of 80.97 in the online evaluation of the pre-RT task of challenge.

For the mid-RT task, the best validation scores are reached much earlier in the training process, and the models with checkpoints at the best validation score outperform the models trained for the whole 1000 epochs in all three configurations according to the aggDSC results of Table 1. Although Table 1 indicates signs of overfitting, the per-patient DSCs in Fig. 2 do not support this observation. Overall, both the aggDSC and the patient-wise DSC evaluations clearly show that the incorporation of the pre-RT data as additional input channels into the basic nnU-Net configuration improves the segmentation performance over the baseline model. We did not find a significant difference between the nnU-Net (c3) and the second stage nnU-Net (c3) configuration, however, the second stage nnU-Net (c3) configuration outperforms the basic nnU-Net (c3) configuration in terms of aggDSC and per-patient DSCs. Accordingly, for the final submission, we selected the two-staged nnU-Net (c3) configuration, where, due to the utilization of the pre-RT input as the first-stage output, only a single second-stage network needs to be trained. Additionally, the second stage nnU-Net (c3) configuration is much faster during inference as the pre-RT mask is also used to guide the sliding window inference. Finally, following our experience in training CNN-based networks for 3D medical image segmentation, where we found that longer training generally still improves distance-based metrics such as the Hausdorff Distance [25,26] as well as the configuration's strong per patient DSC, we also trained the final model for the mid-RT task for 1000 epochs. Ultimately, the submitted second stage nnU-Net (c3) model achieves an aggDSC of 69.84 in the mid-RT task of the HNTS-MRG 2024 challenge. Table 2 presents the aggDSC of the online evaluation of the hidden test set from the Grand Challenge platform.

Table 2. Final aggDSC results based on the online test set evaluation for the pre-RT and the mid-RT tasks.

	Task	GTVp	GTVn	avg
nnU-Net-final	pre-RT	76.71	85.22	80.97
2ndst_nnU-Net (c3)-final	mid-RT	53.86	85.82	69.84

In conclusion, our results show that the medium sized nnU-Net architecture with residual encoders and a parameter count of only 141 million is still a competitive choice for the given complex medical image segmentation tasks, especially when only limited amounts of training data is available. Our two-staged training approach, where the first stage output is used to guide the sampling process for the second stage, is a straightforward and efficient method to improve the segmentation performance, however we found that the performance gains do not outweigh the training of a second nnU-Net. Nonetheless, the incorporation of prior timepoint data as input to a second-stage only network significantly improves the training and inference times as well as the segmentation performance of the mid-RT task, which is the major contribution of our work. Overall we can highlight that it is possible to achieve competitive performance using current consumer hardware and inexpensive-to-train standalone networks.

A compelling continuation in the spirit of this work would be to investigate the potential of patch curriculum learning, recently introduced by Fischer et al. [8]. The method involves training a network using progressively larger image patches, which the authors demonstrated could substantially reduce the number of training epochs needed to attain competitive results. Integrating this approach with our two-stage training method could enhance segmentation performance by guiding the initial, smaller patches to focus on relevant image regions. The combination may further bridge the performance gap between the pre-RT and mid-RT tasks.

Disclosure of Interests. The authors declare to have no competing interests.

References

1. Alterio, D., Marvaso, G., Ferrari, A., Volpe, S., Orecchia, R., Jereczek-Fossa, B.A.: Modern radiotherapy for head and neck cancer. Seminars in Oncology, pp. 233–245 (2019). https://doi.org/10.1053/j.seminoncol.2019.07.002
2. Andrearczyk, V., et al.: Overview of the HECKTOR challenge at MICCAI 2020: Automatic head and neck tumor segmentation in PET/CT. In: Andrearczyk, V., Oreiller, V., Depeursinge, A. (eds.) Head and Neck Tumor Segmentation, pp. 1–21. Springer (2021). https://doi.org/10.1007/978-3-030-67194-5_1
3. Andrearczyk, V., et al.: Overview of the HECKTOR challenge at MICCAI 2021: Automatic head and neck tumor segmentation and outcome prediction in PET/CT images. In: Andrearczyk, V., Oreiller, V., Hatt, M., Depeursinge, A. (eds.) Head and Neck Tumor Segmentation and Outcome Prediction, pp. 1–37. Springer (2022). https://doi.org/10.1007/978-3-030-98253-9_1
4. Andrearczyk, V., et al.: Overview of the HECKTOR challenge at MICCAI 2022: Automatic head and neck tumor segmentation and outcome prediction in PET/CT. In: Andrearczyk, V., Oreiller, V., Hatt, M., Depeursinge, A. (eds.) Head and Neck Tumor Segmentation and Outcome Prediction, pp. 1–30. Springer Nature Switzerland (2023). https://doi.org/10.1007/978-3-031-27420-6_1
5. Andrearczyk, V., Oreiller, V., Jreige, M., Castelli, J., Prior, J.O., Depeursinge, A.: Segmentation and classification of head and neck nodal metastases and primary tumors in PET/CT. In: Annual International Conference of the IEEE Engineering in Medicine and Biology Society (EMBC), pp. 4731–4735 (2022). https://doi.org/10.1109/EMBC48229.2022.9871907
6. Botev, A., Lever, G., Barber, D.: Nesterov's accelerated gradient and momentum as approximations to regularised update descent. In: 2017 International Joint Conference on Neural Networks (IJCNN), pp. 1899–1903 (2017). https://doi.org/10.1109/IJCNN.2017.7966082

7. Özgün Çiçek, Abdulkadir, A., Lienkamp, S.S., Brox, T., Ronneberger, O.: 3D U-net: Learning dense volumetric segmentation from sparse annotation. Proceedings of the Conference on Medical Image Computing and Computer-Assisted Intervention, pp. 424–432 (2016). https://doi.org/10.1007/978-3-319-46723-8_49
8. Fischer, S.M., et al.: Progressive growing of patch size: Resource-efficient curriculum learning for dense prediction tasks. In: Medical Image Computing and Computer Assisted Intervention – MICCAI 2024, pp. 510–520 (2024). https://doi.org/10.1007/978-3-031-72114-4_49
9. Gu, A., Dao, T.: Mamba: Linear-time sequence modeling with selective state spaces (2024). https://arxiv.org/abs/2312.00752
10. He, K., Zhang, X., Ren, S., Sun, J.: Deep residual learning for image recognition. In: IEEE Conference on Computer Vision and Pattern Recognition, pp. 770–778 (2015). https://doi.org/10.1109/CVPR.2016.90
11. Hindocha, S., et al.: Artificial intelligence for radiotherapy auto-contouring: current use, perceptions of and barriers to implementation. Clin. Oncol. **35**(4), 219–226 (2023). https://doi.org/10.1016/j.clon.2023.01.014
12. Isensee, F., Jaeger, P.F., Kohl, S.A., Petersen, J., Maier-Hein, K.H.: nnU-Net: a self-configuring method for deep learning-based biomedical image segmentation. Nat. Methods **18**, 203–211 (2021). https://doi.org/10.1038/s41592-020-01008-z
13. Isensee, F., Wald, T., Ulrich, C., Baumgartner, M., Roy, S., Maier-Hein, K., Jaeger, P.F.: nnU-Net Revisited: A Call for Rigorous Validation in 3D Medical Image Segmentation (2024). https://arxiv.org/abs/2404.09556
14. Kiser, K.J., Smith, B.D., Wang, J., Fuller, C.D.: Après mois, le déluge: preparing for the coming data flood in the MRI-Guided radiotherapy era. Front. Oncol. **9**, 983 (2019). https://doi.org/10.3389/fonc.2019.00983
15. Lin, T.Y., Goyal, P., Girshick, R., He, K., Dollar, P.: Focal loss for dense object detection. In: Proceedings of the IEEE International Conference on Computer Vision (ICCV) (2017). https://doi.org/10.1109/ICCV.2017.324
16. Luo, X., et al.: Segrap2023: A benchmark of organs-at-risk and gross tumor volume segmentation for radiotherapy planning of nasopharyngeal carcinoma (2023). https://arxiv.org/abs/2312.09576
17. Milletari, F., Navab, N., Ahmadi, S.A.: V-Net: Fully convolutional neural networks for volumetric medical image segmentation. In: Conference on 3D Vision, pp. 565–571 (2016). https://doi.org/10.1109/3DV.2016.79
18. Myronenko, A.: 3D MRI brain tumor segmentation using autoencoder regularization (2018). https://arxiv.org/abs/1810.11654
19. Patterson, D., et al.: Carbon emissions and large neural network training (2021). https://arxiv.org/abs/2104.10350
20. Podobnik, G., et al.: HaN-Seg: The head and neck organ-at-risk CT and MR segmentation challenge. Radiotherapy and Oncology **198** (2024). https://doi.org/10.1016/j.radonc.2024.110410
21. Pollard, J.M., Wen, Z., Sadagopan, R., Wang, J., Ibbott, G.S.: The future of image-guided radiotherapy will be MR guided. British J. Radiol., 20160667 (2017). https://doi.org/10.1259/bjr.20160667
22. Ronneberger, O., Fischer, P., Brox, T.: U-net: Convolutional networks for biomedical image segmentation. In: Conference on Medical Image Computing and Computer-Assisted Intervention, pp. 234–241 (2015). https://doi.org/10.1007/978-3-319-24574-4_28
23. Roy, S., et al.: Mednext: transformer-driven scaling of convnets for medical image segmentation. In: Medical Image Computing and Computer Assisted Intervention

– MICCAI 2023, pp. 405–415. Springer, Switzerland (2023). https://doi.org/10.1007/978-3-031-43901-8_39

24. Schmidt, V., et al.: Codecarbon: estimate and track carbon emissions from machine learning computing (2021). https://doi.org/10.5281/zenodo.4658424

25. Tappeiner, E., et al.: Multi-organ segmentation of the head and neck area: an efficient hierarchical neural networks approach. Int. J. Comput. Assist. Radiol. Surg. **14**, 745–754 (2019). https://doi.org/10.1007/s11548-019-01922-4

26. Tappeiner, E., Welk, M., Schubert, R.: Tackling the class imbalance problem of deep learning based head and neck organ segmentation. Int. J. Comput. Assisted Radiol. Surgery, 2103–2111 (2022). https://doi.org/10.1007/s11548-022-02649-5

27. Vaswani, A., et al.: Attention is all you need (2023). https://arxiv.org/abs/1706.03762

28. Wahid, K., Dede, C., Naser, M., Fuller, C.: Training Dataset for HNTSMRG 2024 Challenge (2024). https://doi.org/10.5281/zenodo.11199559

29. Wang, C.Y., Bochkovskiy, A., Liao, M.: Yolov7: trainable bag-of-freebies sets new state-of-the-art for real-time object detectors. In: 2023 IEEE/CVF Conference on Computer Vision and Pattern Recognition (CVPR), pp. 7464–7475 (2023). https://doi.org/10.1109/CVPR52729.2023.00721

30. Warfield, S.K., Zou, K.H., Wells, W.M.: Simultaneous truth and performance level estimation (STAPLE): an algorithm for the validation of image segmentation. IEEE Trans. Med. Imaging **23**, 903–921 (2004). https://doi.org/10.1109/TMI.2004.828354

31. Wu, Z., Shen, C., van den Hengel, A.: Bridging category-level and instance-level semantic image segmentation (2016). https://arxiv.org/abs/1605.06885

32. Ye, Y., Xie, Y., Zhang, J., Chen, Z., Xia, Y.: Uniseg: a prompt-driven universal segmentation model as well as a strong representation learner. In: Medical Image Computing and Computer Assisted Intervention – MICCAI 2023, pp. 508–518 (2023). https://doi.org/10.1007/978-3-031-43898-1_49

33. Zhou, H.Y., et al.: nnFormer: volumetric medical image segmentation via a 3D transformer. IEEE Trans. Image Process. **32**, 4036–4045 (2023). https://doi.org/10.1109/TIP.2023.3293771

Deep Learning for Longitudinal Gross Tumor Volume Segmentation in MRI-Guided Adaptive Radiotherapy for Head and Neck Cancer

Xin Tie$^{(\boxtimes)}$ ⓘ, Weijie Chen ⓘ, Zachary Huemann ⓘ, Brayden Schott ⓘ, Nuohao Liu ⓘ, and Tyler J. Bradshaw ⓘ

University of Wisconsin, Madison, WI, USA
xtie@wisc.edu

Abstract. Accurate segmentation of gross tumor volume (GTV) is essential for effective MRI-guided adaptive radiotherapy (MRgART) in head and neck cancer. However, manual segmentation of the GTV over the course of therapy is time-consuming and prone to interobserver variability. Deep learning (DL) has the potential to overcome these challenges by automatically delineating GTVs. In this study, our team, *UW LAIR*, tackled the challenges of both pre-radiotherapy (pre-RT) (Task 1) and mid-radiotherapy (mid-RT) (Task 2) tumor volume segmentation. To this end, we developed a series of DL models for longitudinal GTV segmentation. The backbone of our models for both tasks was SegResNet with deep supervision. For Task 1, we trained the model using a combined dataset of pre-RT and mid-RT MRI data, which resulted in the improved aggregated Dice similarity coefficient (DSC_{agg}) on a hold-out internal testing set compared to models trained solely on pre-RT MRI data. In Task 2, we introduced mask-aware attention modules, enabling pre-RT GTV masks to influence intermediate features learned from mid-RT data. This attention-based approach yielded slight improvements over the baseline method, which concatenated mid-RT MRI with pre-RT GTV masks as input. In the final testing phase, the ensemble of 10 pre-RT segmentation models achieved an average DSC_{agg} of 0.794, with 0.745 for primary GTV (GTVp) and 0.844 for metastatic lymph nodes (GTVn) in Task 1. For Task 2, the ensemble of 10 mid-RT segmentation models attained an average DSC_{agg} of 0.733, with 0.607 for GTVp and 0.859 for GTVn, leading us to achieve 1st place. In summary, we presented a collection of DL models that could facilitate GTV segmentation in MRgART, offering the potential to streamline radiation oncology workflows.

Keywords: MRI-guided Adaptive Radiotherapy · Longitudinal Imaging · Deep Learning · Segmentation

1 Introduction

Radiation therapy (RT) is a cornerstone of cancer treatment, particularly for head and neck cancer (HNC). Over the past decades, the treatment of HNC with RT has seen significant advancements, evolving from 3D conformal radiation therapy (CRT) into

© The Author(s) 2025
K. A. Wahid et al. (Eds.): HNTS-MRG 2024, LNCS 15273, pp. 99–111, 2025.
https://doi.org/10.1007/978-3-031-83274-1_7

intensity-modulated radiation therapy (IMRT) [1]. IMRT enables improved targeting of tumors while sparing normal tissues. However, this conformality also poses a critical challenge: anatomical changes during treatment, such as tumor shrinkage or weight loss, can drastically alter the dose delivered to both tumor and surrounding organs-at-risk. To address this, adaptive radiation therapy (ART), which involves re-planning during treatment in response to changes taking place in the patient's body, has been developed with the goal of improving target coverage and reducing normal tissue toxicity. Recently, MRI-guided adaptive radiotherapy (MRgART) has emerged as a promising approach for treating HNC patients due to the superior soft tissue contrast provided by MRI, allowing for more accurate tumor delineation [2]. However, manual segmentation of gross tumor volume (GTV) on pre- and mid-treatment MRI scans is usually time-consuming and subject to inter-observer variability. Artificial intelligence (AI), especially deep learning (DL), has the potential to streamline this process, facilitating timely and accurate adjustments to treatment plans.

Extensive studies have focused on segmenting tumors across various imaging modalities using DL [3–7]. Despite these advances, there remains a gap in the development of DL tools for segmenting tumors on multi-time-point imaging data, which is crucial for ART. The annual Medical Image Computing and Computer Assisted Intervention Society (MICCAI) Head and Neck Tumor Segmentation for MR-Guided Applications (HNTS-MRG) 2024 challenge addressed this gap by releasing high-quality annotated MRI data and promoting the development of DL models capable of segmenting GTVs on MRI at different treatment stages. This challenge includes two tasks: Task 1 focuses on automatically segmenting tumor volumes on pre-RT MRI, while Task 2 targets the segmentation of tumor volumes on mid-RT MRI.

In this work, we introduced and validated a series of DL methods designed for longitudinal GTV segmentation in MRgART. Notably, model predictions on current timepoint scans are guided by previous timepoint scan information (when applicable) via specialized attention mechanisms. For each task, we reported the results for ablations studies to understand the impact of each carefully selected component.

2 Datasets and Methods

2.1 Imaging Datasets

The retrospective training dataset consists of 150 patients, each having pre- and mid-RT T2-weighted MRI scans and corresponding labels for primary GTV (GTVp) and metastatic lymph nodes (GTVn). All the cases were manually segmented by multiple physicians independently. The final ground truth contours were combined via the STAPLE algorithm [8] and verified by experienced radiation oncologists. For Task 2, all the pre-RT data has been deformably registered to mid-RT MRI for better spatial alignment. The training data was provided in Neuroimaging Informatics Technology Initiative (NIfTI) format.

2.2 Model Architecture

For both segmentation tasks, we employed SegResNet [9] with deep supervision (Fig. 1) as the backbone of our models. This architecture has demonstrated consistently high performance in previous challenges [10], making it a reliable choice for our study.

SegResNet is a convolutional encoder-decoder model initially designed for brain tumor segmentation on MRI. The encoder is composed of multiple stages, each containing several convolutional blocks with residual connections [11]. Our model's architecture starts with a single residual block in Level 1, followed by progressively deeper configurations of 2, 2, 4, 4, and 4 residual blocks in the subsequent levels. Each residual block is a stack of two units, where each unit includes instance normalization, ReLU activation, and a $3 \times 3 \times 3$ convolution. To effectively capture multi-scale contextual information, we downsample the feature maps by a factor of 2 at each level, while simultaneously increasing the number of feature channels.

The decoder mirrors the encoder's structure, and each level contains a single convolutional block. To reconstruct the pixel-wise segmentation masks, we upsample the feature maps using transposed convolutions and reduced the number of feature channels between levels. Before passing these features into the decoder's convolutional block at each level, we fuse them with the output features from the encoder at the same spatial level. The last layer is a $1 \times 1 \times 1$ convolution, which reduces the channels to three output channels, followed by a softmax function to estimate the probability of each pixel belonging to each class (i.e., background, GTVp and GTVn).

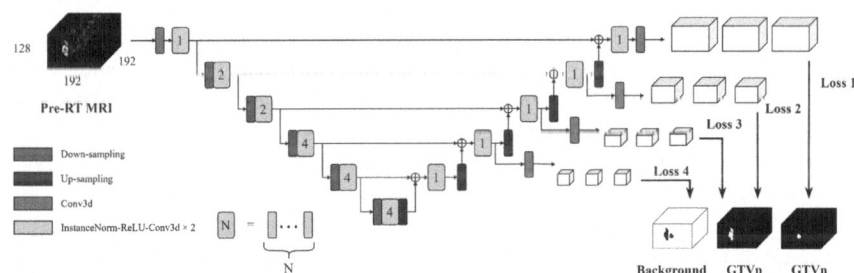

Fig. 1. Diagram of the SegResNet architecture with deep supervision.

For Task 1, the model input is an MRI image from a single time point. For Task 2, the model takes a mid-RT MRI image along with corresponding pre-RT GTVp and GTVn masks as inputs. To allow prior information from pre-RT data to influence intermediate features in the mid-RT model, we integrated a mask-aware attention module at each level of the encoder (Fig. 2). This module is an adaptation of the convolutional block attention module (CBAM) [12].

Similar to CBAM, the mask-aware attention module first applies channel attention by performing global averaging pooling and global max pooling across the spatial dimensions, feeding pooled features into a shared multi-layer perceptron and generating channel attention weights using a sigmoid activation function. The input feature map is multiplied by these attention weights to emphasize important channels. The channel

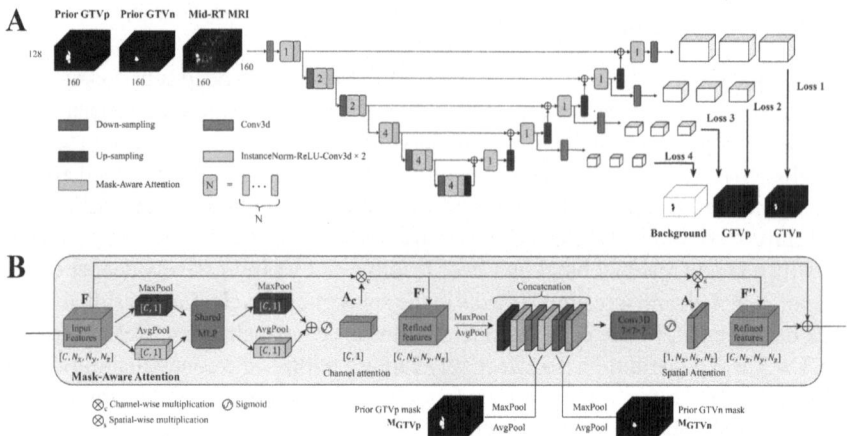

Fig. 2. (**A**) shows the SegResNet architecture augmented with mask-aware attention modules. (**B**) illustrates the inner workings of mask-aware attention.

attention sub-module can be summarized mathematically as follows:

$$A_c = \sigma(MLP(AvgPool_s(F)) + MLP(MaxPool_s(F))) \tag{1}$$

$$F' = A_c \otimes_c F \tag{2}$$

where F denotes the input feature, $AvgPool_s$ and $MaxPool_s$ are the pooling operations along the spatial dimensions, MLP is the multi-layer perceptron, σ denotes the sigmoid function, A_c denotes the channel attention weights, $\otimes_c$ represents channel-wise multiplication, and $F\prime$ represents the refined feature.

Next, the mask-aware attention applies spatial attention by performing averaging pooling and max pooling along the channel axis, then concatenating the resulting feature maps with the masks derived from pre-RT GTVp and GTVn masks. To align the spatial dimensions of the features and pre-RT masks at intermediate layers, we applied max pooling and average pooling to both GTVp and GTVn masks progressively along each spatial axis with a kernel size of $3 \times 3 \times 3$ and a stride of 2. The spatial attention sub-module involves the following operations:

$$A_s = \sigma\left(f_{7\times7\times7}\begin{bmatrix} AvgPool_c(F'), & MaxPool_c(F'), \\ AvgPool_s^{(i)}(M_{GTV_p}), & MaxPool_s^{(i)}(M_{GTV_p}), \\ AvgPool_s^{(i)}(M_{GTV_n}), & MaxPool_s^{(i)}(M_{GTV_n}) \end{bmatrix}\right) \tag{3}$$

$$F'' = A_s \otimes_s F' \tag{4}$$

where F' represents the channel attention-refined feature, $AvgPool_c$ and $MaxPool_c$ are the pooling operations along the channel axis, M_{GTV_p} and M_{GTV_n} are the binary masks of GTVp and GTVn in pre-RT MRI images, $AvgPool_s^{(i)}$ and $MaxPool_s^{(i)}$ indicate i iterations of pooling operations to match the output GTV masks with the feature maps in the spatial dimensions, [·] indicates feature concatenation, $f_{7\times7\times7}$ denotes a $7 \times 7 \times 7$ convolution with the number of filters equal to 6, A_s denotes the spatial attention weights, $\otimes_s$ represents spatial-wise multiplication. The final refined feature, F'', are then added to the input feature F:

$$F_{out} = F + F'' \tag{5}$$

This modification at the intermediate layers of SegResNet allows the mid-RT segmentation model to better focus on important regions informed by the pre-RT data. To distinguish it from the original SegResNet, we refer to this architecture as Mask-Aware SegResNet (MA-SegResNet).

2.3 Implementation

For both tasks, we resampled the MRI images and corresponding GTV masks to a fixed isotropic voxel size of 1 mm. The MRI intensity was then rescaled using z-score normalization, where only non-zero values were used to compute the mean and standard deviation for each MRI volume. The entire dataset was randomly split into 5 folds, with each fold containing 30 patients (60 scans total, as each patient had both pre-RT and mid-RT imaging data). When we ran fivefold cross-validation, data from four folds was combined for model training, while the remaining fold was used for validation.

In Task 1, the SegResNet model operates on $192 \times 192 \times 128$ patches cropped from MRI images centered on the foreground classes with probabilities of 0.45 for GTVp and 0.45 for GTVn (0.1 for background). Both pre-RT and mid-RT MRI data were used for model training, but only pre-RT data in the validation set was used for model selection.

In Task 2, the MA-SegResNet model processes $160 \times 160 \times 128$ patches centered on the foreground classes with the same probabilities as in Task 1. For both tasks, the optimal in-plane patch size was selected based on tuning across 192×192, 160×160 and 128×128. In addition to mid-RT MRI images, registered pre-RT GTV masks were included as input. The original GTV masks for pre-RT MRI images, deformably registered to mid-RT MRI images, were first converted to one-hot encoded masks, and only the masks corresponding to GTVp and GTVn were retrieved and used as additional input channels. When training the mid-RT segmentation model, we also included paired pre-RT data as additional training samples. Since no prior information is available for pre-RT MRI images, we set the input GTVp and GTVn masks to zeros. For model selection, only mid-RT data pairs in the validation set were used.

To alleviate the model overfitting problem, we applied the same data augmentation strategies across both tasks, including random affine transformation (rotation between −25 and 25 degrees, axis flip for all dimensions, zoom between 0.8 to 1.2), Gaussian noise, and Gaussian blur. The loss function for both tasks is a compound loss, comprised of cross-entropy and Dice loss:

$$L = \sum_{k=1}^{4} \frac{1}{2^{k-1}} \sum_{j=1}^{N} \left[L_{CE}\left(y_j^{(k)}, \hat{y}_j^{(k)}\right) + L_{DICE}\left(y_j^{(k)}, \hat{y}_j^{(k)}\right) \right] \tag{6}$$

where $\hat{y}_j^{(k)}, y_j^{(k)}$ denote the prediction and the ground truth for the j-th sample at the deep supervision level k, L_{CE} represents the cross-entropy loss and L_{DICE} represents the Dice loss. When $k = 1$, the loss is computed at the same spatial level as the input images. For $k > 1$, the loss is computed at a spatial level with dimensions reduced to $\frac{1}{2^{k-1}}$ of the input image dimension, and the ground truth masks are interpolated using nearest-neighbor interpolation to match the reduced dimensions.

Both pre-RT and mid-RT segmentation models were trained using the AdamW optimizer [13], with an initial learning rate of 10^{-4}, weight decay regularization of 10^{-5}, and a cosine annealing scheduler. We randomly sampled 2 patches from each sample and set the batch size to 3 (6 patches in a single batch). The model was trained for 400 epochs on a single NVIDIA A100 GPU. Batch size and number of epochs were optimized via grid search (batch sizes of 2, 3, and 4; epochs of 300, 400, and 500). The learning environment requires the following Python (3.8.8) libraries: PyTorch (1.13.0), MONAI (1.3.0) [14].

For both tasks, we used three different random seeds for model training in each training/validation split. The model with the highest aggregated Dice similarity coefficient (DSC_{agg}) in the validation set was selected for each fold, resulting in 5 models for fivefold cross-validation. We repeated the experiments twice with different random seeds to generate the five folds, yielding a total of 10 models. The final submission for each task is an ensemble of these 10 models.

2.4 Inference

For each case, we employed the sliding window method with an overlap rate of 0.625 and blended outputs of overlapping patches using Gaussian weighting. Then we averaged the probability maps estimated by 10 individual models. The class label (0: background, 1: GTVp, 2: GTVn) with the highest averaged probability across all three classes was then assigned to each voxel. In Task 1, we removed any small regions with a volume below $0.5\,\text{cm}^3$ using connected component analysis. In Task 2, we implemented a technique that excludes predicted contours on mid-RT MRI scans that have no overlapping voxels with registered pre-RT GTV contours since the mid-RT GTV is expected to shrink and remain confined within the pre-RT GTV contours throughout the course of treatment. This technique, known as "mask propagation through deformable registration" (or MPDR), has been used in previous literature [7] to reduce false positives.

2.5 Ablation Studies

We conducted ablation studies to identify key factors that contribute to improved segmentation accuracy. First, we held out 30 cases for testing and altered various settings during model development. The training dataset, comprising 120 cases in total, was divided into five folds; in each iteration, four folds were used for training and one fold for validation. Models trained on these five folds were then evaluated on the hold-out testing set to assess the impact of different configurations on performance.

In Task 1, we investigated whether including paired mid-RT data could enhance the performance of pre-RT segmentation models. In Task 2, we focused on evaluating different combinations of input data, the usefulness of pre-RT data for mid-RT GTV segmentation, the impact of model architectures, and whether applying the MPDR method to the model predictions could further improve performance.

For statistical analysis, we employed the bootstrap resampling technique. In each iteration, we randomly sampled 30 cases from the internal test set with replacement and calculated the DSC_{agg} for each model. The DSC_{agg} values were then averaged across the five models to yield a single metric for each bootstrap trial. This process was repeated 10,000 times. The difference between two model configurations was considered statistically significant at the 0.05 level if the metric value computed for one configuration exceeded that of the other in 95% of trials.

2.6 Model Availability

The code and model weights have been made available in an open-source project: https://github.com/xtie97/HNTS-MRG24-UWLAIR.

3 Results

3.1 Quantitative Results

Table 1 presents the results of fivefold cross-validation for Task 1, reported as DSC_{agg} for different fivefold splits (split 1 and split 2 were created using different random seeds for data partitioning). The average DSC_{agg} for split 1 is 0.814, and for split 2, it is 0.816. The overall average DSC_{agg} for the 10 models used in the submission is 0.815 across the validation sets. In the final testing phase, the ensemble of these 10 models achieved an average DSC_{agg} of 0.794 on the hold-out 50 cases, with 0.745 for GTVp and 0.844 for GTVn. The drop in DSC_{agg} is primarily attributed to the decreased performance in the GTVp segmentation.

Table 1. Cross validation results for pre-RT GTV segmentation (Task 1)

	Fivefold Split 1			Fivefold Split 2		
	DSC_{agg} (GTVp)	DSC_{agg} (GTVn)	Average DSC_{agg}	DSC_{agg} (GTVp)	DSC_{agg} (GTVn)	Average DSC_{agg}
Fold 1	0.789	0.837	0.813	0.746	0.838	0.792
Fold 2	0.815	0.844	0.829	0.792	0.882	0.837
Fold 3	0.812	0.881	0.846	0.783	0.853	0.818
Fold 4	0.718	0.820	0.769	0.799	0.860	0.829
Fold 5	0.763	0.861	0.812	0.802	0.812	0.807
Average	0.779 ± 0.040	0.849 ± 0.023	0.814 ± 0.029	0.784 ± 0.023	0.849 ± 0.026	0.816 ± 0.018

Table 2 shows the cross-validation results for Task 2 with different fivefold splits. The average DSC_{agg} of the 10 mid-RT models is 0.754 across the validation sets. In the final testing phase, the ensemble of these models attained an average DSC_{agg} of 0.733 on the hold-out mid-RT data, with 0.607 for GTVp and 0.859 for GTVn. Similar to Task 1, the slight decrease in DSC_{agg} is mainly due to the performance drop of the GTVp segmentation.

Table 2. Cross validation results for mid-RT GTV segmentation (Task 2)

	Fivefold Split 1			Fivefold Split 2		
	DSC$_{agg}$ (GTVp)	DSC$_{agg}$ (GTVn)	Average DSC$_{agg}$	DSC$_{agg}$ (GTVp)	DSC$_{agg}$ (GTVn)	Average DSC$_{agg}$
Fold 1	0.620	0.834	0.727	0.599	0.823	0.711
Fold 2	0.714	0.834	0.774	0.679	0.868	0.774
Fold 3	0.597	0.851	0.724	0.685	0.860	0.772
Fold 4	0.657	0.822	0.739	0.702	0.844	0.773
Fold 5	0.728	0.877	0.802	0.651	0.841	0.746
Average	0.663 ± 0.057	0.843 ± 0.021	0.753 ± 0.034	0.663 ± 0.040	0.847 ± 0.018	0.755 ± 0.027

Figure 3 and Fig. 4 display the example cases from pre-RT MRI and mid-RT MRI scans, respectively. The ground truth and predicted GTV contours are both overlaid on the MRI images. For mid-RT cases, the registered pre-RT scans are also included to reference the initial tumor locations.

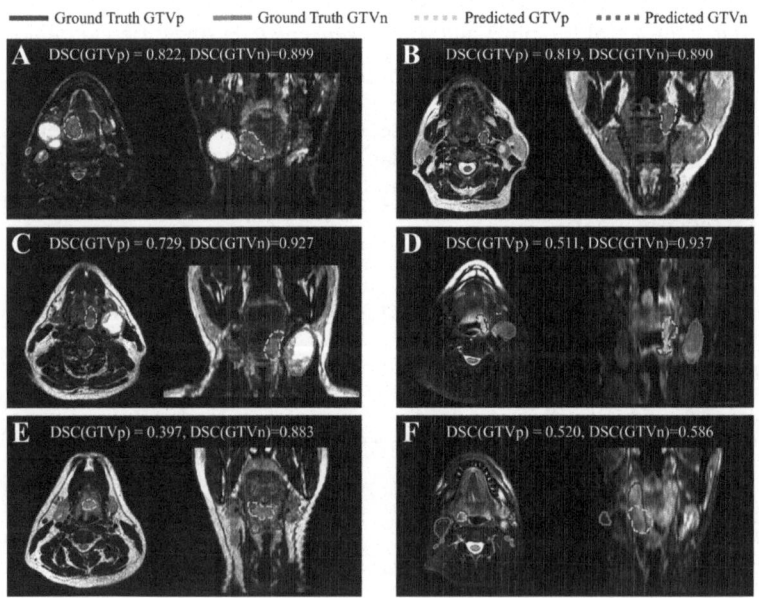

Fig. 3. Pre-RT segmentation results for six sample cases. Dice similarity coefficients (DSCs) for both GTVp and GTVn are reported above each example.

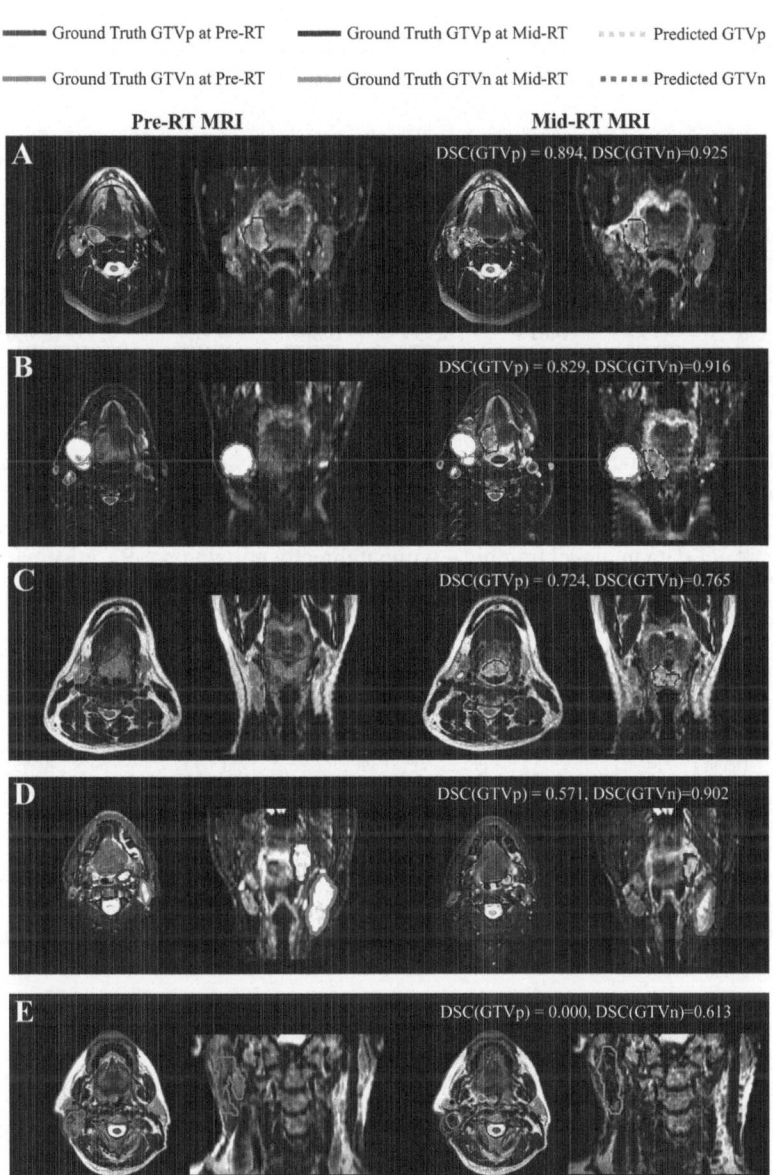

Fig. 4. Mid-RT segmentation results for five sample cases. Dice similarity coefficients (DSCs) for both GTVp and GTVn in midRT MRI scans are labeled above each example. Registered pre-RT MRI images are provided to reference initial tumor sites.

3.2 Ablation Studies

The internal testing results in terms of DSC_{agg} for Task 1 are presented in Table 3. When paired mid-RT data was included in training, the DSC_{agg} averaged over the five folds was 0.799, which is significantly higher (P = 0.004) than the DSC_{agg} achieved by models trained solely on pre-RT data (0.776). This finding supports our decision to train models on data from both time points while only using pre-RT data for model selection.

Table 3. Ablation study investigating the impact of training data on the pre-RT GTV segmentation (Task 1)

Training Data	Fold 1	Fold 2	Fold 3	Fold 4	Fold 5	Average
Pre-RT only	0.770	0.789	0.778	0.774	0.770	0.776 ± 0.008
Mid-RT & Pre-RT	**0.793**	**0.814**	**0.799**	**0.807**	**0.784**	**0.799 ± 0.012**

Table 4 shows the results of ablation studies conducted for Task 2. When no prior information from pre-RT scans was incorporated into mid-RT GTV segmentation, the DSC_{agg} on the internal testing set averaged over the five folds was 0.588. Including paired pre-RT data in model training – without directly utilizing pre-RT information for mid-RT segmentation – resulted in consistent improvements ($P = 0.001$) in DSC_{agg}.

Table 4. Ablation study investigating the impact of training data, model architectures, post-processing techniques on the mid-RT GTV segmentation (Task 2)

Input Data	Training Data	Model Architecture	Post-Processing	Fold 1	Fold 2	Fold 3	Fold 4	Fold 5	Average
MRI images	Mid-RT only	SegResNet	None	0.570	0.558	0.608	0.604	0.602	0.588 ± 0.023
MRI images	Mid-RT & Pre-RT	SegResNet	None	0.621	0.635	0.619	0.646	0.645	0.633 ± 0.013
Mid-RT MRI & Pre-RT MRI & Pre-RT GTV masks	Mid-RT only	SegResNet	None	0.696	0.662	0.713	0.698	0.694	0.693 ± 0.019
MRI images & Prior GTV masks	Mid-RT only	SegResNet	None	0.700	0.713	0.714	0.707	0.686	0.704 ± 0.012
MRI images & Prior GTV masks	Mid-RT & Pre-RT	SegResNet	None	**0.723**	0.685	0.721	0.714	0.719	0.712 ± 0.016

(*continued*)

Table 4. (*continued*)

Input Data	Training Data	Model Architecture	Post-Processing	Fold 1	Fold 2	Fold 3	Fold 4	Fold 5	Average
MRI images & Prior GTV masks	Mid-RT & Pre-RT	MA-SegResNet	None	0.717	0.723	0.746	0.726	0.713	0.725 ± 0.013
MRI images & Prior GTV masks	Mid-RT & Pre-RT	MA-SegResNet	MPDR	0.722	**0.733**	**0.751**	**0.730**	**0.731**	**0.733 ± 0.011**

The most significant performance increase from the baseline configuration (i.e., using only mid-RT data for model development) occurred when pre-RT GTV masks were integrated. In this setup, simply concatenating mid-RT MRI images with pre-RT GTV masks as model inputs improved the DSC_{agg} from 0.588 to 0.704 (P < 0.001). Building on it, further improvements were observed by including pre-RT scans as additional training data (with prior GTV masks set to zeros) and replacing SegResNet model with MA-SegResNet, leading to a DSC_{agg} of 0.725 (from 0.704, P = 0.039). Lastly, applying MPDR to the predictions of the MA-SegResNet model consistently increased the DSC_{agg} across all five folds. The configuration in the last row of Table 4 was the setup used in our final submission for Task 2.

4 Discussion

In this work, we addressed both pre-RT and mid-RT MRI segmentation tasks using the SegResNet architecture as the backbone. Our results demonstrated that including paired data from different time points in the treatment as additional training samples enhanced segmentation performance across both tasks. For mid-RT GTV segmentation, integrating prior information from pre-RT scans significantly improved accuracy. Moreover, architectural modifications and post-processing techniques led to further improvements without adding excessive computational complexity and inference time.

There are several limitations in our study. First, we did not incorporate co-registered pre-RT MRI images as an additional input channel for mid-RT GTV segmentation. Adding this information may lead to further improvements. Second, in Task 2, we did not investigate whether the mask-aware attention modules could enhance performance with backbone architectures other than SegResNet. Third, the applicability of our findings and approaches to other longitudinal tumor segmentation tasks in radiotherapy remains uncertain. Lastly, although our method shows clear advancements, its impact on clinical efficiency is unknown. Prospective studies are essential to rigorously evaluate any AI models intended for clinical use.

In conclusion, we developed a series of DL models for automatic GTV contouring in MRgART, demonstrating potential to streamline radiation oncology workflows for treating HNC patients.

Acknowledgments. We acknowledge the organizers of the HNTS-MRG 24 Challenge for releasing high-quality, well-annotated data and for holding such a great challenge to advance the field of image-guided adaptive radiotherapy. We also thank the Center for High Throughput Computing (CHTC) at University of Wisconsin-Madison for providing GPU resources. Dr. Tyler Bradshaw is currently funded by the National Institute Of Biomedical Imaging And Bioengineering of the National Institutes of Health under Award Number R01EB033782.

Disclaimer: The content is solely the responsibility of the authors and does not necessarily represent the official views of the National Institutes of Health.

Disclosure of Interests. The authors have no competing interests to declare that are relevant to this article.

References

1. Morgan, H.E., Sher, D.J.: Adaptive radiotherapy for head and neck cancer. Cancers Head Neck **5**, 1 (2020). https://doi.org/10.1186/s41199-019-0046-z
2. Benitez, C.M., Chuong, M.D., Künzel, L.A., Thorwarth, D.: MRI-guided adaptive radiation therapy. Semin. Radiat. Oncol. **34**(1), 84–91 (2024). https://doi.org/10.1016/j.semradonc.2023.10.013
3. Huemann, Z., Tie, X., Hu, J., Bradshaw, T.J.: ConTEXTual Net: a multimodal vision-language model for segmentation of pneumothorax. J. Imaging Inform. Med., 1–12, March 2024. https://doi.org/10.1007/s10278-024-01051-8
4. Yousefirizi, F., et al.: TMTV-Net: fully automated total metabolic tumor volume segmentation in lymphoma PET/CT images — a multi-center generalizability analysis. Eur. J. Nucl. Med. Mol. Imaging (2024). https://doi.org/10.1007/s00259-024-06616-x
5. Zhang, S., et al.: Deep learning for automatic gross tumor volumes contouring in esophageal cancer based on contrast-enhanced computed tomography images: a multi-institutional study. Int. J. Radiat. Oncol. (2024). https://doi.org/10.1016/j.ijrobp.2024.02.035
6. Wahid, K.A., et al.: Evaluation of deep learning-based multiparametric MRI oropharyngeal primary tumor auto-segmentation and investigation of input channel effects: Results from a prospective imaging registry. Clin. Transl. Radiat. Oncol. **32**, 6–14 (2022). https://doi.org/10.1016/j.ctro.2021.10.003
7. Tie, X., et al.: Automatic Quantification of Serial PET/CT Images for Pediatric Hodgkin Lymphoma Patients Using a Longitudinally-Aware Segmentation Network. ArXiv, p. arXiv:2404.08611v1, April 2024
8. Warfield, S.K., Zou, K.H., Wells, W.M.: Simultaneous Truth and Performance Level Estimation (STAPLE): an algorithm for the validation of image segmentation. IEEE Trans. Med. Imaging **23**(7), 903–921 (2004). https://doi.org/10.1109/TMI.2004.828354
9. Myronenko, A.: 3D MRI brain tumor segmentation using autoencoder regularization, 19 November 2018, arXiv: arXiv:1810.11654. Accessed 19 Feb 2024. http://arxiv.org/abs/1810.11654
10. Andrearczyk, V., et al.: Automatic Head and Neck Tumor segmentation and outcome prediction relying on FDG-PET/CT images: Findings from the second edition of the HECKTOR challenge. Med. Image Anal. **90**, 102972, December 2023. https://doi.org/10.1016/j.media.2023.102972
11. He, K., Zhang, X., Ren, S., Sun, J.: Deep Residual Learning for Image Recognition, Dec. 10, 2015, arXiv: arXiv:1512.03385. Accessed 19 Feb 2024. http://arxiv.org/abs/1512.03385

12. Woo, S., Park, J., Lee, J.-Y., Kweon, I.S.: CBAM: Convolutional Block Attention Module, Jul. 18, 2018, arXiv: arXiv:1807.06521. Accessed 24 Oct 2023. http://arxiv.org/abs/1807.06521

13. Loshchilov, I., Hutter, F.: Decoupled Weight Decay Regularization, Jan. 04, 2019, arXiv: arXiv:1711.05101. Accessed 31 Aug 2023. http://arxiv.org/abs/1711.05101

14. Cardoso, M.J., et al.: MONAI: An open-source framework for deep learning in healthcare, Nov. 04, 2022, arXiv: arXiv:2211.02701. Accessed 01 Mar 2023. http://arxiv.org/abs/2211.02701

Enhancing Head and Neck Tumor Segmentation in MRI: The Impact of Image Preprocessing and Model Ensembling

Mehdi Astaraki[1,2(✉)] and Iuliana Toma-Dasu[1,2]

[1] Department of Medical Radiation Physics, Stockholm University, Stockholm, Sweden
[2] Department of Oncology-Pathology, Karolinska Institutet, Solna, Sweden
mehdi.astaraki@ki.se

Abstract. The adoption of online adaptive MR-guided radiotherapy (MRgRT) for Head and Neck Cancer (HNC) treatment faces challenges due to the complexity of manual HNC tumor delineation. This study focused on the problem of HNC tumor segmentation and investigated the effects of different preprocessing techniques, robust segmentation models, and ensembling steps on segmentation accuracy to propose an optimal solution . We contributed to the MICCAI Head and Neck Tumor Segmentation for MR-Guided Applications (HNTS-MRG) challenge which contains segmentation of HNC tumors in Task1) pre-RT and Task2) mid-RT MR images. In the internal validation phase, the most accurate results were achieved by ensembling two models trained on maximally cropped and contrast-enhanced images which yielded average volumetric Dice scores of (0.680, 0.785) and (0.493, 0.810) for (GTVp, GTVn) on pre-RT and mid-RT volumes. For the final testing phase, the models were submitted under the team's name of "Stockholm_Trio" and the overall segmentation performance achieved aggregated Dice scores of (0.795, 0.849) and (0.553, 0.865) for pre- and mid-RT tasks, respectively. The developed models are available at https://github.com/Astarakee/miccai24

Keywords: head-neck tumor · MR-guided radiotherapy · GTV · segmentation

1 Introduction

Head and neck cancer (HNC) encompasses a diverse range of malignancies arising within the anatomical structures of the head and neck, including the oral cavity, nasopharynx, oropharynx, larynx, and hypopharynx [1, 2].

Approximately half of all cancer patients undergo radiation therapy (RT). Successful treatment hinges on the accurate and precise delivery of radiation to the targeted areas while minimizing damage to adjacent healthy tissue. Intensity-modulated radiation therapy as the most common treatment modality for HNC enabled conformal dose delivery through technological advancements. Magnetic Resonance (MR) imaging linear accelerators combine these two technologies to provide improved soft tissue contrast in real-time. MR-guided RT (MRgRT) has a number of advantages over conventional RT, including superior soft tissue contrast compared to CT, continuous intrafraction MR

K. A. Wahid et al. (Eds.): HNTS-MRG 2024, LNCS 15273, pp. 112–122, 2025.
https://doi.org/10.1007/978-3-031-83274-1_8

imaging, and the ability to adapt treatment plans in real-time. Online adaptive MRgRT, performed while the patient remains in the treatment position, takes into account real-time anatomical changes like soft-tissue deformity and volume changes. Clinically, these technical advantages result in the ability to safely escalate the dose while reducing toxicity and treatment time. As a result, the use of MRgRT techniques has increased over time, particularly in the treatment of pancreatic, prostate, lung, and liver cancers [3, 4].

HNC tumors present significant challenges for manual delineation due to their complex anatomical structures. Furthermore, the sheer volume of data acquired for MRgRT planning makes manual segmentation almost impractical. Thus, developing accurate and automated segmentation methods for HNC tumors is crucial for streamlining the delineation process and ensuring greater consistency in contouring. The advancements in Deep Learning (DL) techniques over the past decade have resulted in the development of sophisticated DL models that demonstrate exceptional performance in various medical imaging applications, notably tumor segmentation [5–8]. Within the context of HNC segmentation, in recent years, three challenges have been established to benchmark algorithmic progress. The Head and Neck Tumor (HECKTOR) challenge [9] focused on evaluating the segmentation of primary and nodal gross tumors in PET-CT images. In this challenge, top performing models achieved Dice scores in the range of 0.77, through the utilization of UNet-based models enhanced with a variety of techniques. These techniques included the incorporation of attention mechanisms, residual connections, image preprocessing, and model ensembling strategies. The segmentation of OARs and GTVs of nasopharyngeal carcinoma for radiotherapy planning (SegRap) challenge [10] assessed the segmentation of both tumors and 45 healthy structures in bi-modal CT images. Top performing models in the GTVs segmentation task utilized preprocessing techniques, notably intensity harmonization and background cropping, prior to model training. Encoder decoder segmentation models including nnU-Net [6] and its modified version, as well as MultiTalent [7] resulted in average Dice scores reaching 0.73. Lastly, the Head and Neck Segmentation (HaN-Seg) challenge [10] aimed to benchmark the performance of automatic segmentation across 30 OARs in multimodal CT-MRIs. Nevertheless, the objective assessments of MR-only tumor segmentation techniques for MRgRT applications have not been investigated thoroughly.

The MICCAI Head and Neck Tumor Segmentation for MR-Guided Applications (HNTS-MRG) 2024 challenge [11] has been focused on the segmentation of HNC tumors for MRI-guided Adaptive RT applications. This paper presents our contribution to this challenge in which we conducted a set of extensive experiments to identify the best workflow for HNC tumor segmentation in MR images.

2   Materials and Methods

The HNTS-MRG challenge focuses on the automated segmentation of gross tumor volumes (GTVs) from images acquired at two distinct time points in two independent tasks: task 1) pre-RT MR images and task 2) mid-RT MR images. The core objective of this challenge is to investigate whether integrating prior timepoint data into the algorithms can improve the segmentation accuracy of primary GTV (GTVp) and metastatic lymph node GTV (GTVn).

2.1 Studied Datasets

All data utilized in this study were collected from patients with histologically confirmed head and neck cancer (HNC), primarily oropharyngeal cancer, who underwent radiotherapy (RT) at The University of Texas MD Anderson Cancer Center. Imaging data included a mixed set of fat-suppressed and non-fat-suppressed T2-weighted (T2w) sequences of the head and neck region.

Both pre- and mid-RT volumes were examined by three to four independent expert physicians, and manual delineation was conducted by reviewing the patient's medical history, including relevant previous imaging such as PET-CT. The obtained segmentation masks were then reviewed by an experienced radiation oncologist before utilizing the STAPLE algorithm to generate the final segmentation masks.

The training dataset comprises 150 T2w MRIs accompanied by their corresponding segmentation masks. The developed algorithms are mandated to be containerized and submitted to an online platform for an objective assessment on an unseen testing set encompassing 50 subjects.

Task 1: Pre-RT
The pre-RT images were acquired one to three weeks prior to the commencement of RT. During the training phase of model development, participants had the option to utilize exclusively pre-RT volumes or to incorporate mid-RT volumes to expand their training dataset. However, the testing phase exclusively involved pre-RT volumes.

Task 2: Mid-RT
The mid-RT volumes were acquired two to four weeks after the initiation of RT. In addition to the mid-RT volumes, for each subject, the pre-RT volumes were included in two distinct geometrical configurations: the original volume and the ones deformably registered to the mid-RT volumes. Essentially, the pre-RT data could function as prior knowledge, and participants had the flexibility to employ any desired combination settings between the pre- and mid-RT volumes.

Figure 1 illustrates examples of the tumor appearance in the examined training subjects.

2.2 Methods

Preprocessing
The provided image volumes by the organizer were already cropped to encompass the region from the superior aspect of the clavicles to the inferior aspect of the nasal septum. This standardized the field of view and removed identifiable facial features.

To further refine the datasets for model training, a two-stage preprocessing pipeline was implemented. In the first stage, to optimize patch extraction and minimize background inclusion, MR volumes were cropped to maximally exclude background regions. Thresholding and connected component analysis were employed to segment the volumes into foreground (body) and background, utilizing the body skin as the delineating boundary. A bounding box was then generated to enclose the widest portion of the volume, ensuring optimal loading of head and neck tissues within each patch while eliminating unnecessary background information. The second preprocessing stage focused on intensity enhancement. The N4 bias correction field algorithm [12] was applied to mitigate intensity inhomogeneities within each MR volume. Subsequently, to address the substantial variability in the dynamic range of MR volume intensities, cases with peak intensities exceeding 700 underwent clipping of values above the 99th percentile. Finally, Z-score standardization was performed as the concluding intensity modification step.

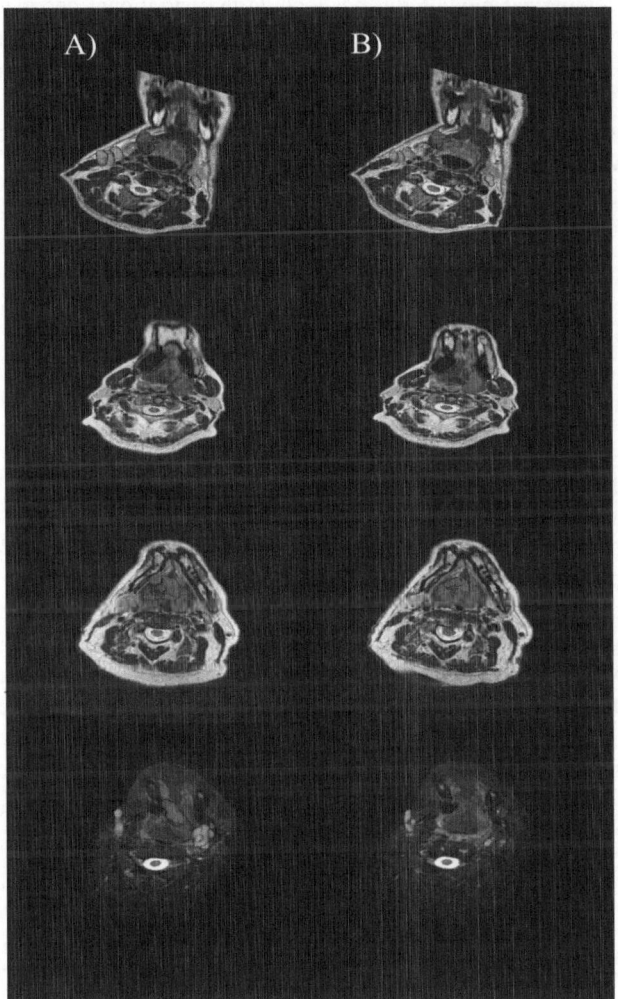

Fig. 1. Illustrative examples of tumor appearance in pre-RT (left column) and mid-RT (right column) images. Primary tumors are depicted in red and metastatic lymph nodes are visualized in green. Pseudo-3D visualization in axial-coronal views was employed to provide a comprehensive context of tumor volumes and their relative positions.

SegResNet Model

The encoder-decoder network architecture of MONAI SegResNet [13] contains ResNet blocks in the encoder part where each block consists of two convolution layers, with normalization and skip connections. The decoder part consists of a single block per spatial level. Each block within the decoder path starts by reducing the depth of feature maps, doubling the spatial dimension, followed by skipping connections. It should be noted that the employed models did not include the variational decoder branch. The models were trained for 500 epochs with the following hyperparameters: variable training and validation iterations per epoch to cover all the samples, batch size of 4, initial learning rate of 1e−4, LeakyReLU as activation function, instance normalization, four levels of deep supervision, and patch sizes were ($96 \times 160 \times 128$). The network encoder consisted of six stages with (1, 3, 4, 4, 6, 6) number of convolutions.

nnU-Net ResENC Model

The nnU-Net model V2 [6] served as the second segmentation pipeline. A ResNet-enhanced U-Net ("nnU-Net ResENCM") with original configurations [14] was employed. Models were trained for 1200 epochs across five folds with 250 training and 50 validation iterations, an initial learning rate of 1e-2, a batch size of 4, enabled deep supervision, and a patch size of ($64 \times 224 \times 160$). The network architecture include six encoder stages with (32, 64, 128, 256, 320, 320) kernels per stage.

MedNeXt Model

MedNeXt [15] is a network architecture composed exclusively of ConvNeXt blocks [16] thereby fully leveraging the latter's design advantages. To preserve contextual information during upsampling and downsampling operations, the architecture replaces standard up-down/sample blocks with Residual Inverted Bottlenecks. To mitigate performance saturation associated with large kernel sizes, training commences with smaller kernels but can be progressively increased through the UpKern technique. In general, MedNeXt is a scalable encoder-decoder framework optimized for 3D segmentation tasks, designed to maximize the potential of ConvNeXt architectures within the constraints of limited datasets. The MedNeXt model was compiled using large-scale network architectures with a kernel size of 3, a batch size of 2, isotropic spacing of 1mm, and a patch size of ($128 \times 128 \times 128$), following the nnU-Net V1 training protocol.

U-Mamba Model

Recent advancements in structured state-space sequence modeling, exemplified by the S4 architecture [17], have established these models as efficient building blocks for complex deep networks. Building upon this foundation, Mamba [18] introduced a selective mechanism to enhance S4's capability to focus on relevant input information. The U-Mamba [19] architecture further innovates by integrating state-space modules into convolutional blocks, enabling a hybrid approach that effectively captures both localized features and long-range dependencies within image data. This model offers a compelling alternative to transformer-based models by providing linear scaling with respect to feature size, in contrast to the quadratic complexity inherent in the self-attention mechanisms of transformers. In this study, we investigated the functionality of the U-Mamba model for both tasks by using the original implementation within the nnU-Net V2 framework.

Model Development and Ensembling

All the described models were optimized with a combination of Cross Entropy and Dice loss. For the pre-RT task, all models were trained exclusively on pre-RT MR volumes to segment the images into GTVp, GTVn, and background. In contrast, for the mid-RT task, an ablation study was conducted to explore different combinations of image data and segmentation masks. These settings included training the model using only mid-RT MR volumes (*no-prior*), incorporating registered pre-RT MRs as a second channel (*preMR-prior*), including pre-RT segmentation masks as the second channel (*preMask-prior*), and integrating registered pre-RT MR and segmentation masks together into a three-channel model (*preMRMask-prior*). Additionally, we dilated the pre-RT masks with $3 \times 3 \times 3$ structure elements, then signed distance maps were derived from the dilated masks and incorporated as prior information to guide the network's attention (*preDistance-prior*). Figure 2 shows examples of distance maps generated from pre-RT masks.

To enhance the overall model's robustness, the complementary roles of the developed models were assessed. To this end, the best-performing models for GTVp and GTVn were combined to generate the final segmentation masks.

3 Results

The segmentation models were trained with a 5-fold cross-validation resampling strategy on the training data set except for the MedNeXt model which was trained with 3-folds. For better readability, only the quantified Dice metrics are reported in this paper. Table 1 shows the segmentation results for pre-RT task over the training set.

Table 1. Quantified conventional volumetric Dice metric for pre-RT task on the training set. The model with overall better performance is marked as bold.

Model	Dice Metric ($\mu \pm \sigma$)			
	Original Data		Preprocessed Data	
	GTVp	GTVn	GTVp	GTVn
SegResNet	0.638 ± 0.060	0.718 ± 0.057	0.642 ± 0.053	0.740 ± 0.048
ResENC	0.627 ± 0.088	0.723 ± 0.062	0.680 ± 0.079	0.755 ± 0.053
MedNeXt	0.639 ± 0.073	0.754 ± 0.051	0.657 ± 0.083	0.785 ± 0.048
U-Mamba	0.619 ± 0.081	0.726 ± 0.059	0.643 ± 0.087	0.739 ± 0.067
Ensemble (ResENC&MedNeXT)	--	--	$\mathbf{0.680 \pm 0.079}$	$\mathbf{0.785 \pm 0.048}$

Table 1 reveals that ResENC and ModNeXt models consistently surpassed other models in performance, particularly when trained on the preprocessed dataset. Consequently, these two models were chosen for the mid-RT ablation experiments. However, due to challenges in optimizing the MedNeXt model for this task, only the nnU-Net ResENC model was ultimately utilized for training. The results of the ablation experiments

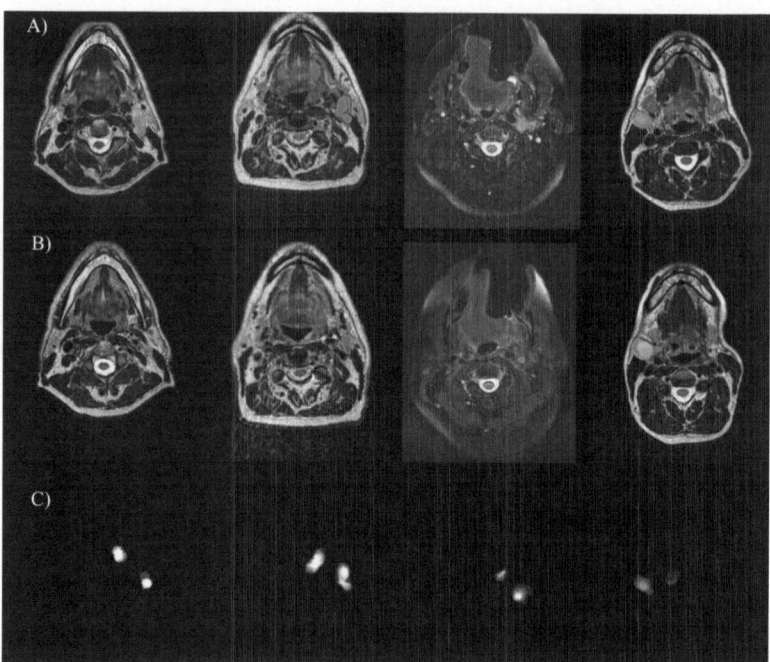

Fig. 2. Examples of preDistance-prior calculated from segmentation masks of first time-point MRIs. Axial slices showcasing preRT (row A), mid-RT (row B), and prior channel (row C).

conducted with various combinations of image and mask data for the mid-RT task are presented in Table 2.

Table 2. Quantified conventional volumetric Dice metric of nnU-Net ResENC model for mid-RT task on the training set with different combinations of prior information. The model with overall better performance is marked as bold.

Data	Dice Metric ($\mu \pm \sigma$)			
	Original Data		Preprocessed Data	
	GTVp	GTVn	GTVp	GTVn
no-prior	0.459 ± 0.021	0.695 ± 0.026	0.470 ± 0.016	0.709 ± 0.023
preMR-prior	0.454 ± 0.069	0.751 ± 0.026	0.471 ± 0.078	0.763 ± 0.019
preMask-prior	0.439 ± 0.063	0.750 ± 0.025	0.459 ± 0.068	0.772 ± 0.023
preMRMask-prior	0.457 ± 0.072	0.766 ± 0.018	0.468 ± 0.063	0.778 ± 0.015
preDistance-prior	$\mathbf{0.473 \pm 0.053}$	$\mathbf{0.796 \pm 0.013}$	$\mathbf{0.493 \pm 0.054}$	$\mathbf{0.810 \pm 0.008}$

The results displayed in Table 2 clearly indicate the superior performance of the *preDistance-prior* settings compared to other models. Figure 3 provides a visual representation of the segmentation performance achieved by the top-performing models for both pre-RT and mid-RT tasks.

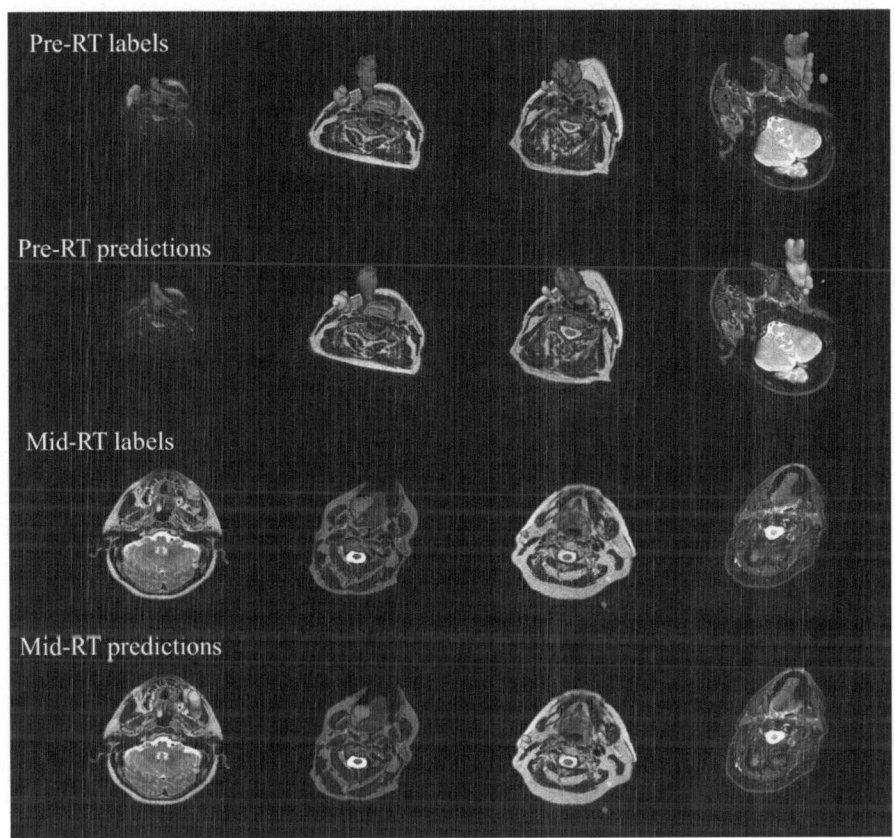

Fig. 3. Displaying predicted segmentation masks for pre-RT and mid-RT tasks. For each task, the upper row presents ground truth masks with GTVp in red and GTVn in green. The lower row illustrates corresponding predicted masks, where GTVp is shown in magenta and GTVn in yellow. Pseudo-3D visualization in axial views was employed to provide a comprehensive context of tumor volumes and their relative positions.

In a nutshell, the empirical results obtained on the training dataset demonstrate that the implemented preprocessing techniques significantly enhanced the segmentation accuracy of the models. The optimal results were attained by ensembling ResENC and MedNeXt models for the pre-RT task and incorporating distance maps as prior knowledge for the mid-RT task. Notably, the inference time for the pre-RT task was under 6 min and approximately 3 min for the mid-RT task, utilizing a local workstation with limited resources (12GB GPU memory, 32GB RAM, and 8 CPU threads). However, the execution time of the Docker image on the evaluation platform surpassed the 20-min limit, resulting in failed job submissions. Consequently, to meet the resource limitations, the final submitted models underwent substantial simplifications. Specifically, the preprocessing pipeline was simplified by removing both bias correction and volume cropping. Furthermore, the ensemble approach initially planned for the pre-RT task, which involved both ResENC and MedNeXT models, was revised to utilize only three trained folds of the MedNeXT model. Similarly, for the mid-RT task, only three folds of the ResENC model were employed. The performance of these simplified submitted models on both the preliminary development phase and final testing datasets is presented in Table 3. Furthermore, the final ranking of the challenge was determined by calculating the average of aggregated Dice metrics (mean DSC_{agg}) over the GTVp and the GTVn. The resulting mean DSC_{agg} values of this contribution were 0.822 and 0.710 for the pre-RT and mid-RT tasks, respectively.

Table 3. Performance of submitted containerized algorithms on testing datasets in terms of aggregated Dice metric.

Testing phase	Aggregated Dice Metric			
	pre-RT		mid-RT	
	GTVp	GTVn	GTVp	GTVn
Preliminary development (n = 2)	0.868	0.929	0.746	0.814
Final (n = 50)	0.795	0.849	0.553	0.865

4 Discussion

Accurate delineation of the GTVs is critical for effective RT planning in HNC. However, the complex morphology of head and neck structures and low target-to-background contrast present remarkable challenges for delineation tasks. In addition, manual delineation is time-consuming and prone to inter-observer variability, particularly in MRgRT applications, where rapid and precise segmentation is essential.

In this study, we developed a pipeline for segmenting GTVp and GTVn in both pre- and mid-RT T2w MR images. Our findings demonstrate that applying simple preprocessing steps to the acquired MR images can significantly improve the performance of segmentation models. Furthermore, the effective integration of pre-RT data as prior knowledge into mid-RT segmentation models can substantially enhance segmentation performance and model robustness.

Specifically, maximal cropping of image data, elimination of background regions, intensity normalization, and bias-field correction improved segmentation accuracy by an average of nearly 3% for pre-RT and 2% for mid-RT tasks. More importantly, our quantitative metrics reveal that directly integrating pre-RT MR volumes or segmentation masks into the mid-RT pipeline does not lead to meaningful improvements. This is likely because tumor response to treatment can alter tumor appearance, resulting in differing characteristics (even significant shrinkage) in mid-RT volumes. Therefore, directly adding first-time point data, where tumors appear intact, can confuse learning algorithms attempting to map the same segmentation masks onto two different tumor appearances. Conversely, representing the first-time point tumor appearance as probabilistic maps introduces a set of non-binary hypotheses to the segmentation model, resulting in significant accuracy enhancement. In practice, this strategy improved the Dice metric of GTVp by 2% and GTVn by up to 11%.

It is worth to note that we anticipated some performance drops in the testing phase due to the simplification of submitted models necessitated by limited computational resources. Nevertheless, the achieved results remain robust enough to validate the potential of the proposed pipeline.

Finally, while our pipeline yielded promising results for both tasks, certain limitations will be addressed in future studies. In particular, we explored integrating prior information into the segmentation network by simply adding distance maps as second input channels. Future investigations will consider alternative strategies, such as late fusion techniques.

Acknowledgments. This study was supported by the Cancer Research Funds of Radiumhemmet. We gratefully acknowledge the Swedish Cancer Society and The Swedish Research Council (2020-04618). Our appreciation also goes to Stockholm Medical Image Laboratory and Education (SMILE) for providing access to their computational resources.

Disclosure of Interests. The authors have no competing interests to declare that are relevant to the content of this article.

References

1. Guo, K., Xiao, W., Chen, X., Zhao, Z., Lin, Y., Chen, G.: Epidemiological trends of head and neck cancer: a population-based study. Biomed. Res. Int. **2021**(1), 1738932 (2021). https://doi.org/10.1155/2021/1738932
2. Gormley, M., Creaney, G., Schache, A., Ingarfield, K., Conway, D.I.: Reviewing the epidemiology of head and neck cancer: definitions, trends and risk factors. British Dental J. **233**(9), 780–786 (2022). https://doi.org/10.1038/s41415-022-5166-x
3. Pollard, J.M., Wen, Z., Sadagopan, R., Wang, J., Ibbott, G.S.: The future of image-guided radiotherapy will be MR guided. British J. Radiol. **90**(1073), May 2017, https://doi.org/10.1259/BJR.20160667/7445697
4. Benitez, C.M., Chuong, M.D., Künzel, L.A., Thorwarth, D.: MRI-guided adaptive radiation therapy. Semin. Radiat. Oncol. **34**(1), 84–91 (2024). https://doi.org/10.1016/J.SEMRADONC.2023.10.013
5. Antonelli, M., et al.: The Medical Segmentation Decathlon, June 2021. https://doi.org/10.48550/arXiv.2106.05735
6. Isensee, F., Jaeger, P.F., Kohl, S.A.A., Petersen, J., Maier-Hein, K.H.: nnU-Net: a self-configuring method for deep learning-based biomedical image segmentation. Nature Methods **18**(2), 203–211 (2020). https://doi.org/10.1038/s41592-020-01008-z
7. Ulrich, C., Isensee, F., Wald, T., Zenk, M., Baumgartner, M., Maier-Hein, K.H.: MultiTalent: a multi-dataset approach to medical image segmentation. Lecture Notes in Computer Science (including subseries Lecture Notes in Artificial Intelligence and Lecture Notes in Bioinformatics), vol. 14222 LNCS, pp. 648–658 (2023). https://doi.org/10.1007/978-3-031-43898-1_62/TABLES/2

8. Hatamizadeh, A., Nath, V., Tang, Y., Yang, D., Roth, H.R., Xu, D.: Swin UNETR: swin transformers for semantic segmentation of brain tumors in MRI images. Lecture Notes in Computer Science (including subseries Lecture Notes in Artificial Intelligence and Lecture Notes in Bioinformatics), vol. 12962. LNCS, pp. 272–284, January 2022. https://doi.org/10.1007/978-3-031-08999-2_22

9. Oreiller, V., et al.: Head and neck tumor segmentation in PET/CT: The HECKTOR challenge. Med. Image Anal. **77**, 102336 (2022). https://doi.org/10.1016/J.MEDIA.2021.102336

10. Luo, X., et al.: SegRap2023: A Benchmark of Organs-at-Risk and Gross Tumor Volume Segmentation for Radiotherapy Planning of Nasopharyngeal Carcinoma, December 2023. https://doi.org/10.7937/K9/TCIA.2017.8oje5q00

11. Head and Neck Tumor Segmentation for MR-Guided Applications (HNTS-MRG) 2024. Accessed 07 Sep 2024. https://hntsmrg24.grand-challenge.org/

12. Tustison, N.J., et al.: N4ITK: improved N3 bias correction. IEEE Trans. Med. Imaging **29**(6), 1310–1320 (2010). https://doi.org/10.1109/TMI.2010.2046908

13. Myronenko, A.: 3D MRI brain tumor segmentation using autoencoder regularization, pp. 311–320. Springer, Cham (2019). https://doi.org/10.1007/978-3-030-11726-9_28

14. Isensee, F., et al.: nnU-Net revisited: a call for rigorous validation in 3D medical image segmentation, April 2024, Accessed 29 July 2024. https://arxiv.org/abs/2404.09556v2

15. Roy, S., et al.: MedNeXt: transformer-driven scaling of ConvNets for medical image segmentation. Lecture Notes in Computer Science (including subseries Lecture Notes in Artificial Intelligence and Lecture Notes in Bioinformatics), vol. 14223. LNCS, pp. 405–415, 2023, https://doi.org/10.1007/978-3-031-43901-8_39/TABLES/2

16. Liu, Z., Mao, H., Wu, C.Y., Feichtenhofer, C., Darrell, T., Xie, S.: A ConvNet for the 2020s. Proceedings of the IEEE Computer Society Conference on Computer Vision and Pattern Recognition, vol. 2022-June, pp. 11966–11976, January 2022. https://doi.org/10.1109/CVPR52688.2022.01167

17. Gu, A., Goel, K., Ré, C.: Efficiently modeling long sequences with structured state spaces. In: ICLR 2022 - 10th International Conference on Learning Representations, October 2021. Accessed 30 Jul 2024. https://arxiv.org/abs/2111.00396v3

18. Gu, A., Dao, T.: Mamba: linear-time sequence modeling with selective state spaces, December 2023. https://arxiv.org/abs/2312.00752v2. Accessed 30 July 2024

19. Ma, J., Li, F., Wang, B.: U-mamba: enhancing long-range dependency for biomedical image segmentation, January 2024. https://arxiv.org/abs/2401.04722v1. Accessed 30 July 2024

UMamba Adjustment: Advancing GTV Segmentation for Head and Neck Cancer in MRI-Guided RT with UMamba and NnU-Net ResEnc Planner

Jintao Ren[1,2]($\boxtimes$)(ID), Kim Hochreuter[1,2](ID), Jesper Folsted Kallehauge[1,2](ID), and Stine Sofia Korreman[1,2,3](ID)

[1] Department of Clinical Medicine, Aarhus University, Nordre Palle Juul-Jensens Blvd. 11, 8200 Aarhus, Denmark
jintaoren@clin.au.dk
[2] Aarhus University Hospital, Danish Centre for Particle Therapy, Palle Juul-Jensens Blvd. 25, 8200 Aarhus, Denmark
[3] Aarhus University, Department of Oncology, Palle Juul-Jensens Blvd. 35, 8200 Aarhus, Denmark

Abstract. Magnetic Resonance Imaging (MRI) plays a crucial role in MRI-guided adaptive radiotherapy for head and neck cancer (HNC) due to its superior soft-tissue contrast. However, accurately segmenting the gross tumor volume (GTV), which includes both the primary tumor (GTVp) and lymph nodes (GTVn), remains challenging. Recently, two deep learning segmentation innovations have shown great promise: UMamba, which effectively captures long-range dependencies, and the nnU-Net Residual Encoder (ResEnc), which enhances feature extraction through multistage residual blocks. In this study, we integrate these strengths into a novel approach, termed 'UMambaAdj'. Our proposed method was evaluated on the HNTS-MRG 2024 challenge test set using pre-RT T2-weighted MRI images, achieving an aggregated Dice Similarity Coefficient (DSC_{agg}) of 0.751 for GTVp and 0.842 for GTVn, with a mean DSC_{agg} of 0.796. This approach demonstrates potential for more precise tumor delineation in MRI-guided adaptive radiotherapy, ultimately improving treatment outcomes for HNC patients. Team: DCPT-Stine's group.

Keywords: Deep learning · Mamba · Tumor segmentation · Head and Neck Cancer · MRI

1 Introduction

Magnetic Resonance Imaging (MRI) plays a pivotal role in radiotherapy (RT), particularly in MRI-guided adaptive radiotherapy, due to its superior soft-tissue contrast compared to other imaging modalities like computed tomography (CT). This soft-tissue contrast enables more accurate delineation, which is especially

© The Author(s) 2025
K. A. Wahid et al. (Eds.): HNTS-MRG 2024, LNCS 15273, pp. 123–135, 2025.
https://doi.org/10.1007/978-3-031-83274-1_9

crucial in head and neck cancer (HNC), where the intricate anatomy and proximity of vital structures, such as the salivary glands, optic nerves, and spinal cord [1,4], make precise tumor targeting critical. MRI's ability to differentiate between tumor tissues and surrounding normal tissues enhances radiation delivery accuracy, reducing the risk of collateral damage to critical structures [3,23,24].

Despite these advantages, accurate delineation of HNC tumors, including both the primary tumor volume (GTVp) and involved nodal metastasis (GTVn), remains challenging. The heterogeneous and diffuse nature of HNC tumors often makes obtaining clear margins difficult [17,28]. Additionally, MRI's lower spatial resolution in the third dimension (through-slice direction) compared to the in-plane resolution can complicate tumor delineation, potentially leading to variability in interpretation among clinicians and contributing to inter-observer variation (IOV).

Given these challenges, there is a growing need for automated, accurate segmentation methods to enhance the consistency and precision of tumor delineation in MRI-guided adaptive radiotherapy. Deep learning-based medical image segmentation has emerged as a promising solution, often building on the classic U-Net architecture, known for its symmetrical encoder-decoder design and skip connections [27]. It plays a crucial role in medical image analysis by identifying and delineating structures such as organs, lesions, tumors, and tissues across various 2D and 3D imaging modalities, including CT, MRI and Positron Emission Tomography (PET), thereby aiding in diagnosis, treatment planning, and prognoses.

Recently, the leading deep learning models for segmentation have shifted between convolutional neural networks (CNNs) and transformer-based architectures. CNNs excel at capturing translational invariances and local features but often face challenges with long-range dependencies [21], such as relationships between distant regions of an image or global structural patterns. In contrast, Vision Transformers (ViTs) [8] effectively capture global context by treating the image as a sequence of patches. However, their self-attention mechanism incurs a quadratic computational cost relative to the number of patches [9], and Transformers tend to be prone to overfitting, especially when working with limited datasets [13,19]. Leveraging the complementary strengths of both architectures, many studies have explored hybrid models that integrate ViTs with CNNs, resulting in architectures such as nnFormer [36], TransUNet [5], UNETR [12], SwinUNETR [11], and UNETR++ [29]. These hybrid models have also gained popularity in HNC GTV segmentation. For instance, Hung Chu et al. [6] demonstrated that the SwinUNETR achieved an average Dice similarity coefficient (DSC) of 0.626 on CT/PET data, while a cross-modal Swin transformer achieved a mean DSC of 0.769 for GTVp using CT/PET modalities [18]. Despite these advancements, the U-Net architecture continues to be a foundational design in all segmentation models.

The field is now advancing with structured state-space models (SSMs), such as Mamba [7,9,10], which offer improved segmentation performance by efficiently modeling long-range dependencies and scaling effectively with sequence length [13,35]. Mamba's ability to capture complex anatomical relationships makes it well-suited for segmenting GTVp and GTVn in HNC, as their locations are often closely correlated with each other. However, the optimal model configuration depends on task-specific factors such as the foreground-to-background ratio, image resolution, and tumor size variability, as each architecture's effectiveness varies with different imaging challenges and anatomical complexities.

Recently, two innovative approaches, UMamba [22] and the new nnU-Net Residual Encoder (ResEnc) planner [15], have gained significant attention in medical image segmentation. The default UMamba encoder incorporates a Mamba layer after each CNN block, which can be computationally expensive, especially at the first level where image features have a large resolution. This design, which includes both a residual encoder and decoder, can be cumbersome to train and provides only limited accuracy improvements in its default configuration [15]. In contrast, the nnU-Net ResEnc enhances feature extraction through multiple blocks of residual CNN encoding and employs only a single CNN layer in the decoder, offering a more efficient solution.

This study focuses on addressing the first task of the HNTS-MRG 2024 challenge, which aims to segment both GTVp and GTVn using pre-RT T2-weighted MRI images. We aim to improve gross tumor volume (GTV) segmentation in T2-weighted MRI for head and neck cancer by integrating the strengths of UMamba and nnU-Net ResEnc. We refer to this integrated approach as **UMambaAdj** in this study.

Our contributions are as follows:

– We optimize UMamba by removing the Mamba layer in the first stage and the residual blocks in the decoder, significantly enhancing computational efficiency while preserving its ability to capture long-range dependencies in deeper stages.
– We combine UMamba's long-range dependency modeling with nnU-Net ResEnc's enhanced residual encoding to improve the accuracy of GTV delineation in the complex anatomy of head and neck cancer.

2   Material and Methods

2.1   Data

The dataset used in this study was provided by the organizers of the HNTS-MRG 2024 challenge task 1, consisting of 150 HNC patients, primarily with oropharyngeal cancer (OPC). Each patient had T2-weighted MRI sequences of the head and neck region, acquired at University of Texas MD Anderson Cancer Center [30]. The images included pre-RT scans taken 1–3 weeks before the start of radiotherapy. For all cases, GTV for the primary tumor (GTVp) and involved lymph nodes (GTVn) were independently segmented by 3 to 4 expert physician observers based on the MRI images. The ground truth segmentation was then generated using the Simultaneous Truth And Performance Level Estimation (STAPLE) algorithm [32].

2.2   Network Architecture

The proposed network architecture is based on a combination of a 3D ResEnc U-Net and Mamba blocks. The CNN part of network architecture was designed according to the new nnU-Net Residual encoder planner (M - median sized). The U-Net consists of 6 stages, each with varying features per stage (32, 64, 128, 256, 320, 320). The network uses 3D convolutional layers with kernel sizes mostly set to (3, 3, 3), except for the first stage where it is (1, 3, 3). The strides vary across stages to enable down-sampling at different levels, with a stride set of (1, 2, 2) between the first and second stages, and (2, 2, 2) for the remaining stages. Each stage contains a different number of residual CNN blocks with counts of (1, 3, 4, 6, 6, 6) in the encoder, and a Mamba layer is appended after each residual CNN block except the first stage. A skip connection with concatenation was applied to connect the Mamba layer and the decoder blocks, while each decoder block consists of only one 3D CNN block. Instance normalization and the Leaky ReLU activation function are used. Additionally, deep supervision is applied at the top four levels of the network outputs. The overall structure of the network can be seen in Fig. 1a.

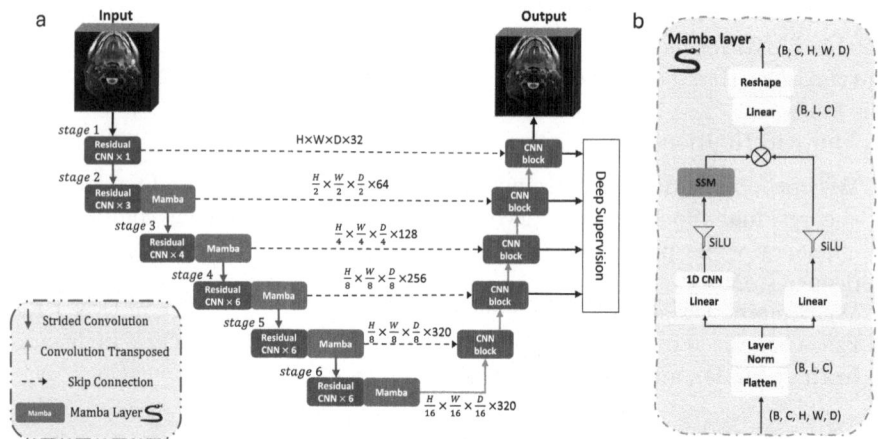

Fig. 1. (a) Overview of the proposed UMamba adjustment (UMambaAdj) network architecture. (b) Details of the Mamba layer.

2.3   Mamba Layer

The Mamba layer, adapted from the UMamba design for capturing long-range dependencies, processes input image feature maps of shape (B, C, H, W, D), where B is the batch size, C the channel, and H, W, D the spatial dimensions. These feature maps are first reshaped and transposed into a flattened representation (B, L, C), where $L = H \times W \times D$, treating all spatial locations as individual patch tokens. The reshaped features are then normalized using Layer Normalization. Following normalization, the features undergo two parallel branches. In the first branch, the features are expanded to $(B, 2L, C)$ through a linear transformation, followed by a 1D convolutional layer and a SiLU activation function, ultimately passing through the SSM layer. In the second branch, a similar expansion process is performed via a linear transformation and a SiLU activation function, but without the convolutional or SSM steps. The outputs from the two branches are then combined using element-wise (Hadamard) multiplication. The resulting features are projected back to the original token dimension, reshaped, and transposed to restore the original input shape of (B, C, H, W, D), maintaining the spatial structure of the image feature maps for further processing. The detailed flow of a Mamba layer can be seen in Fig. 1b.

2.4   Training Parameters

Training was conducted with a batch size of 4, using a patch size of (48, 192, 192) and Z-score normalization for data preprocessing. The median image size in voxels was (123, 512, 511), with spacing set at (1.199, 0.5, 0.5). Resampling spline interpolation functions were employed to adjust both image and segmentation data, using an interpolation order of 3 for images and an order of 1 (linear) for masks. The training was performed using the SGD optimizer with a PolyLR scheduler (exponent = 0.9), starting with a learning rate of 0.01. The adoption of the Mamba layer often led to gradient vanishing or explosion during training, especially when using mixed-precision (fp16) with automatic casting. To address this, normalized gradients were clipped with a value of 1.

The 150 patients were randomly divided into 5 folds, with each fold comprising 120 patients for training and 30 for validation. Each model was trained for a maximum of 1,000 epochs, and the final models from the last epoch were saved for prediction. For the final challenge submission, predictions on the test set were generated using an ensemble of all models trained across the five folds. In line with reproducibility and verification guidelines [14], all source code, predicted masks, training logs and trained weights have been made publicly available on GitHub[1].

2.5   Evaluation

The aggregated Dice Similarity Coefficient (DSC_{agg}) [2] was used as the primary evaluation metric in accordance with the guidelines of the HNTS-MRG 2024 challenge. Additionally, we employed the mean 95th percentile Hausdorff Distance (HD_{95}), and the mean surface distance (MSD) as supplementary metrics to further evaluate the segmentation performance for both GTVp and GTVn. Hausdorff Distance (HD) was used for case study.

To evaluate the performance of the proposed method, we compared both the segmentation accuracy and training epoch times of the default nnU-Net,

---

[1] https://github.com/Aarhus-RadOnc-AI/UMambaAdj.

nnU-Net Residual Encoder (ResEnc), UMamba Encoder (UMambaEnc), and the proposed UMambaAdj using the DSC_{agg} metric. Since the Mamba block involves multiple tensor shape manipulations that are not fully represented by FLOPs, we measured the stable epoch time after the first epoch as a direct indicator of model efficiency. All compared groups were trained with a batch size of 4.

2.6 System Environment

The experiments were conducted on a system equipped with dual AMD Ryzen Threadripper 3990X 64-core processors (128 threads) and 256GB of system memory. An NVIDIA RTX A6000 GPU with 48GB VRAM was used for training. The software environment included Python 3.12.4, PyTorch 2.4.0, CUDA 12.6 and nnU-Net 2.5.1. Distance metrics were calculated using SimpleITK 2.4.0.

3   Results

3.1   Cross-Validation Performance

Table 1. GTVp performance on 5-Fold cross-validation

Metric	Fold 0	Fold 1	Fold 2	Fold 3	Fold 4	Average
DSC_{agg}	0.804	0.742	0.804	0.774	0.776	0.779
HD_{95} [mm]	5.7	8.1	6.8	6.5	10.4	7.5
MSD [mm]	2.6	2.8	1.9	1.9	4.2	2.68

Table 2. GTVn performance on 5-Fold cross-validation

Metric	Fold 0	Fold 1	Fold 2	Fold 3	Fold 4	Average
DSC_{agg}	0.874	0.849	0.751	0.875	0.885	0.847
HD_{95} [mm]	15.2	15.4	24.5	21.2	17.6	18.78
MSD [mm]	3.1	2.6	4.2	3.6	3.3	3.36

The performance metrics for GTVp and GTVn across 5-fold cross-validation are summarized in Table 1 and Table 2. For GTVp, DSC_{agg} ranged from 0.742 to 0.804 across different folds, with an average of 0.779. The HD_{95} varied from 5.7 mm to 10.4 mm, yielding an average of 7.5 mm. MSD ranged between 1.9 mm and 4.2 mm, with an average of 2.68 mm. For GTVn, DSC_{agg} ranged from 0.751 to 0.885, with an average of 0.847. The HD_{95} showed a wider range from 15.2 mm to 24.5 mm, resulting in an average of 18.78 mm. The MSD values spanned from 2.6 mm to 4.2 mm, with an average of 3.36 mm.

3.2   Comparison of nnU-Net, ResEnc, UmambaEnc and UmambaAdj

Table 3. Performance comparison between nn-UNet default, ResEnc, UMambaEnc, and the proposed UMambaAdj.

Methods	GTVp			GTVn		
	DSC_{agg}	HD_{95}	MSD	DSC_{agg}	HD_{95}	MSD
nnUNet default	0.78	14.5	3.7	0.859	28.3	4.4
nnUNet ResEnc	0.803	10.5	5.5	0.872	15.8	3.1
UMambaEnc	0.794	8.7	3.5	**0.880**	**10.5**	**2.2**
UMambaAdj	**0.804**	**5.7**	**2.6**	0.874	15.2	3.1

Bold numbers indicate the best performance for each metric

Table 3 summarizes the performance of various models for GTVp and GTVn segmentation. For GTVp, the proposed UMambaAdj achieved the highest DSC_{agg} (0.804) and the best results in terms of HD_{95} (5.7 mm) and MSD (2.6 mm). In the case of GTVn, UMambaEnc achieved the highest DSC_{agg} (0.880) and outperformed others with the lowest HD_{95} (10.5 mm) and MSD (2.2 mm).

Two cases were selected for illustration in Fig. 2. For patient (a), all methods except UMambaAdj predicted a significantly smaller GTVp (DSC range 0.277–0.584), with the lower part missing, whereas UMambaAdj achieved a higher DSC of 0.703, despite all methods failing to capture the upper part. Additionally, the default nnU-Net model incorrectly identified a lymph node as GTVn, resulting in an HD of 75.2 mm, compared to 3.2–3.5 mm for the other methods. For patient (b), all methods except UMambaAdj made false positive predictions of GTVp in the same location. Moreover, all methods except UMambaEnc incorrectly predicted a lymph node as positive bilaterally, leading to an HD of 60 mm.

The training epoch time for the nnU-Net default model was 116 s, while the nnU-Net ResEnc took 127 s. The UMambaEnc required 400 s, and UMambaAdj took 199 s. Mixed-precision (16-bit floating point) autocast was enabled for all layers, except the Mamba layer, to ensure training stability.

3.3   Final Test Score

We submitted our trained UMambaAdj models in a Docker container to the HNTS-MRG 2024 challenge on the grand challenge platform. Predictions were made using an ensemble of all five models trained across the 5-folds. Our model was evaluated on the test set using pre-RT T2-weighted MRI images, achieving an DSC_{agg} of 0.751 for GTVp and 0.842 for GTVn, resulting in an overall mean DSC_{agg} of **0.796**.

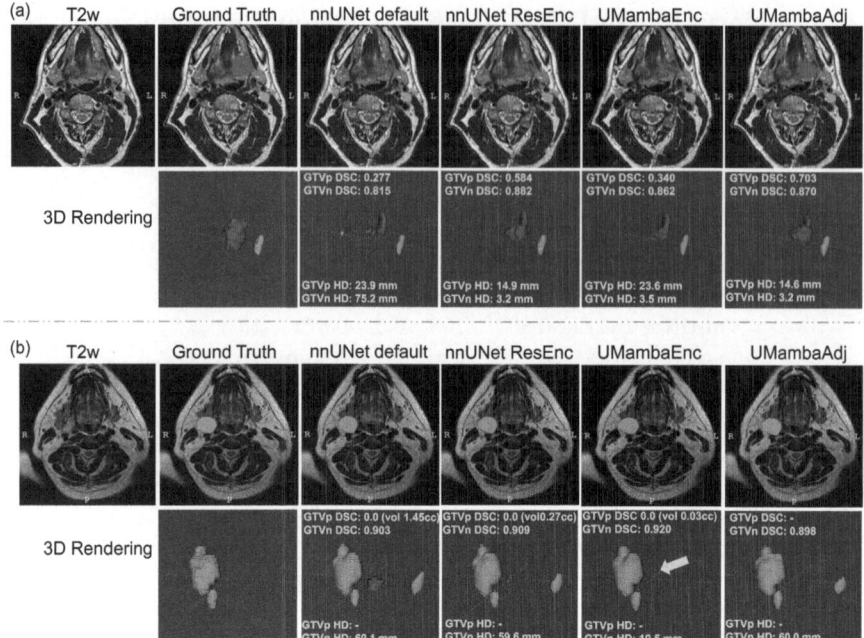

Fig. 2. Two patients (a and b) were selected for illustration. For each patient, the first row (left to right) displays the original T2-weighted MRI image, the ground truth overlaid on the image, and the segmentation results from all compared methods overlaid on the image. The second row shows the 3D renderings of the delineations and segmentations. Red represents the GTVp, and Green represents the GTVn. Segmentation metrics, including DSC and HD, are shown on the rendering subfigures. A yellow arrow in patient (b) indicates a nearly invisible small false-positive GTVp segmentation predicted by UMambaEnc. (Color figure online)

4   Discussion

In this study, we developed a customized network that integrates features from both UMamba and the nnU-Net Residual Encoder for T2-weighted MRI head and neck tumor segmentation. The aim was to combine the feature extraction strength of the residual encoder with the long-range dependency capabilities of Mamba blocks. Compared to the original UMambaEnc, the proposed UMambaAdj demonstrated comparable segmentation accuracy with reduced training and inference time, and outperformed UMambaEnc for GTVp. It also achieved significantly better HD_{95} and MSD while maintaining similar DSC_{agg} compared to nnU-Net ResEnc. All recent methods outperformed the default nnU-Net model across all metrics, confirming the complementary strengths of UMamba and nnU-Net ResEnc in the proposed approach.

The cross-validation and final test results revealed a notable performance gap between the segmentation accuracy of the primary tumor (GTVp) and the nodal disease (GTVn). Although the DSC_{agg} for GTVn was substantially higher than for GTVp, the HD_{95} and MSD metrics were significantly larger for GTVp. This discrepancy suggests that the model struggled more with accurately delineating the nodal boundaries or even on detecting the nodes, often due to falsely predicted lymph nodes. These false predictions greatly influenced the distance based metrics.

Our experiments demonstrated that models incorporating the UMamba block achieved significant improvements in distance-based metrics (HD_{95} and MSD), underscoring the value of long-range dependencies provided by the Mamba. This capability is crucial for capturing the intricate structures of HNC tumors, where understanding dependencies between primary tumors and metastatic lymph nodes is vital. Notably, the proposed UMambaAdj model, which excludes the Mamba block from the first stage, matched the ResEnc model's DSC_{agg} performance while achieving HD_{95} and MSD metrics similar to UMambaEnc. This suggests UMambaAdj effectively balances volumetric overlap and boundary delineation, although GTVn results indicate a need for the Mamba layer in the first stage.

The evaluation results show that while HD_{95} and MSD metrics vary significantly among the methods, DSC_{agg} remains relatively consistent. This difference is due to the metrics' sensitivities: DSC_{agg}, being a global overlap measure, is less affected by minor boundary discrepancies or isolated false predictions. In contrast, HD_{95} is highly sensitive to boundary inaccuracies, making it more responsive to small over-segmentation, under-segmentation, or isolated false predictions. This sensitivity makes these metrics more reflective of clinically relevant errors, where even small false positives or negatives would be unacceptable. This observation underscores the importance of employing multiple evaluation metrics for a comprehensive assessment of segmentation performance. Although DSC_{agg} might be suitable as a single ranking metric in public challenges due to its straightforward interpretation, relying solely on it can be misleading. The best DSC_{agg} method does not always correspond to the best overall segmentation performance, particularly in accurately capturing the boundaries or avoiding false predictions-a critical factor in real-world clinical applications.

Furthermore, we observed that the increased computational demands of the UMamba block were reflected in extended training times rather than memory usage. This was not fully explained by its computational complexity (as FLOPs showed only a marginal increase) but was largely due to tensor shape manipulations, such as permute, transpose, and reshaping, which became more significant with larger tensor sizes.

Despite the promising performance of the proposed UMambaAdj model, its effectiveness must be validated on external datasets. Our current validation was limited to a single fold from a single institutional dataset, and thus, further testing on public datasets or other private datasets is essential to confirm the generalizability and robustness of our approach. This need for broader validation is especially relevant given the rapid emergence of various Mamba-based segmentation models since Mamba's initial publication.

Recent adaptations, such as the "Swin" feature with Mamba [20], the tri-oriented vision Mamba approach [34], and the Visual Mamba U-Net [31], have shown that integrating Mamba with existing CNN blocks can lead to notable segmentation accuracy, all claiming achieved state-of-the-art (SOTA). However, some of these models still require significant modifications to match the 3D segmentation performance of nnU-Net. Our adjustments based on UMamba indicate that Mamba holds particular promise for HNC tumor segmentation, especially for GTVn.

Nevertheless, even with these advances, head and neck cancer tumor segmentation remains a challenging task that is far from being a "solved" problem. Fully automatic segmentation methods often face limitations that necessitate human intervention to ensure accurate treatment planning. In our study, despite achieving decent DSC_{agg} scores, even SOTA models struggled with accurately identifying tumor locations, highlighting the persistent difficulties in this domain. The complex anatomy of the head and neck region, coupled with the challenge of distinguishing tumors in T2-weighted images (as evidenced in the GTVp cases from Fig. 2), reinforces these challenges. Therefore, incorporating complementary information such as FDG-PET imaging [16,26], biopsy data, patient reports [25], or human interaction [33] may be crucial for improving the accuracy and reliability of HNC GTV segmentation.

5   Conclusion

In conclusion, our customized UMambaAdj model successfully combines the strengths of long-range dependencies from UMamba blocks with the feature encoding capabilities of the nnU-Net Residual encoder, offering a balanced solution for GTV segmentation in HNC. The model showed promise in achieving accurate segmentations with a more efficient architecture, demonstrating comparable or improved performance over existing methods. However, further validation on diverse datasets and incorporating complementary tumor information with human-in-the-loop strategies will be necessary to advance the application of automatic segmentation in clinical practice for MRI guided adaptive RT.

References

1. Ahmed, M., et al.: The value of magnetic resonance imaging in target volume delineation of base of tongue tumours-a study using flexible surface coils. Radiother. Oncol. **94**(2), 161–167 (2010)
2. Andrearczyk, V., et al.: Overview of the hecktor challenge at MICCAI 2021: automatic head and neck tumor segmentation and outcome prediction in PET/CT images. In: 3D Head and Neck Tumor Segmentation in PET/CT Challenge, pp. 1–37. Springer (2021)
3. Benitez, C.M., Chuong, M.D., Künzel, L.A., Thorwarth, D.: MRI-guided adaptive radiation therapy. In: Seminars in Radiation Oncology, vol. 34, pp. 84–91. Elsevier (2024)

4. Brouwer, C.L., et al.: Ct-based delineation of organs at risk in the head and neck region: DAHANCA, EORTC, GORTEC, HKNPCSG, NCIC CTG, NCRI, NRG oncology and TROG consensus guidelines. Radiother. Oncol. **117**(1), 83–90 (2015)

5. Chen, J., et al.: Transunet: transformers make strong encoders for medical image segmentation. arXiv preprint arXiv:2102.04306 (2021)

6. Chu, H., et al.: Swin UNETR for tumor and lymph node segmentation using 3D PET/CT imaging: a transfer learning approach. In: 3D Head and Neck Tumor Segmentation in PET/CT Challenge, pp. 114–120. Springer (2022)

7. Dao, T., Gu, A.: Transformers are SSMS: generalized models and efficient algorithms through structured state space duality. arXiv preprint arXiv:2405.21060 (2024)

8. Dosovitskiy, A.: An image is worth 16x16 words: transformers for image recognition at scale. arXiv preprint arXiv:2010.11929 (2020)

9. Gu, A., Dao, T.: Mamba: linear-time sequence modeling with selective state spaces. arXiv preprint arXiv:2312.00752 (2023)

10. Gu, A., Goel, K., Ré, C.: Efficiently modeling long sequences with structured state spaces. arXiv preprint arXiv:2111.00396 (2021)

11. Hatamizadeh, A., Nath, V., Tang, Y., Yang, D., Roth, H.R., Xu, D.: Swin UNETR: swin transformers for semantic segmentation of brain tumors in MRI images. In: International MICCAI Brainlesion Workshop, pp. 272–284. Springer (2021)

12. Hatamizadeh, A., et al.: UNETR: transformers for 3D medical image segmentation. In: Proceedings of the IEEE/CVF Winter Conference on Applications of Computer Vision, pp. 574–584 (2022)

13. Heidari, M., et al.: Computation-efficient era: a comprehensive survey of state space models in medical image analysis. arXiv preprint arXiv:2406.03430 (2024)

14. Hurkmans, C., et al.: A joint ESTRO and AAPM guideline for development, clinical validation and reporting of artificial intelligence models in radiation therapy. Radiother. Oncol. **197**, 110345 (2024)

15. Isensee, F., et al.: nnU-Net revisited: a call for rigorous validation in 3D medical image segmentation. arXiv preprint arXiv:2404.09556 (2024)

16. Jensen, K., et al.: Imaging for target delineation in head and neck cancer radiotherapy. In: Seminars in Nuclear Medicine, vol. 51, pp. 59–67. Elsevier (2021)

17. Jensen, K., et al.: The danish head and neck cancer group (DAHANCA) 2020 radiotherapy guidelines. Radiother. Oncol. **151**, 149–151 (2020)

18. Li, G.Y., Chen, J., Jang, S.I., Gong, K., Li, Q.: Swincross: cross-modal swin transformer for head-and-neck tumor segmentation in PET/CT images. Med. Phys. **51**(3), 2096–2107 (2024)

19. Lin, T., Wang, Y., Liu, X., Qiu, X.: A survey of transformers. AI Open **3**, 111–132 (2022)

20. Liu, J., et al.: Swin-umamba: mamba-based unet with imagenet-based pretraining. arXiv preprint arXiv:2402.03302 (2024)

21. Luo, W., Li, Y., Urtasun, R., Zemel, R.: Understanding the effective receptive field in deep convolutional neural networks. In: Advances in Neural Information Processing Systems, vol. 29 (2016)

22. Ma, J., Li, F., Wang, B.: U-mamba: enhancing long-range dependency for biomedical image segmentation. arXiv preprint arXiv:2401.04722 (2024)

23. McDonald, B.A., Dal Bello, R., Fuller, C.D., Balermpas, P.: The use of MR-guided radiation therapy for head and neck cancer and recommended reporting guidance. In: Seminars in Radiation Oncology, vol. 34, pp. 69–83. Elsevier (2024)

24. Mohamed, A.S., et al.: Prospective in silico study of the feasibility and dosimetric advantages of MRI-guided dose adaptation for human papillomavirus positive oropharyngeal cancer patients compared with standard IMRT. Clin. Transl. Radiat. Oncol. **11**, 11–18 (2018)

25. Rajendran, P., et al.: Large language model-augmented auto-delineation of treatment target volume in radiation therapy. arXiv preprint arXiv:2407.07296 (2024)

26. Ren, J., Eriksen, J.G., Nijkamp, J., Korreman, S.S.: Comparing different CT, PET and MRI multi-modality image combinations for deep learning-based head and neck tumor segmentation. Acta Oncol. **60**(11), 1399–1406 (2021)

27. Ronneberger, O., Fischer, P., Brox, T.: U-net: convolutional networks for biomedical image segmentation. In: Medical Image Computing and Computer-Assisted Intervention–MICCAI 2015: 18th International Conference, Munich, Germany, 5–9 October 2015, Proceedings, part III 18, pp. 234–241. Springer (2015)

28. Rühle, A., Nicolay, N.H.: Head and neck cancer. In: Grosu, A.L., Nieder, C., Nicolay, N.H. (eds.) Target Volume Definition in Radiation Oncology, pp. 91–114. Springer, Cham (2023). https://doi.org/10.1007/978-3-031-45489-9_5

29. Shaker, A.M., Maaz, M., Rasheed, H., Khan, S., Yang, M.H., Khan, F.S.: UNETR++: delving into efficient and accurate 3d medical image segmentation. IEEE Trans. Med. Imaging (2024)

30. Wahid, K., Dede, C., Naser, M., Fuller, C.: Training dataset for HNTSMRG 2024 challenge. https://doi.org/10.5281/zenodo.11199559 (2024). [Data set]

31. Wang, Z., Zheng, J.Q., Zhang, Y., Cui, G., Li, L.: Mamba-UNet: UNet-like pure visual mamba for medical image segmentation. arXiv preprint arXiv:2402.05079 (2024)

32. Warfield, S.K., Zou, K.H., Wells, W.M.: Simultaneous truth and performance level estimation (staple): an algorithm for the validation of image segmentation. IEEE Trans. Med. Imaging **23**(7), 903–921 (2004)

33. Wei, Z., Ren, J., Korreman, S.S., Nijkamp, J.: Towards interactive deep-learning for tumour segmentation in head and neck cancer radiotherapy. Phys. Imaging Radiat. Oncol. **25**, 100408 (2023)

34. Xing, Z., Ye, T., Yang, Y., Liu, G., Zhu, L.: Segmamba: long-range sequential modeling mamba for 3D medical image segmentation. arXiv preprint arXiv:2401.13560 (2024)

35. Xu, R., Yang, S., Wang, Y., Cai, Y., Du, B., Chen, H.: Visual mamba: a survey and new outlooks (2024)

36. Zhou, H.Y., Guo, J., Zhang, Y., Yu, L., Wang, L., Yu, Y.: nnformer: interleaved transformer for volumetric segmentation. arXiv preprint arXiv:2109.03201 (2021)

Comparative Analysis of nnUNet and MedNeXt for Head and Neck Tumor Segmentation in MRI-Guided Radiotherapy

Nikoo Moradi[1,2(✉)], André Ferreira[2,3,4,5,6], Behrus Puladi[5,6],
Jens Kleesiek[2,7,8,9], Emad Fatemizadeh[1], Gijs Luijten[2,10,11], Victor Alves[3],
and Jan Egger[2,4,7,10,11]

[1] Department of Electrical Engineering, Sharif University of Technology, Tehran, Iran
nikoo.moradi@ee.sharif.edu
[2] Institute for AI in Medicine (IKIM), University Medicine Essen, Girardetstraße 2,
45131 Essen, Germany
[3] Center Algoritmi/LASI, University of Minho, 4710-057 Braga, Portugal
[4] Computer Algorithms for Medicine Laboratory, Graz, Austria
[5] Department of Oral and Maxillofacial Surgery, University Hospital RWTH Aachen,
Aachen, Germany
[6] Institute of Medical Informatics, University Hospital RWTH Aachen, Aachen,
Germany
[7] Cancer Research Center Cologne Essen (CCCE), University Medicine Essen (AöR),
Essen, Germany
[8] German Cancer Consortium (DKTK), Partner Site Essen, Essen, Germany
[9] Department of Physics, TU Dortmund University, Dortmund, Germany
[10] Center for Virtual and Extended Reality in Medicine (ZvRM), University
Medicine Essen, Essen, Germany
[11] Institute of Computer Graphics and Vision (ICG), Graz University of Technology,
Inffeldgasse 16/II, 8010 Graz, Austria

Abstract. Radiation therapy (RT) is essential in treating head and neck cancer (HNC), with magnetic resonance imaging(MRI)-guided RT offering superior soft tissue contrast and functional imaging. However, manual tumor segmentation is time-consuming and complex, and therefore remains a challenge. In this study, we present our solution as team TUMOR to the HNTS-MRG24 MICCAI Challenge which is focused on automated segmentation of primary gross tumor volumes (GTVp) and metastatic lymph node gross tumor volume (GTVn) in pre-RT and mid-RT MRI images. We utilized the HNTS-MRG2024 dataset, which consists of 150 MRI scans from patients diagnosed with HNC, including original and registered pre-RT and mid-RT T2-weighted images with corresponding segmentation masks for GTVp and GTVn. We employed two state-of-the-art models in deep learning, nnUNet and MedNeXt. For Task 1, we pretrained models on pre-RT registered and mid-RT images, followed by fine-tuning on original pre-RT images. For Task 2, we combined registered pre-RT images, registered pre-RT segmentation masks,

K. A. Wahid et al. (Eds.): HNTS-MRG 2024, LNCS 15273, pp. 136–153, 2025.
https://doi.org/10.1007/978-3-031-83274-1_10

and mid-RT data as a multi-channel input for training. Our solution for **Task 1** achieved 1st place in the final test phase with an aggregated Dice Similarity Coefficient of **0.8254**, and our solution for **Task 2** ranked 8th with a score of **0.7005**. The proposed solution is publicly available at Github Repository.

Keywords: HNTS-MRG24 · MICCAI24 · nnUNet · MedNeXt

1   Introduction

Radiation therapy (RT) is a fundamental treatment modality for various malignancies, with head and neck cancer (HNC) being a primary beneficiary. Traditional RT planning has largely relied on computed tomography (CT) imaging. However, recent advancements have driven significant interest in magnetic resonance imaging (MRI)-guided RT. MRI offers superior soft tissue contrast compared to CT and enables functional imaging through multiparametric sequences, such as diffusion-weighted imaging. Additionally, MRI-guided RT facilitates daily adaptive treatment using MRI-Linac devices, optimizing tumor destruction while minimizing adverse effects. These advantages suggest that MRI-guided adaptive RT has the potential to revolutionize clinical practice for HNC [1,2].

Despite these benefits, MRI-guided RT planning generates extensive data, making manual tumor segmentation by physicians-the current clinical standard-a time-consuming and impractical process. This challenge is intensified by the complex anatomy of head and neck (H&N) tumors, which are notoriously difficult to delineate accurately. As a result, there is a growing interest in leveraging artificial intelligence (AI) to automate and improve the segmentation process.

Deep learning (DL), a subset of AI, has shown remarkable success in medical image segmentation, particularly in challenging domains like HNC. Various public challenges, such as the HECKTOR [3] and SegRap [4] challenges, have driven advancements in this field by providing datasets and benchmarks for AI model development. However, no large-scale, publicly available datasets for MRI-guided RT in HNC exist, highlighting the need for community-driven efforts to develop AI tools for clinical translation.

The HNTS-MRG24 challenge[1] addresses this gap by focusing on the segmentation of H&N tumors in MRI-guided adaptive RT. The challenge is divided into two tasks:

Task 1: Segmentation of primary gross tumor volume (GTVp) and metastatic node gross tumor volume (GTVn) on pre-RT MRI images.

Task 2: Extends this to mid-RT MRI images. In this task mid-RT image, pre-RT image with segmentation, and registered pre-RT image with registered segmentation are all available and can be used as input.

---

[1] https://hntsmrg24.grand-challenge.org.

A unique aspect of this challenge is its exploration of whether incorporating prior time point data (pre-RT and mid-RT) into segmentation algorithms can enhance performance in RT applications.

Given the potential of AI to streamline and enhance MRI-guided RT planning, the development of robust, automated segmentation algorithms could significantly impact clinical workflows, reducing the burden on clinicians and improving patient outcomes.

This paper presents our approach to the HNTS-MRG24 challenge, utilizing state-of-the-art DL models, i.e., nnUNet [5] and MedNeXt [6,7], and ensemble techniques to achieve accurate and reliable segmentation of H&N tumors.

State of the Art

Recent advancements in AI-based segmentation of HNC have demonstrated significant potential, particularly in the context of RT planning. Various studies have utilized DL models to automate and improve the segmentation of tumors and organs at risk, addressing the challenges posed by the complex anatomy of the H&N region.

Li et al. (2020) [8] proposed a semi-supervised framework for medical image segmentation using deep convolutional neural networks. Their method combines labeled and unlabeled data to improve segmentation accuracy by generating pseudo labels and iteratively refining the model. The framework incorporates ensemble learning to reduce errors from poor-quality pseudo labels. The approach was evaluated on the ISIC 2018 dataset for skin lesion segmentation [9,10], and it demonstrated superior performance compared to fully supervised models and earlier semi-supervised methods.

Astaraki et al. (2023) [11] focused on nasopharyngeal carcinoma, a subset of HNC, in the SegRap 2023 challenge. They developed a fully automated segmentation framework using a standard 3D U-Net model, which was effective in segmenting both GTVs and organs at risk from CT images. Their approach achieved first place in the second task of SegRap 2023 challenge, underscoring the robustness of the U-Net architecture for this task.

Myronenko et al. (2022) [12] presented a fully automated solution for H&N tumor segmentation using positron emission tomography (PET)/CT images in the HECKTOR 2022 [12] challenge. They employed the SegResNet architecture from MONAI, a semantic segmentation network optimized for 3D medical imaging. Their approach included 5-fold cross-validation, image normalization, and model ensembling, which helped achieve first place in the challenge.

These studies collectively highlight the ongoing efforts to refine DL techniques for HNC segmentation, with a particular focus on improving accuracy, handling data imbalance, and integrating multi-modal imaging data. The continued development of AI-driven segmentation tools holds promise for enhancing the precision and efficiency of RT planning, ultimately improving patient outcomes.

The remainder of this paper is structured as follows: Sect. 2 covers the materials and methods used in our approach, Sect. 3 presents the results and evaluation of our models, Sect. 4 discusses the findings, and Sect. 5 concludes with a summary and future directions.

2   Materials and Methods

2.1   Dataset

HNTS-MRG24 Dataset: For this study, we utilized the dataset provided by the HNTS-MRG24 challenge, which focuses on the segmentation of H&N tumors for MRI-guided adaptive RT. The dataset comprises MRI images from 150 patients diagnosed with HNC, collected at The University of Texas MD Anderson Cancer Center. It includes both pre-RT and mid-RT T2-weighted (T2w) MRI scans, with corresponding segmentation masks for GTVp and GTVn.

An important aspect of the dataset is the pre-registration of pre-RT images to mid-RT images, which was performed by the challenge organizers. This registration process was designed to align the images spatially, facilitating more accurate comparison and analysis between the different time points. Detailed parameters for this registration process can be found in the challenge's official GitHub repository[2].

For Task 1, we utilized the mid-RT and pre-RT registered images to pretrain our models, followed by fine-tuning on the original pre-RT images. During this phase, all 150 cases were included, and no cases were discarded.

For Task 2, the input consisted of a multi-channel format combining mid-RT, pre-RT registered images, and the corresponding pre-RT registered segmentation. All 150 cases were used to train nnUNet, but we encountered issues while training MedNeXt. To resolve these issues, cases with zero ground truth for either label 1, label 2, or both were discarded, leaving 115 samples after removing 35 cases. An example of a pre-RT image with its segmentation is shown in Fig. 1. All the visualizations in this paper were created using 3D Slicer [13].

BraTS24 Meningioma Radiotherapy Dataset: In addition to the challenge data, external public datasets were permitted. However, finding datasets that matched the challenge criteria (T2w MRI, at least two segmentation labels, H&N region) proved challenging. We experimented with the BraTS 2024 Meningioma RT Segmentation Challenge dataset [14], which contains 500 samples and focuses on the segmentation of GTV for meningiomas in brain MRI scans. This dataset uses 3D postcontrast T1w images and preserves extracranial structures through defacing techniques.

We used all 500 cases of the BraTS dataset specifically for Task 1 of the challenge and applied it as pretraining data for our models.

2.2   Networks

The effectiveness of two state-of-the-art models in DL, nnUNet and MedNeXt, was tested in both tasks of the HNTS-MRG24 challenge. Their performance

---

[2] https://github.com/kwahid/HNTSMRG_2024.

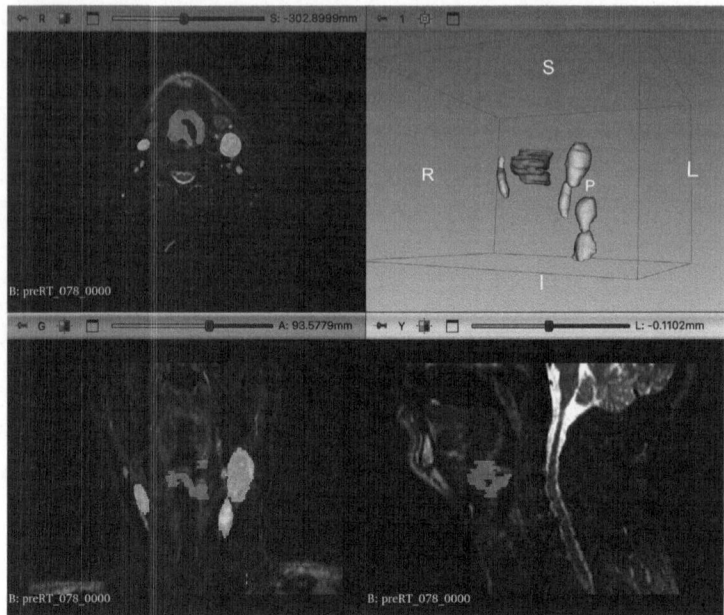

Fig. 1. A sample pre-RT image (Case 78) with its corresponding segmentation. The green label represents GTVp (label = 1), while the yellow label represents GTVn (label = 2). The images show axial, coronal, and sagittal views, along with a 3D rendering of the segmented tumors. (Color figure online)

was compared across different architectures and training strategies. A detailed description of the architectures used for each task will follow.

NnUNet: nnUNet is an automated DL framework that self-configures based on the properties of the input data. It has proven effective across a wide range of medical image segmentation tasks due to its adaptability and robust performance [5]. Specifically, we employed the following architectures, each for 5 folds:

- 3D Full Resolution (FullRes) U-Net with the default planner
- 3D U-Net Cascade with the default planner
- 3D FullRes U-Net with Large Residual Encoder (ResEnc) Presets (nnUNet-PlannerResEncLPlans)

For further information about nnUNet different configurations and planners, please refer to its documentation[3].

MedNeXt: MedNeXt is a DL architecture tailored for medical image analysis, particularly effective in handling varying image modalities [6,7]. We utilized the following architectures, each for 5 folds:

---
[3] https://github.com/MIC-DKFZ/nnUNet/tree/master/documentation.

– Small model with $3 \times 3 \times 3$ kernel size
– Small model with $5 \times 5 \times 5$ kernel size
– Large model with $3 \times 3 \times 3$ kernel size
– Large model with $5 \times 5 \times 5$ kernel size

When referring to small and large MedNeXt models, the terms relate to the compound scaling of the model. This refers to the simultaneous scaling of depth (number of layers), width (number of channels), and receptive field (kernel size). The small functional design (MedNeXt-S) utilizes 32 channels, an expansion ratio of 2, and a block count of 2. The largest architecture (MedNeXt-L) consists of 62 MedNeXt blocks and uses high values of both expansion ratio and block count [6].

The decision to explore these configurations (kernel sizes $3 \times 3 \times 3$ and $5 \times 5 \times 5$ for small and large models) is supported by findings in the MedNeXt study [6]. Smaller kernels, such as $3 \times 3 \times 3$, provide a robust baseline with balanced computational efficiency and performance. Larger kernels, such as $5 \times 5 \times 5$, leverage the ability of MedNeXt to learn long-range spatial dependencies, which are particularly useful for medical images with complex anatomical structures.

The MedNeXt study also demonstrates that MedNeXt-L outperforms or is competitive with smaller variants across tasks involving heterogeneous datasets (brain and kidney tumors, organs), varying modalities (CT, MRI), and diverse training set sizes. On the other hand MedNeXt-S can be more computationally efficient and data-efficient, which are important considerations in medical image segmentation where computational resources may be limited and datasets are often small [6].

For further information about different configurations of MedNeXt, please visit its documentation[4].

2.3   Ensemble Strategy

To enhance the performance of our segmentation models, we implemented a multi-level ensemble strategy. First, we applied the default ensembling methods for each model framework. For nnUNet models, we used `nnUNetv2_ensemble` and for MedNeXt models, we used `MedNeXtv1_ensemble` to combine the predictions from different architectures of each model. `nnUNetv2_ensemble` and `MedNeXtv1_ensemble` are the commands provided by the frameworks, which are average ensemble.

After obtaining the ensembled predictions from nnUNet and MedNeXt separately, average ensemble method was applied on these outputs to produce the final segmentation result. However, due to differences in the size of the probability maps generated by MedNeXt and nnUNet, additional preprocessing was required. Specifically, we padded the probability maps from MedNeXt to match the size of the original compressed nifti images. Once the probability maps were

---

[4] https://github.com/MIC-DKFZ/MedNeXt.

aligned, we computed average and converted the averaged probability map into a segmentation image.

2.4   Methodology

The training and ensemble strategies, as well as the incorporation of external data, are described below.

Task 1: For Task 1, different configurations of nnUNet and MedNeXt (as described in Sect. 2.2) were pre-trained on mid-RT and pre-RT registered images as individual inputs, and then fine-tuned on the original pre-RT images. After training all the models, ensembling strategies were applied to improve performance. Initially, every possible combination of nnUNet models (FullRes, Cascade, and ResEnc) was aggregated. Following this, the outputs of MedNeXt were combined with the best-performing combination of nnUNet models to evaluate possible improvements.

To explore the effectiveness of the BraTS dataset for the HNC segmentation task, nnUNet FullRes was first pretrained on the BraTS dataset (500 samples) and then fine-tuned on the original pre-RT images. For this experiment, to handle the transition from the BraTS dataset (a single-channel output) to the HNTS-MRG24 dataset requiring segmentation of two channels (GTVp and GTVn), two labels were assigned instead of one during the pretraining phase. This adjustment ensured that the output layer of the pretrained model had two channels, and it is compatible with fine-tuning on the HNTS-MRG24 dataset without further modification to the output layer.

Additionally, nnUNet FullRes was pretrained on a combined dataset of BraTS and mid-RT and pre-RT registered images (800 samples) to assess the benefit of combining external data with challenge-specific data. Following this, fine-tuning was performed on the original pre-RT images.

Task 2: To evaluate the potential value of pre-RT images and their segmentation masks for mid-RT segmentation, nnUNet FullRes and MedNeXt small model with kernel size 3 were trained on four different datasets:

- Mid-RT images only, referred to as Dataset 504.
- Mid-RT and registered pre-RT images as a multi-channel input, referred to as Dataset 505.
- Mid-RT images, registered pre-RT images, and pre-RT segmentation masks as a multi-channel input, referred to as Dataset 506.
- Mid-RT and registered pre-RT segmentation masks as a multi-channel input, referred to as Dataset 507.

After identifying the best dataset, all model configurations mentioned in Sect. 2.2 were trained on it. The same ensembling strategy as in Task 1 was applied, first aggregating combinations of nnUNet models (FullRes, Cascade, ResEnc) and then combining the best-performing combination of nnUNet models outputs with MedNeXt predictions to evaluate possible improvements.

2.5 Evaluation

The validation set was evaluated using the Aggregated Dice Similarity Coefficient (DSC_{agg}) [15] , same as the challenge evaluation standards.

$$DSC_{agg} = \frac{2 \sum_i |A_i \cap B_i|}{\sum_i (|A_i| + |B_i|)} \qquad (1)$$

In this context, A_i and B_i represent the ground truth and predicted segmentations for image i, respectively, where i ranges across the entire test set.

Additionally, DSC was calculated for each label (GTVp, GTVn) on a persample basis, for each model [16].

$$DSC_i = \frac{2|A_i \cap B_i|}{|A_i| + |B_i|} \qquad (2)$$

In this context, A_i represents the ground truth and B_i represents the predicted segmentations for image i.

For cases with zero ground truth ($|A_i| = 0$), the predictions were checked to determine whether the model produced a true empty segmentation (no tumor predicted for a sample with zero ground truth). In such cases, the DSC was assigned a value of 1. Conversely, if the model produced a non-empty segmentation for a sample with zero ground truth, the DSC was assigned a value of 0. After addressing these scenarios, the mean and standard deviation (STD) of the DSC were calculated across all samples to further assess model performance.

3 Results

The experiments were conducted using the cluster node of the Institute for Artificial Intelligence in Medicine (IKIM) in Essen, Germany. The node has 6 NVIDIA RTX 6000, 48 GB of VRAM, 1024 GB of RAM, and AMD EPYC 7402 24-Core Processor. The software environment included Python 3.9.19, PyTorch version 2.3.1+cu121, nnUNet version 2.5, and MedNeXt version 1.7.0.

Task 1:

The performance of different configurations of the nnUNet and MedNeXt models was evaluated using DSC_{agg}. While we intended to train all configurations of both models, we encountered issues with certain MedNeXt configurations. The small model with kernel size 3 was the only one that trained successfully and other configurations kept collapsing after a few epochs. On the other hand, nnUNet was successfully trained in all three configurations: FullRes, ResEnc, and Cascade.

For nnUNet, we applied its default ensembling strategy (nnUNetv2_ ensemble) to combine predictions from the FullRes, ResEnc, and Cascade configurations. Every possible combination of these models was ensembled. We also attempted an additional step of averaging the best predictions from nnUNet (ensemble of Cascade and ResEnd) and MedNeXt. An overview of these results is given in Table 1.

In addition to DSC_{agg}, mean DSC and STD were also calculated the for each predicted label (GTVp, GTVn) across all samples for each model configurations to further investigate model stability and robustness under different conditions. These results are shown in Table 2.

Figure 2 shows a comparison between the best-performing MedNeXt prediction, the worst-performing average ensemble of nnUNet and MedNeXt, and the ground truth segmentation.

Table 1. DSC_{agg} for each model configuration for Task 1

Model	GTVp DSC_{agg}	GTVn DSC_{agg}	Mean
nnUNet FullRes	0.7772	0.8517	0.8144
nnUNet ResEnc	0.7873	0.8586	0.8230
nnUNet Cascade	0.7847	0.8550	0.8198
nnUNet Cascade + FullRes	0.7846	0.8573	0.8210
nnUNet Cascade + ResEnc	0.7919	0.8633	0.8276
nnUNet FullRes + ResEnc	0.7896	0.8601	0.8249
nnUNet All Ensembled	0.7889	0.8618	0.8254
MedNeXt Small (Kernel 3)	**0.8066**	**0.8710**	**0.8388**
nnUNet + MedNeXt (Average)	0.7931	0.8166	0.8049

Table 2. Mean DSC $\pm$ STD over all cases for each model configuration for Task 1

Model	GTVp DSC	GTVn DSC
nnUNet FullRes	0.6101 $\pm$ 0.3453	0.7488 $\pm$ 0.2761
nnUNet ResEnc	0.6066 $\pm$ 0.3509	0.7619 $\pm$ 0.2750
nnUNet Cascade	0.6272 $\pm$ 0.3331	0.7431 $\pm$ 0.2772
nnUNet Cascade + FullRes	0.6247 $\pm$ 0.3392	0.7502 $\pm$ 0.2762
nnUNet Cascade + ResEnc	0.6359 $\pm$ 0.3402	0.7701 $\pm$ 0.2700
nnUNet FullRes + ResEnc	0.6283 $\pm$ 0.3483	0.7765 $\pm$ 0.2611
nnUNet All	0.6265 $\pm$ 0.3405	0.7682 $\pm$ 0.2630
MedNeXt Small (Kernel 3)	**0.6940 $\pm$ 0.2982**	**0.8010 $\pm$ 0.2341**
nnUNet + MedNeXt (Average)	0.6636 $\pm$ 0.3147	0.7654 $\pm$ 0.2520

Additionally inference of pretrained models compared to fine-tuned models was tested, see Table 1 and 3. Interestingly, these pretrained models performed

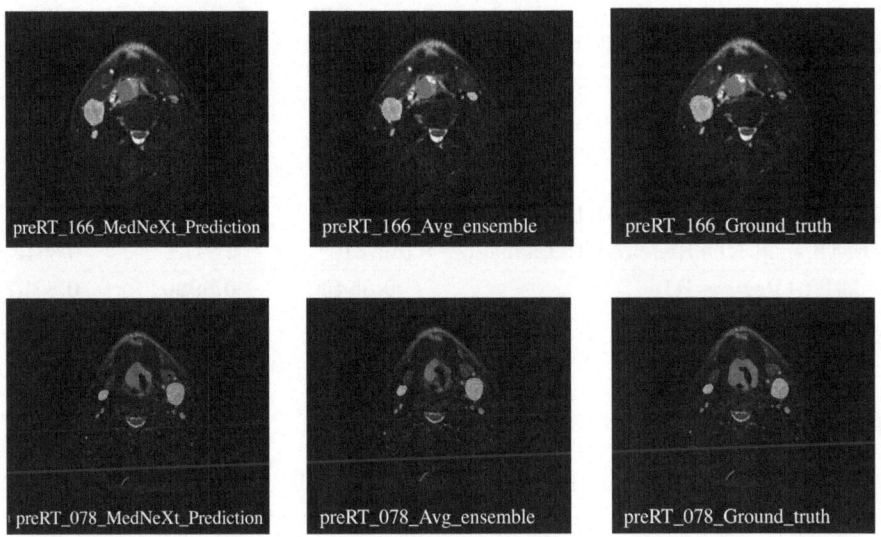

Fig. 2. Comparison of predicted segmentations of two pre-RT samples (Case 78 and 166) for Task 1. The left image shows the prediction from MedNeXt with the best DSCagg, the middle image shows the prediction from the average ensemble of nnUNet and MedNeXt, which had the lowest DSCagg, and the right image shows the ground truth segmentation. The green label represents GTVp (label = 1), and the yellow label represents GTVn (label = 2). (Color figure online)

better on the original pre-RT images compared to the models that were fine-tuned on the original pre-RT images.

Table 3. DSC_{agg} for Pretrained Models on Pre-RT Registered and Mid-RT Images

Model	GTVp DSC_{agg}	GTVn DSC_{agg}	Mean
Pretrained nnUNet FullRes	0.8545	0.8930	0.8737
Pretrained nnUNet Cascade	0.8701	0.8942	0.8822
Pretrained nnUNet ResEnc	**0.8936**	**0.9197**	**0.9066**
Pretrained MedNeXt Small (Kernel 3)	0.8748	0.8994	0.8871

We experimented with using the BraTS dataset as external data. It was used in the pretraining step, either alone or combined with mid-RT and pre-RT registered images, and then fine-tuned on the original pre-RT images. The results of these experiments are presented in Table 4.

For Task 1, the final submission used the MedNeXt small model with kernel size 3. It achieved a DSC_{agg} of **0.8728** for GTVn and **0.7780** for GTVp, with an overall mean DSC_{agg} of **0.8254** in the final test phase on the 50 test patients.

Table 4. DSC$_{agg}$ for nnUNet FullRes Pretrained on Different Datasets and Fine-Tuned on Original Pre-RT for Task 1

Pretrain on	Fine-tune on	GTVp DSC$_{agg}$	GTVn DSC$_{agg}$	Mean
BraTS	_	0.0904	0.0000	0.0452
BraTS	Original pre-RT	0.6871	0.7605	0.7238
BraTS+mid-RT+Reg pre-RT	_	0.8347	0.8740	0.8544
BraTS+mid-RT+Reg pre-RT	Original pre-RT	0.7611	0.8473	0.8042
mid-RT+Reg pre-RT	_	**0.8545**	**0.8930**	**0.8737**
mid-RT+Reg pre-RT	Original pre-RT	0.7772	0.8517	0.8144

Task 2:

For this task, nnUNet FullRes and MedNeXt small model with kernel size 3 were trained on four datasets with different combinations of mid-RT images, registered pre-RT images, and their segmentation masks (see Sect. 2.4).

nnUNet FullRes was successfully trained for all experiments. However, while training MedNeXt small with kernel size 3 on Dataset 506, the model collapsed after a few hundred epochs for some folds. Table 5 outlines the number of epochs each fold was trained for. Despite the incomplete training, we proceeded with inference for MedNeXt small model with kernel size 3 on Dataset 506.

Table 5. Training Epochs for MedNeXt (Small with Kernel size 3) on Dataset 506 Across Folds

	fold 0	fold 1	fold 2	fold 3	fold 4
Num of epochs	344	1000	1000	626	672

To address the training issues with MedNeXt on dataset 506, we discarded 35 samples where either label 1, label 2, or both were zero in the segmentation mask of the registered pre-RT. This resulted in 115 samples (dataset 516), which were then used to train MedNeXt. The results are presented in Table 6.

After experimenting on different datasets and finding the best one, we used the same architectures for Task 2 as in Task 1 to compare different models and configurations and ensemble strategies. All configurations of nnUnet trained successfully. Other MedNeXt architectures (small model with kernel size 5, large model with kernel size 3, and large model with kernel size 5) collapsed after only a few epochs. As with Task 1, we ensembled all possible combinations of nnUNet predictions using default nnUNet ensembling (`nnUNetv2_ensemble`), and then ensembled the best nnUNet predictions (ensemble of Cascade and FullRes) with MedNeXt predictions (from dataset 506, despite incomplete training on some folds) using the average ensembling method. An overview is given in Table 7.

Table 6. DSC_{agg} for nnUNet FullRes and MedNeXt Small with Kernel size 3 Trained on Different Datasets for Task 2

Model	Trained on	GTVp DSC_{agg}	GTVn DSC_{agg}	Mean
nnUNet FullRes	Dataset 504	0.5000	0.8085	0.6542
nnUNet FullRes	Dataset 505	0.4608	0.8025	0.6316
nnUNet FullRes	Dataset 506	**0.6100**	**0.8508**	**0.7304**
nnUNet FullRes	Dataset 507	0.6007	0.8513	0.7260
MedNeXt Small (Kernel 3)	Dataset 504	0.5676	0.8162	0.6919
MedNeXt Small (Kernel 3)	Dataset 505	0.5678	0.8091	0.6884
MedNeXt Small (Kernel 3)	Dataset 506	0.6099	0.8306	0.7202
MedNeXt Small (Kernel 3)	Dataset 507	**0.6231**	**0.8446**	**0.7339**
MedNeXt Small (Kernel 3)	Dataset 516	0.5904	0.8275	0.7089

In addition to DSC_{agg}, we also calculated the mean DSC and STD for each predicted label (GTVp, GTVn) across all samples for each model configurations to further evaluate model robustness and performance consistency under varying conditions. These results are provided in Table 8.

Figure 3 presents a comparison between the best-performing nnUNet model (ensemble of Cascade and FullRes) and the worst-performing average ensemble of nnUNet and MedNeXt model trained on Dataset 506.

Table 7. DSC_{agg} for nnUNet and MedNeXt trained on Dataset 506 for Task2

Model	GTVp DSC_{agg}	GTVn DSC_{agg}	Mean
nnUNet FullRes	0.6100	0.8508	0.7304
nnUNet Cascade	0.6105	0.8521	0.7313
nnUNet ResEnc	0.5742	0.8293	0.7018
nnUNet All Ensembled	0.6159	0.8544	0.7351
nnUNet Cascade + FullRes Ensembled	**0.6173**	**0.8543**	**0.7358**
nnUNet Cascade + ResEnc Ensembled	0.6031	0.8472	0.7251
nnUNet FullRes + ResEnc Ensembled	0.6022	0.8463	0.7242
MedNeXt (Small with Kernel 3)	0.6099	0.8306	0.7202
nnUNet + MedNeXt (Average)	0.6049	0.7735	0.6892

For Task 2, the final submission used an nnUNet ensemble of FullRes and Cascade models. It achieved a DSC_{agg} of **0.8519** for GTVn and **0.5491** for GTVp, with an overall mean DSC_{agg} of **0.7005** in the final test phase on the 50 test patients.

Table 8. Mean DSC ± STD over all cases for nnUNet and MedNeXt trained on Dataset 506 for Task2

Model	GTVp DSC	GTVn DSC
nnUNet FullRes	0.5148 ± 0.3374	0.8124 ± 0.1690
nnUNet Cascade	0.5134 ± 0.3324	0.8096 ± 0.1797
nnUNet ResEnc	0.4896 ± 0.3295	0.7869 ± 0.1886
nnUNet All Ensembled	0.5068 ± 0.3328	**0.8167 ± 0.1619**
nnUNet Cascade + FullRes Ensembled	0.5094 ± 0.3350	0.8139 ± 0.1719
nnUNet Cascade + ResEnc Ensembled	0.4992 ± 0.3342	0.8001 ± 0.1830
nnUNet FullRes + ResEnc Ensembled	0.4957 ± 0.3315	0.7996 ± 0.1834
MedNeXt (Small with Kernel 3)	**0.5577 ± 0.3202**	0.7804 ± 0.2150
nnUNet + MedNeXt (Average)	0.5147 ± 0.3525	0.7158 ± 0.3026

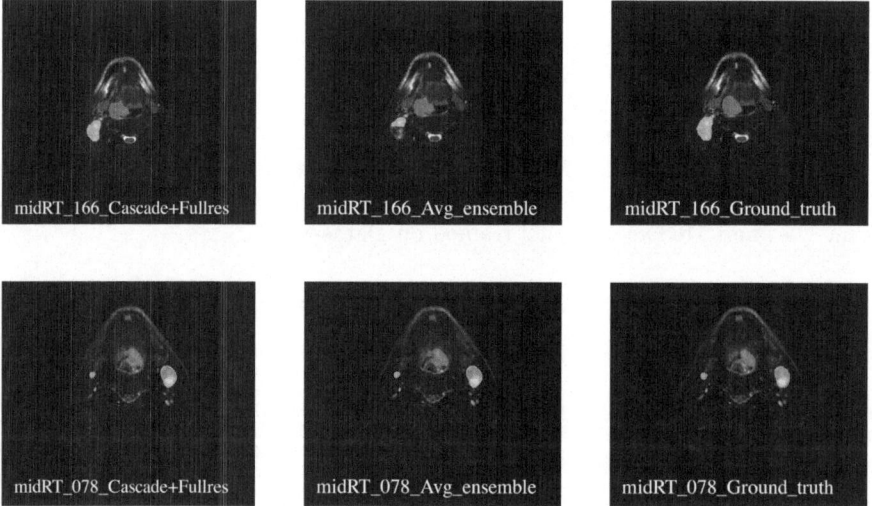

Fig. 3. Comparison of segmentation predictions of sample mid-RT (Case 78 and 166) for Task 2. The left image shows the prediction from ensemble of nnUNet Cascade and FullRes with the best DSCagg, the middle image shows the prediction from the average ensemble of nnUNet and MedNeXt, which had the lowest DSCagg, and the right image shows the ground truth segmentation. The green label represents GTVp (label = 1), and the yellow label represents GTVn (label = 2). (Color figure online)

4   Discussion

The results for Task 1 and Task 2 provide insights into the performance of various configurations of nnUNet and MedNeXt models, as well as the impact of pretraining with external datasets and ensembling strategies.

For **Task 1**, the MedNeXt small model with kernel size 3 achieved the best performance with the highest DSC_{agg}, outperforming all configurations of nnUNet, see Table 1. As a result, we chose the MedNeXt small kernel size 3 configuration as our final submission for Task 1.

Despite the fact that MedNeXt small model with kernel size 3 outperformed all nnUNet models, other MedNeXt architectures faced stability issues while training. Specifically, MedNeXt small model with kernel size 5, large model with kernel size 3, and large model with kernel size 5 repeatedly collapsed too early in training-after only a few epochs-so these models were not trained for enough epochs to be used effectively. Consequently, we were unable to fully compare the performance of different MedNeXt architectures and use the `mednextv1_ensemble` method to aggregate predictions from various configurations of MedNeXt, as originally planned.

Interestingly, the average ensemble of nnUNet and MedNeXt predictions led to a lower DSC_{agg} than using either model independently, suggesting that while both models have strengths, averaging their predictions may have introduced inconsistencies that reduced performance. Specifically, the average ensembling approach increased the number of false negatives and false positives, while also decreasing the number of true positives. This imbalance likely contributed to the overall drop in DSC_{agg}.

The comparison of mean DSC $\pm$ STD values shows that MedNeXt consistently achieved a higher mean DSC with less variability, for both GTVp and GTVn, see Table 2. This indicates greater robustness in its segmentation performance across different samples and superior overall performance. The nnUNet models had higher variability and lower mean DSC, particularly in GTVp predictions. In conclusion, MedNeXt proved to be a stronger candidate for reliable segmentation compared to nnUNet.

Pretrained models on the registered pre-RT and mid-RT images had higher DSC_{agg} than those that were fine-tuned on original pre-RT images, see Table 3. This observation can be attributed to the fact that the pretraining process used all available input samples, without a separate validation set. Since the original pre-RT images and the registered pre-RT images are highly similar, it can be expected that the pretrained models, which were trained on the registered images, would perform better on a data that is quite similar to what was already seen. In other words, the evaluation on these pretrained models was not a reliable measure of their true performance. In contrast, the fine-tuned models likely had lower DSC_{agg} because they were fine-tuned using a 5-fold cross-validation setup. Finally, we concluded that fine-tuned models would most likely generalize better on unseen test data.

When we experimented with pretraining nnUNet using the BraTS dataset, the results were mixed. Pretraining on BraTS alone led to poor performance, particularly for GTVn segmentation, see Table 4. Several factors likely contributed to this outcome. First, the BraTS dataset consists of images of the brain, while our challenge data includes the more anatomically complex H&N region. Second, the BraTS dataset uses T1w MRI images, whereas our challenge data is T2w, which may have caused discrepancies in the features learned during pretraining. Finally, the BraTS dataset contains only one label (tumor region), whereas our challenge requires segmentation of two distinct labels (GTVp and GTVn). These differences likely inhibited the ability of the pretrained model to generalize well to the challenge-specific data. However, when BraTS was combined with mid-RT and pre-RT registered images, there was a notable improvement, although it still did not surpass the models trained solely on challenge-specific data. This dataset was less aligned with the challenge's specific needs, and thus its contribution to the final model performance was limited.

For **Task 2**, the impact of including registered pre-RT images and their segmentation masks in the multi-channel input was evaluated. It was observed that using registered pre-RT images alone, without their segmentation masks, did not provide useful information for segmenting mid-RT images. Using only the segmentation masks of registered pre-RT images along with the mid-RT images resulted in a significant improvement. However, including both registered pre-RT images and their segmentation masks further improved the performance of nnUNet FullRes, and it achieved the highest DSC_{agg} for the mid-RT segmentation task when it was trained on dataset 506 (see Table 6).

MedNeXt faced stability issues when trained on dataset 506, collapsing after a few hundred epochs for several folds, see Table 5. Despite incomplete training, the MedNeXt model trained on Dataset 506 outperformed models trained on Datasets 505 and 504. Interestingly, the MedNeXt model trained on Dataset 507, achieved the best performance among MedNeXt models for Task 2. The observation that MedNeXt performed better on Dataset 507 than on Dataset 506, in contrast to nnUNet, is likely due to its inability to successfully complete 1000 epochs for all folds on Dataset 506. To address these training challenges, the dataset was refined by discarding samples with zero ground truth for either label, resulting in a stable training process for MedNeXt on Dataset 516. Nevertheless, MedNeXt showed its best performance when trained on Dataset 507.

Both nnUNet and MedNeXt models trained on datasets which included segmentation masks of registered pre-RT images (506 and 507) performed better than those trained on dataset 505 and 504. This further demonstrates the importance of including segmentation masks in the input data. This improvement can be attributed to the fact that the primary difference between mid-RT and pre-RT images lies in the size of the GTVn and GTVp, as mid-RT images are taken after some RT treatments. By incorporating pre-RT images and their segmentation, into the model's input, the model can better understand the region of the GTVn and GTVp in the mid-RT images, and can be more accurate in localization and segmentation.

Other MedNeXt architectures, specifically, MedNeXt small model with kernel size 5, large model with kernel size 3, and large model with kernel size 5 faced stability issues while training in Task 2 as well. They repeatedly collapsed too early in training-after only a few epochs-so these models were not trained for enough epochs to be used effectively. As a result, it was not possible to compare the performance of different MedNeXt architectures and use the `mednextv1_ensemble` method to aggregate predictions from various configurations of MedNeXt, as originally planned.

The comparison of different models highlights that nnUNet, particularly the nnUNet ensemble of FullRes and Cascade model, outperformed MedNeXt in terms of DSC_{agg}, which is the primary ranking metric for the challenge, see Table 7. As a result, the nnUNet ensemble of FullRes and Cascade was chosen as the final model for Task 2.

The ensemble of nnUNet and MedNeXt predictions for Task 2, using the average ensembling method, resulted in a lower DSC_{agg} than using either model independently. This suggests that averaging predictions between models may not have fully captured their individual strengths and could have introduced inconsistencies in the final segmentation. Specifically, the average ensembling approach increased the number of false negatives and false positives, while reducing the number of true positives, ultimately lowering the DSC_{agg} and overall performance.

The comparison of mean DSC $\pm$ STD values shows that nnUNet ensemble of all generally achieved a higher mean DSC with less variability, especially for GTVn, see Table 8. This indicates greater robustness in its segmentation performance across samples. For GTVp, MedNeXt outperformed nnUNet models. However, MedNeXt had higher variability and lower mean DSC, in GTVn predictions. This suggests that its stability issues and variability in results and lower DSC_{agg} made it less reliable compared to the more robust and stable nnUNet ensembles.

For practitioners looking to implement these approaches, the choice of model depends on the task requirements and available resources. MedNeXt is recommended for its robustness and high performance in GTVp and GTVn segmentation for Task 1. However, it is important to carefully monitor training stability when using larger configurations or datasets with imbalances. For Task 2, the nnUNet is suggested due to its consistent performance and ability to handle multi-channel inputs effectively.

The computational requirements for the two models are also different. MedNeXt needs more GPU memory and longer training time due to its complex design. Training a MedNeXt model takes about 180 s per epoch on an NVIDIA RTX 6000 GPU. In comparison, nnUNet takes about 60 s per epoch under the same conditions. nnUNet's lower resource usage and faster training make it better for environments with limited resources, while MedNeXt is more suitable for tasks that require detailed spatial modeling and where sufficient resources are available.

5   Conclusion

In this study, we sought to address the issue of segmenting tumor volumes in HNC using MRI data for the purpose of adaptive RT planning. Two DL models, nnUNet and MedNeXt, were tested, with an investigation of diverse architectural configurations, ensemble methodologies, and pretraining on external dataset.

In conclusion, the nnUNet model, particularly when ensemble predictions are leveraged, demonstrated high efficacy in the segmentation tasks. MedNeXt also demonstrated potential, particularly in Task 1, but encountered challenges with stability during training for Task 2. Pretraining with domain-specific data proved to be a crucial step in Task 1, and the incorporation of registered pre-RT segmentation masks proved beneficial for enhancing the performance of both models for Task 2. This finding addresses the key aspect of this challenge, which explored whether incorporating prior time point data (pre-RT and mid-RT) into segmentation algorithms could enhance performance in RT applications. The results clearly show that using both registered pre-RT images and their segmentation masks significantly improves the model's ability to accurately segment mid-RT images. Future work could focus on exploring more effective ensemble methods, such as weighted average ensemble, to better combine the strengths of both models. Additionally, addressing the stability issues faced by MedNeXt, particularly in Task 2, may involve adjusting the training process to prevent collapse during training. Improving the use of external datasets through more sophisticated domain adaptation could also enhance the effectiveness of pretraining.

Acknowledgments. This work is supported by the Plattform für KI-Translation Essen (KITE) from the REACT-EU initiative (EFRE-0801977, https://kite.ikim.nrw/) and "NUM 2.0" (FKZ: 01KX2121) and FWF enFaced 2.0 (grant number: KLI-1044, https://enfaced2.ikim.nrw/). André Ferreira thanks the Fundação para a Ciência e Tecnologia (FCT) Portugal for the grant 2022.11928.BD.

References

1. Kiser, K.J., Smith, B.D., Wang, J., Fuller, C.D.: Après mois, le déluge: preparing for the coming data flood in the MRI-guided radiotherapy era. Front. Oncol. **9**, 983 (2019)
2. Pollard, J.M., Wen, Z., Sadagopan, R., Wang, J., Ibbott, G.S.: The future of image-guided radiotherapy will be MR guided. Br. J. Radiol. **90**(1073), 20160667 (2017)
3. Andrearczyk, V., Oreiller, V., Abobakr, M., Akhavanallaf, A., et al.: Overview of the hecktor challenge at MICCAI 2022: automatic head and neck tumor segmentation and outcome prediction in PET/CT. In: Head and Neck Tumor Segmentation and Outcome Prediction. HECKTOR 2022. LNCS, vol. 13626, pp. 1–30. Springer, Cham (2023)
4. Luo, X., et al.: Segrap2023: a benchmark of organs-at-risk and gross tumor volume segmentation for radiotherapy planning of nasopharyngeal carcinoma. In: MICCAI SegRap 2023 (2023)
5. Isensee, F., Jaeger, P.F., Kohl, S.A., Petersen, J., Maier-Hein, K.H.: nnU-Net: a self-configuring method for deep learning-based biomedical image segmentation. Nat. Methods **18**(2), 203–211 (2021)
6. Roy, S., et al.: Mednext: transformer-driven scaling of convnets for medical image segmentation. In: International Conference on Medical Image Computing and Computer-Assisted Intervention (MICCAI) (2023)
7. Isensee, F., Jaeger, P.F., Kohl, S.A., Petersen, J., Maier-Hein, K.H.: nnU-Net: a self-configuring method for deep learning-based biomedical image segmentation. Nat. Methods 1–9 (2020)

8. Li, R., Auer, D., Wagner, C., Chen, X.: A generic ensemble based deep convolutional neural network for semi-supervised medical image segmentation. arXiv preprint arXiv:2004.07995 (2020)
9. Codella, N., et al.: Skin lesion analysis toward melanoma detection 2018: a challenge hosted by the international skin imaging collaboration (ISIC). arXiv preprint arXiv:1902.03368 (2019)
10. Tschandl, P., Rosendahl, C., Kittler, H.: The ham10000 dataset, a large collection of multi-source dermatoscopic images of common pigmented skin lesions. Sci. Data **5**, 180161 (2018)
11. Astaraki, M., Bendazzoli, S., Toma-Dasu, I.: Fully automatic segmentation of gross target volume and organs-at-risk for radiotherapy planning of nasopharyngeal carcinoma. arXiv:2310.02972 (2023)
12. Myronenko, A., Siddiquee, M.M.R., Yang, D., He, Y., Xu, D.: Automated head and neck tumor segmentation from 3D PET/CT: HECKTOR 2022 challenge report. arXiv preprint arXiv:2209.10809 (2022)
13. Pieper, S., Halle, M., Kikinis, R.: 3D slicer. In: 2004 2nd IEEE International Symposium on Biomedical Imaging: Nano to Macro (IEEE Cat No. 04EX821), vol. 1, pp. 632–635 (2004)
14. LaBella, D., et al.: Brain tumor segmentation (BRATS) challenge 2024: meningioma radiotherapy planning automated segmentation. arXiv preprint arXiv:2405.18383 (2024)
15. Andrearczyk, V., Oreiller, V., Jreige, M., Castelli, J., Prior, J.O., Depeursinge, A.: Segmentation and classification of head and neck nodal metastases and primary tumors in PET/CT. In: 2022 44th Annual International Conference of the IEEE Engineering in Medicine & Biology Society (EMBC), pp. 4731–4735 (2022)
16. Dice, L.R.: Measures of the amount of ecologic association between species. Ecology **26**(3), 297–302 (1945)

A Coarse-to-Fine Framework
for Mid-Radiotherapy Head and Neck Cancer
MRI Segmentation

Jing Ni[1], Qiulei Yao[1], Yanfei Liu[2(✉)], and Haikun Qi[1(✉)]

[1] ShanghaiTech University, Shanghai, China
qihk@shanghaitech.edu.cn
[2] Shenzhen United Imaging Research Institute of Innovative Medical Equipment, Shenzhen, China
yanfei.liu@cri-united-imaging.com

Abstract. Radiotherapy is the preferred treatment modality for head and neck cancer (HNC). During the treatment, adaptive radiation therapy (ART) technology is commonly employed to account for changes in target volume and alterations in patient anatomy. This adaptability ensures that treatment remains precise and effective despite these physiological variations. Magnetic resonance imaging (MRI) provides higher-resolution soft tissue images, making it valuable in target delineation of HNC treatment. The delineation in ART should adhere to the same principles as those used in the initial delineation. Consequently, the contouring performed on MR images during ART should reference the earlier delineations for consistency and accuracy. To address this, we proposed a coarse-to-fine cascade framework based on 3D U-Net to segment mid-radiotherapy HNC from T2-weighted MRI. The model consists of two interconnected components: a coarse segmentation network and a fine segmentation network, both sharing the same architecture. In the coarse segmentation phase, different forms of prior information were used as input, including dilated pre-radiotherapy masks. In the fine segmentation phase, a resampling operation based on a bounding box focuses on the region of interest, refining the prediction with the mid-radiotherapy image to achieve the final segmentation. In our experiment, the final results were achieved with an aggregated Dice Similarity Coefficient (DSC) of 0.562, indicating that the prior information plays a crucial role in enhancing segmentation accuracy. (Team name: TNL_skd)

Keywords: Head and neck cancer · 3D segmentation · Coarse-to-Fine · Magnetic Resonance Image

1 Introduction

Head and neck cancer (HNC) is among the most common types of cancer globally; almost 880,000 patients are diagnosed every year [1]. Radiation therapy plays an important role in HNC treatment, but it relies heavily on the accuracy of delineation. A precise manual contouring process of a HNC patient often takes the clinician lots of time and

K. A. Wahid et al. (Eds.): HNTS-MRG 2024, LNCS 15273, pp. 154–165, 2025.
https://doi.org/10.1007/978-3-031-83274-1_11

the average reported time taken from 2.7 to 3.0 h [2], therefore an effective auto segmentation method is necessary. Due to the higher soft tissue resolution of magnetic resonance images compared to computed tomography images in the head and neck region, magnetic resonance guided radiotherapy (MRgRT) has become increasingly popular in HNC treatment. However, MR-guided ART may require multiple contouring for one patient, making the need for automatic contouring urgent.

In recent years, an increasing number of research has the advantage of deep learning in medical image segmentation [3]. Using positron emission tomography (PET) and computed tomography (CT) to segment head and neck cancer was explored in the HECTOR challenge, and the algorithms from participants showed great results [4–6]. Notably, there is also a growing interest in utilizing MRI for this purpose. Schouten et al. [7] introduced an automated segmentation pipeline for head and neck squamous cell cancer using a multi-view convolutional neural network (MV-CNN) with multimodal MRI sequences, achieving moderate results. Bielak et al.'s [8] method utilizes multiple MRI contrasts at different time points to train CNNs for lesion segmentation in head and neck cancer. The contribution of each contrast is assessed by comparing a reference CNN with all contrasts to CNNs where one input channel is excluded. A method proposed by Korte et al. involved the development of three convolutional neural network-based auto-segmentation architectures [9]. Their results also demonstrated improved geometric accuracy in the segmentation process. However, these methods, which directly delineate MR images, are limited in their ability to account for inherent structural and spatial changes that may occur during treatment. This presents a challenge for ART, where reference to initial images or contours is necessary for accurate treatment adaptation. Wahid et al. [10] proposed utilizing multiparametric MRI for oropharyngeal cancer primary gross tumor volume auto-segmentation, leveraging additional channel combinations to improve segmentation performance. Their work highlights the importance of incorporating multiple MRI contrasts to achieve more robust segmentation.

Due to the lack of publicly available datasets for evaluating model performance in this area, the HNTS-MRG2024 has been organized, offering a platform to assess and explore adaptive radiotherapy (ART) in real-world clinical scenarios.

ART has similarities with future frame prediction, and both techniques rely heavily on prior information to guide future outcomes (Fig. 1). In ART, previous CT or MRI scans provide a reference for adjusting treatment plans when the patient's anatomy changes. In the same way, during future frame prediction, past frames are used to predict subsequent frames in natural images. Inspired by Z. Gao et al. [11], they proposed a simpler yet effective CNN model for video prediction and achieved state-of-the-art results without introducing any complex modules, strategies, and tricks. Similarly, UNet3D-based next-frame prediction models have also been explored [12], and they have achieved better performance compared to CNN-LSTM and Convolutional LSTMs. In the ref-based segmentation, we can consider it as a combination of one-frame prediction and segmentation.

Our work focuses on task 2 mid-radiotherapy segmentation of the HNTS-MRG2024 challenge. We present an end-to-end UNet-based model that utilizes prior knowledge, specifically through pre-radiotherapy masks, to improve segmentation performance in

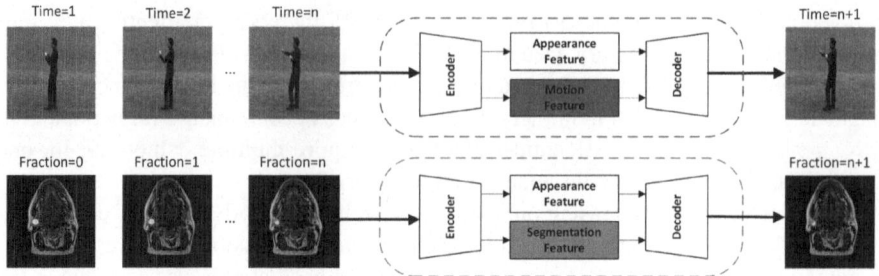

Fig. 1. The workflows of ART and future frame prediction.

adaptive radiotherapy. To assess the impact of varying levels of prior information, we conducted three comparative experiments. For the first, no prior information was incorporated. In the second experiment, pre-radiotherapy masks were used as the only source of prior information. In contrast, the third experiment combined both pre-radiotherapy masks and pre-radiotherapy images to evaluate their joint effect on segmentation accuracy.

2   Method

2.1   Dataset

The HNTS-MRG 2024 challenge training dataset contains 150 patients. The dataset consists of T2-weighted (T2w) anatomical sequences of the head and neck region, collected at the University of Texas MD Anderson Cancer Center. The dataset includes both fat-suppressed and non-fat-suppressed images, with all patients being immobilized during the scans. The mask includes three categories: background, primary gross tumor volumes (GTVp), and metastatic lymph nodes (GTVn). For each case, there are 3 sets of images, including the original pre-radiotherapy T2w MRI, the original mid-radiotherapy T2w MRI, and the registered pre-radiotherapy T2w MRI. Corresponding segmentation masks are also provided for each of these, including the pre-RT segmentation mask, the mid-radiotherapy segmentation, and the registered pre-radiotherapy segmentation mask.

2.2   Data Preprocessing

Normalization. In the data preprocessing stage, we used the registered pre-images and masks provided by the official dataset for all 150 training cases. For each image, we selected the area corresponding to the mask and calculated the 0.5th and 99.5th percentiles within this region. These calculated low and high values were then applied to clip the entire image. After that, we used Z-score normalization.

Image Cropping. In the pre-phase, the maximum xy dimensions, referring to the cross-sectional plane, are 768×768, and the minimum is 512×512, with the maximum z-plane being 162. In the mid-radiotherapy, the maximum xy dimensions are 768×768, and the minimum is 480×480, with the maximum z-plane being 168. Additionally, the majority of images have a resolution of $0.5 \times 0.5 \times 2$ mm, with a shape of $512 \times 512 \times z$. To meet the input requirement of the UNet structure, we resampled all images to a fixed size of $336 \times 336 \times 160$ with a resolution of $0.5 \times 0.5 \times 2$ mm.

Mask Dilated. We used pre-radiotherapy masks as prior information to segment the mid-radiotherapy images. To adequately overlap with the mid-radiotherapy mask, the radius of the pre-radiotherapy masks was dilated along the x, y, and z axes. In our experiment, we tested dilation radii ranging from 1 to 10, finally selecting 6 pixels for the xy-plane and 3 pixels for the z-plane, represented as $(6, 6, 3)$. This choice was based on the spacing resolution of the imaging data and the practical 3 mm clinical margin used by physicians to account for uncertainties in tumor position and treatment delivery. After applying this dilation, the overlap ratio of the mid-radiotherapy mask reached 99.12% for GTVp and 99.71% for GTVn, calculated across all cases by excluding voxels with zero values. These values represent the mean overlap ratios for individual cases after removing zero-value results from the Statistic. By using dilated pre-radiotherapy masks, we could incorporate prior information to guide the model toward more accurate localization, capturing contextual details that improve segmentation performance.

Post Processing. In the inference stage, post-processing was used to refine the results and improve segmentation accuracy. We mainly used connected component analysis and region merging. Connected component analysis was applied to identify different regions, while region merging helped ensure accurate segmentation boundaries and removed smaller irrelevant components. In some cases, the segmentation results may show an overlap between GTVp and GTVn, which does not happen in the clinical reality.

2.3 Network Architecture

The end-to-end framework is built on encoder-decoder architecture. The network operates on a set of inputs and a corresponding set of outputs. Inspired by the success of Jiang et al. [13] in BraTS Challenge2019 and Sun et al. [5] in HECTOR Challenge2022, we propose an end-to-end coarse-to-fine segmentation network (Fig. 2).

Target Location. Due to the anatomical complexity of the region for head and neck cancer, precise location is the key to delineation. Clinical experience has shown that tumors can both shrink and shift during radiotherapy. GTVp may temporarily swell due to inflammation or the tumor's biological response, but it often shrinks as treatment progresses [14]. In contrast, GTVn typically decreases in volume throughout radiotherapy. However, positional shifts can occur, influenced by factors such as patient weight loss or changes in positioning during treatment sessions [15]. To accommodate potential tumor movement in ART, we dilated the pre-radiotherapy mask to create a sufficient margin around the tumor. This margin ensures that even with slight anatomical variations during treatment, the tumor remains within the pre-radiotherapy masks, maintaining accurate targeting throughout radiotherapy. We dilated registered pre-radiotherapy

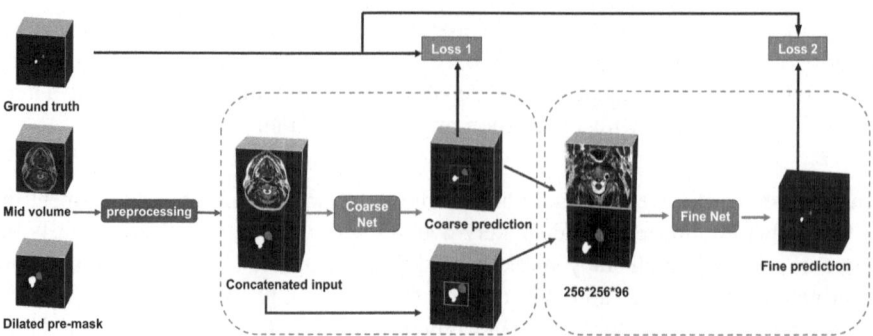

Fig. 2. The proposed end-to-end coarse-to-fine segmentation network.

masks as prior information and concatenated it with mid-radiotherapy image as input to the model.

Coarse Segmentation. The coarse network is based on the 3D UNet implemented using the MONAI framework (Fig. 3), which adopts an encoder-decoder architecture with symmetric skip connections to preserve spatial information. The encoder path consists of five stages, with channel dimensions progressively increasing from 16, 32, 64, 128 to 256. At each stage in the encoder, residual units with two convolutional layers are used. Those layers with kernel size $3 \times 3 \times 3$ and stride 2 perform downsampling, while batch normalization (BN) and PReLU activation enhance stability. The decoder mirrors the encoder, using transposed convolutions (Conv3DTranspose) to upsample the feature maps. Each upsampling operation doubles the spatial resolution while halving the number of channels. Skip connections between the encoder and decoder are used to concatenate high-resolution details from earlier layers, facilitating better localization in the final segmentation. In the final layer of the decoder, a $1 \times 1 \times 1$ convolution reduces the feature maps to 3 channels, aligning with the output requirement for segmentation classes. This streamlined architecture makes it suitable for segmentation tasks.

Bounding Box. Before performing the fine segmentation, we cropped it again to reduce redundant information. After getting coarse output, we calculated its bounding box relative to the dilated pre-radiotherapy mask. In addition, we respectively calculated the bounding box of the original pre-radiotherapy mask and dilated pre-radiotherapy mask for comparison. The results are shown in Table 1. Therefore, a fixed size of $256 \times 256 \times 96$ was selected, centered on the combined bounding box of the two.

Fine Segmentation. The fine segmentation network architecture is the same as the coarse segmentation network, while they have different inputs. For the fine network, the cropped coarse output was concatenated with the cropped mid-radiotherapy image and used as the input.

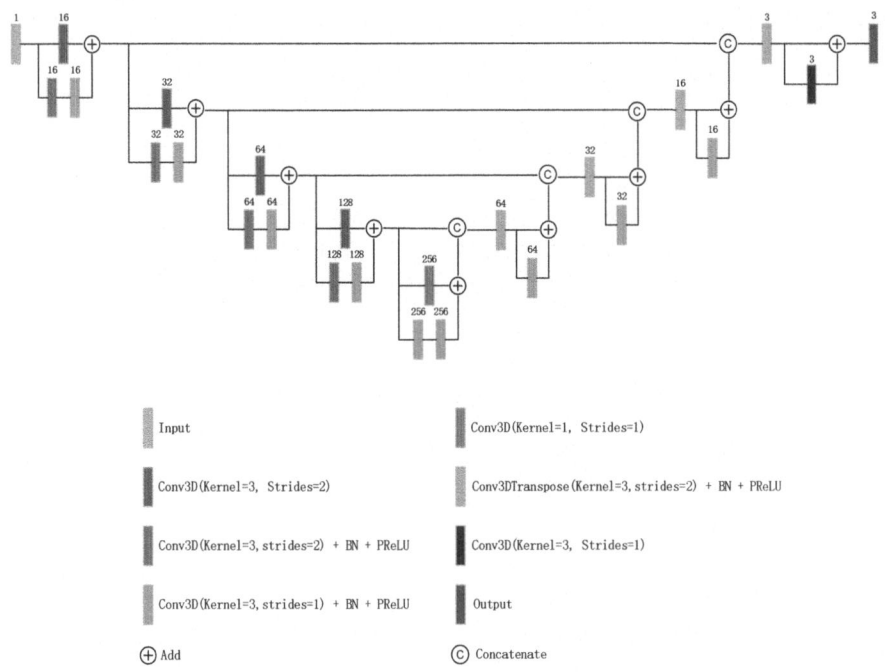

Fig. 3. An illustration of the coarse model based on 3D Unet. Here, we show the case with a single-channel input. In our experiments, the input will be adjusted according to different situations

2.4 Training Details

The training dataset consists of 150 samples, which were split into 80% for training, 10% for validation, and 10% for testing. We used 3-fold cross-validation during training, and the model was trained for 200 epochs in each fold using the DiceCE loss function and the AdamW optimizer [16]. To evaluate the prediction of accuracy, we used the aggregated Dice Similarity Coefficient (DSCagg) [17]. Besides, the CyclicLR schedule was used to adapt the learning rate and prevent overfitting. All the operations were conducted on an NVIDIA GeForce RTX 3090 with 24GB of memory.

Data Augmentation. For data augmentation, we relied on the MONAI platform [18]. We applied zooming, rotation, and affine transformations to both images and masks, while RandAdjustContrastd and RandGaussianNoised were applied only to images. A summary of the training details is as follows (Table 2):

Loss Function. The loss function we chose is DiceCE, which combines the advantages of multi-class dice loss [19] and cross-entropy loss. We applied the loss function to both the coarse and fine networks. Rather than a simple summation, the total loss is weighted dynamically at each stage. In the coarse segmentation (loss 1), the weight

Table 1. Bounding box measurements for pre-radiotherapy mask and dilated pre-radiotherapy mask

Measurement	pre-RT mask	dilated pre-RT mask
x range	(27, 247)	(33, 253)
x mean	119.74	126.99
y range	(23, 198)	(33, 222)
y mean	91.12	105.48
z range	(5, 69)	(11, 75)
z mean	29.13	39.64

Table 2. Training methodology and hyper-parameters.

Component	Value
Epochs	200
Batch size	1
Learning rate	$3e^{-5}$
Loss function	DiceCE
Optimizer	AdamW
Scheduler	CyclicLR
Data augmentation	RandRotate90d, RandZoomd[0.9, 1.1], RandAffined (scale range[0.1, 0.1, 0.1]), RandAdjustContrastd (gamma[0.9, 1.1]), RandGaussianNoised

decreases with epoch from 0.9 to 0.1, while in the fine segmentation (loss 2), the weight increases from 0.1 to 0.9. To ensure that the coarse segmentation results are as accurate as possible in the early stages of training, we need to implement effective measures. This will contribute to improving the overall performance and effectiveness of subsequent model training.

Evaluation Metrics. The official evaluation metric is DSCagg [20], which minimizes the impact of isolated false negatives or false positives on the overall performance evaluation, ensuring a more robust and balanced assessment.

$$\mathrm{DSC}_{agg} = \frac{2 \sum_i |A_i \cap B_i|}{\sum_i |A_i| + |B_i|}$$

where A_i and B_i are the ground truth and predictions for image i.

The surface dice similarity coefficient (surface DSC) evaluates the agreement between the segmentation surface and the reference surface [21]. It quantifies the proportion of points on the automatically segmented surface that lie within a specified tolerance distance τ from the reference surface. When evaluating the Surface DSC, it is necessary to define a threshold within which variation is clinically acceptable.

Specifically, a surface S is defined as the boundary of a mask ∂M, with its area represented by $|S| = \int_S d\sigma$, where σ is a point on the surface. The mapping from this point on the surface to a point in three-dimensional space is represented by $\xi(\sigma)$. Using this mapping, the border region $\mathcal{B}_i^{(\tau)}$ for a surface S_i at a given tolerance τ is defined as the set of points within a distance of τ from the surface S_i.

$$\mathcal{B}_i^{(\tau)} = \left\{ \mathbf{x} \in \mathbb{R}^3 \,\middle|\, \exists \sigma \in S_i, \, \|\mathbf{x} - \xi(\sigma)\| \leq \tau \right\},$$

The surface DSC metric, denoted as $R_i^{(\tau)}$, is then calculated by measuring the overlap of each surface with the border region of the other, normalized by the sum of the surface areas.

$$R_{i,j}^{(\tau)} = \frac{|S_i \cap \mathcal{B}_j^{(\tau)}| + |S_j \cap \mathcal{B}_i^{(\tau)}|}{|S_i| + |S_j|},$$

The Hausdorff Distance (HD) is an important geometric metric used to measure the similarity between two sets of points [22]. It defines the maximum of the minimum distances from points in one set to the closest points in the other set, capturing the worst-case deviation between the two sets. Given two point sets A and B, the Hausdorff Distance $H(A, B)$ is defined as:

$$H(A, B) = \max(h(A, B), h(B, A))$$

where

$$h(A, B) = \max_{a \in A} \min_{b \in B} \|a - b\|$$

To exclude unreasonable distances caused by some outliers and maintain overall numerical stability, we use the 95% Hausdorff Distance (95% HD).

3   Results

We evaluated the effect of incorporating different types of prior information on the segmentation performance. Specifically, we compared three different input combinations: mid-radiotherapy (mid-RT) image alone, pre-radiotherapy (pre-RT) mask + mid-radiotherapy (mid-RT) image, pre-radiotherapy image + pre-radiotherapy mask + mid-radiotherapy image. Tables 3, 4, 5 presents the segmentation results across these input combinations for GTVp and GTVn. The test process for both models utilized the same architecture and parameters.

The results indicate that using mid-radiotherapy images alone provided a baseline performance while adding the pre-radiotherapy image to the mid-radiotherapy image resulted in noticeable improvements in both GTVp and GTVn segmentation accuracy. Furthermore, using the pre-radiotherapy mask in combination with the mid-radiotherapy image demonstrated the highest mean accuracy in fold 1. Among the results, GTVn consistently outperformed GTVp. These results highlight the positive impact of incorporating prior information to enhance segmentation performance.

We also tested the results of training the coarse and fine networks separately but found it inferior to the end-to-end model. This suggests that the coherence of information is crucial for segmentation tasks.

Table 3. Comparison of Results for Different Inputs using DSCagg

Input	Fold	GTVp	GTVn	Mean
	fold1	0.012	0.292	0.152
mid-RT image	fold2	0.021	0.229	0.125
	fold3	0.003	0.345	0.173
	fold1	0.536	0.702	**0.619**
pre-RT mask mid-RT image	fold2	0.528	0.704	0.616
	fold3	0.477	0.650	0.564
pre-RT image	fold1	0.426	0.600	0.513
pre-RT mask	fold2	0.471	0.745	0.608
mid-RT image	fold3	0.409	0.609	0.509

Table 4. Comparison of Results for Different Inputs using surface DSC

Input	Fold	GTVp	GTVn	Mean
	fold1	0.037	0.233	0.135
mid-RT image	fold2	0.107	0.225	0.166
	fold3	0.022	0.238	0.130
	fold1	0.417	0.613	0.515
pre-RT mask mid-RT image	fold2	0.365	0.646	0.506
	fold3	0.391	0.569	0.478
pre-RT image	fold1	0.420	0.604	0.520
pre-RT mask	fold2	0.452	0.667	**0.559**
mid-RT image	fold3	0.431	0.599	0.515

Table 5. Comparison of Results for Different Inputs using 95%HD

Input	Fold	GTVp	GTVn	Mean
	fold1	91.685	100.119	96.505
mid-RT image	fold2	59.064	57.349	58.257
	fold3	65.849	58.355	62.102
	fold1	9.423	9.008	**9.242**
pre-RT mask mid-RT image	fold2	16.924	9.282	13.740
	fold3	15.849	11.906	13.972
pre-RT image	fold1	12.788	35.082	19.476
pre-RT mask	fold2	11.460	15.793	13.265
mid-RT image	fold3	17.142	17.054	17.112

In the final submission phase, we selected the model using a pre-radiotherapy mask and mid-radiotherapy image. The final test results were: GTVp = 0.500, GTVn = 0.625, mean = 0.562.

4 Discussion

The findings suggest that the inclusion of different types of prior information can significantly enhance segmentation performance. The highest performance was achieved when combining pre-radiotherapy masks with mid-radiotherapy images, demonstrating that integrating temporal information along with spatial priors contributes to more robust segmentation. This implies that the model benefits not only from knowing the current anatomy but also from the historical context provided by pre-radiotherapy images, which might help in anticipating deformations or other changes over time.

Our results align with the findings of Wahid et al. [10], where the segmentation of oropharyngeal primary tumors using different types of data (five mpMRI input channels: T2, T1, ADC, Ktrans, Ve) showed that adding all data types did not yield the best performance. Instead, the combination of T1 and T2 achieved the best results. Similarly, in our study, adding different prior information did not always result in better outcomes, but the strategic combination of pre-radiotherapy mask and mid-radiotherapy image was most effective.

Based on the competition results, GTVp was consistently harder to segment than GTVn. Our results for GTVp segmentation were comparable to other participants, which demonstrates that our simple model has a certain level of capability in handling the segmentation of complex structures. However, the GTVn segmentation results showed a larger gap compared to other participants. We attribute this discrepancy to the following reasons: First, we did not use additional datasets, and all training and validation were conducted using the official training set only. Incorporating additional multi-modal training could help the model learn more comprehensive features. Second, we did not introduce modules or techniques to enhance the model's performance, as our goal was to explore whether a simple model could effectively predict and segment in the ART field. Lastly, our model might have experienced overfitting during training, which could have potentially impacted its ability to generalize effectively.

5 Conclusion

In our experiment, the proposed simple cascade model still demonstrates a certain ability to segment and predict the location of mid-radiotherapy cancer targets. This study aims to investigate the impact of prior knowledge on the performance of ART segmentation tasks. We explored how prior knowledge influences segmentation accuracy by incorporating prior information through a dilated pre-radiotherapy mask during the coarse segmentation step and refining the results based on a fixed region determined by the bounding box.

More relevant datasets need to be used for training in future work, allowing the model to learn more effectively. Additionally, incorporating different modalities of data from the same patient as prior knowledge could be investigated to determine if it further improves model performance. Finally, adding small spatial prediction modules to the network might be explored as a potential improvement, which could better serve adaptive radiotherapy by enhancing the accuracy of tumor deformation and positional change predictions.

References

1. Bray, F., Ferlay, J., Soerjomataram, I., Siegel, R.L., Torre, L.A., Jemal, A.: Global cancer statistics 2018: Globocan estimates of incidence and mortality worldwide for 36 cancers in 185 countries. CA Cancer J. Clin. **68**(6), 394–424 (2018)
2. Kosmin, M., et al.: Rapid advances in auto-segmentation of organs at risk and target volumes in head and neck cancer. Radiother. Oncol. **135**, 130–140 (2019)
3. Singh, S.P., Wang, L., Gupta, S., Goli, H., Padmanabhan, P., Gulyás, B.: 3D deep learning on medical images: a review. Sensors **20**(18), 5097 (2020)
4. Myronenko, A., Siddiquee, M.M.R., Yang, D., He, Y., Xu, D.: Automated head and neck tumor segmentation from 3D PET/CT HECKTOR 2022 challenge report. In: 3D Head and Neck Tumor Segmentation in PET/CT Challenge, pp. 31–37. Springer (2022)
5. Sun, X., An, C., Wang, L.: A coarse-to-fine ensembling framework for head and neck tumor and lymph segmentation in CT and PET images. In: 3D Head and Neck Tumor Segmentation in PET/CT Challenge, pp. 38–46. Springer (2022)
6. Jiang, H., Haimerl, J., Gu, X., Lu, W.: A general web-based platform for automatic delineation of head and neck gross tumor volumes in PET/CT images. In: 3D Head and Neck Tumor Segmentation in PET/CT Challenge, pp. 47–53. Springer (2022)
7. Schouten, J.P.E., et al.: Automatic segmentation of head and neck primary tumors on MRI using a multi-view CNN. Cancer Imaging **22**(1), 8 (2022)
8. Bielak, L., et al.: Convolutional neural networks for head and neck tumor segmentation on 7-channel multiparametric MRI: a leave-one-out analysis. Radiat. Oncol. **15**, 1–9 (2020)
9. Korte, J.C., Hardcastle, N., Ng, S.P., Clark, B., Kron, T., Jackson, P.: Cascaded deep learning-based auto-segmentation for head and neck cancer patients: organs at risk on T2-weighted magnetic resonance imaging. Med. Phys. **48**(12), 7757–7772 (2021)
10. Wahid, K.A., et al.: Evaluation of deep learning-based multiparametric MRI oropharyngeal primary tumor auto-segmentation and investigation of input channel effects: results from a prospective imaging registry. Clin. Transl. Radiat. Oncol. **32**, 6–14 (2022)
11. Gao, Z., Tan, C., Wu, L., Li, S.Z.: Simvp: simpler yet better video prediction. In: Proceedings of the IEEE/CVF Conference on Computer Vision and Pattern Recognition, pp. 3170–3180 (2022)
12. Akbacak, E.: Unet3d based next frame prediction. In: 2024 8th International Artificial Intelligence and Data Processing Symposium (IDAP), pp. 1–4. IEEE (2024)
13. Jiang, Z., Ding, C., Liu, M., Tao, D.: Two-stage cascaded u-net: 1st place solution to brats challenge 2019 segmentation task. In: Brainlesion: Glioma, Multiple Sclerosis, Stroke and Traumatic Brain Injuries: 5th International Workshop, BrainLes 2019, Held in Conjunction with MICCAI 2019, Shenzhen, China, 17 October 2019, Revised Selected Papers, Part I 5, pp. 231–241. Springer (2020)
14. Suarez-Gironzini, V., Khoo, V.: Imaging advances for target volume definition in radiotherapy. Curr. Radiol. Rep. **3**, 1–9 (2015)
15. Aly, F., Miller, A.A., Jameson, M.G., Metcalfe, P.E.: A prospective study of weekly intensity modulated radiation therapy plan adaptation for head and neck cancer: improved target coverage and organ at risk sparing. Australas. Phys. Eng. Sci. Med. **42**, 43–51 (2019)

16. Loshchilov, I.: Decoupled weight decay regularization. arXiv preprint arXiv:1711.05101 (2017)
17. Andrearczyk, V., et al.: Overview of the HECKTOR challenge at MICCAI 2021: automatic head and neck tumor segmentation and outcome prediction in PET/CT images. In: 3D Head and Neck Tumor Segmentation in PET/CT Challenge, pp. 1–37. Springer (2021)
18. Cardoso, M.J., et al.: MONAI: an open-source framework for deep learning in healthcare (2022)
19. Sudre, C.H., Li, W., Vercauteren, T., Ourselin, S., Jorge Cardoso, M.: Generalised dice overlap as a deep learning loss function for highly unbalanced segmentations. In: Deep Learning in Medical Image Analysis and Multimodal Learning for Clinical Decision Support: Third International Workshop, DLMIA 2017, and 7th International Workshop, ML-CDS 2017, Held in Conjunction with MICCAI 2017, Québec City, QC, Canada, September 14, Proceedings 3, pp. 240–248. Springer (2017)
20. Andrearczyk, V., Oreiller, V., Jreige, M., Castelli, J., Prior, J.O., Depeursinge, A.: Segmentation and classification of head and neck nodal metastases and primary tumors in PET/CT. In: 2022 44th Annual International Conference of the IEEE Engineering in Medicine & Biology Society (EMBC), pp. 4731–4735. IEEE (2022)
21. Nikolov, S., et al.: Clinically applicable segmentation of head and neck anatomy for radiotherapy: deep learning algorithm development and validation study. J. Med. Internet Res. **23**(7), e26151 (2021)
22. Huttenlocher, D.P., Klanderman, G.A., Rucklidge, W.J.: Comparing images using the hausdorff distance. IEEE Trans. Pattern Anal. Mach. Intell. **15**(9), 850–863 (1993)

Assessing Self-supervised xLSTM-UNet Architectures for Head and Neck Tumor Segmentation in MR-Guided Applications

Abdul Qayyum[1]([✉]), Moona Mazher[2], and Steven A. Niederer[1]

[1] National Heart and Lung Institute, Faculty of Medicine, Imperial College London, London, UK
a.qayyum@imperial.ac.uk

[2] Centre for Medical Image Computing, Department of Computer Science, University College London, London, UK

Abstract. Radiation therapy (RT) plays a pivotal role in treating head and neck cancer (HNC), with MRI-guided approaches offering superior soft tissue contrast and daily adaptive capabilities that significantly enhance treatment precision while minimizing side effects. To optimize MRI-guided adaptive RT for HNC, we propose a novel two-stage model for Head and Neck Tumor Segmentation. In the first stage, we leverage a Self-Supervised 3D Student-Teacher Learning Framework, specifically utilizing the DINOv2 architecture, to learn effective representations from a limited unlabeled dataset. This approach effectively addresses the challenge posed by the scarcity of annotated data, enabling the model to generalize better in tumor identification and segmentation. In the second stage, we fine-tune an xLSTM-based UNet model that is specifically designed to capture both spatial and sequential features of tumor progression. This hybrid architecture improves segmentation accuracy by integrating temporal dependencies, making it particularly well-suited for MRI-guided adaptive RT planning in HNC. The model's performance is rigorously evaluated on a diverse set of HNC cases, demonstrating significant improvements over state-of-the-art deep learning models in accurately segmenting tumor structures. Our proposed solution achieved an impressive mean aggregated Dice Coefficient of 0.81 for pre-RT segments and 0.65 for mid-RT segments, underscoring its effectiveness in automated segmentation tasks. This work advances the field of HNC imaging by providing a robust, generalizable solution for automated Head and Neck Tumor Segmentation, ultimately enhancing the quality of care for patients undergoing RT. Our team name is DeepLearnAI (CEMRG). The code for this work is available at https://github.com/RespectKnowledge/SSL-based-DINOv2_Vision-LSTM_Head-and-Neck-Tumor_Segmentation.

Keywords: xLSTM-UNet architecture · Student-teacher SSL-based models · Self-supervised learning · UxLSTM · Head and neck cancer segmentation · MRI-guided radiotherapy

K. A. Wahid et al. (Eds.): HNTS-MRG 2024, LNCS 15273, pp. 166–178, 2025.
https://doi.org/10.1007/978-3-031-83274-1_12

1   Introduction

Radiation therapy (RT) continues to play a vital role in cancer treatment, offering an effective method for managing a broad spectrum of malignancies. Head and neck cancer (HNC) is one area where RT is highly beneficial, as precise targeting is crucial to achieving effective tumor control while minimizing damage to nearby healthy tissues. Traditionally, CT-based planning has been the standard approach in RT for identifying and mapping tumors and surrounding anatomical structures. However, recent advancements have sparked growing interest in MRI-guided RT planning, which provides several notable advantages over CT. MRI provides enhanced soft tissue contrast for improved tumor visualization and enables advanced multiparametric sequences such as diffusion-weighted imaging (DWI), which adds functional imaging capabilities to the planning process. Furthermore, MRI-Linac systems support daily adaptive RT by integrating real-time imaging with treatment delivery [1]. These capabilities together enhance the precision of tumor localization and the overall efficacy of RT.

The introduction of MRI-guided adaptive RT marks a significant leap forward in cancer therapy. This approach allows for dynamic adjustments to treatment plans based on real-time MRI scans taken during RT, enabling more tailored and responsive treatments. By continuously adapting the treatment to account for changes in the tumor and surrounding tissues, MRI-guided adaptive RT aims to enhance tumor elimination while sparing healthy tissue, potentially revolutionizing treatment practices for HNC [2].

Despite its transformative potential, several challenges accompany the widespread adoption of MRI-guided RT in clinical practice. The process generates large amounts of data, especially for HNC, where accurate tumor delineation is particularly crucial. The task of manual segmentation by clinicians, which remains the clinical standard, is labor-intensive and time-consuming [3]. Additionally, the complexity of head and neck tumors makes them some of the most difficult anatomical structures to accurately segment [4]. These hurdles have prompted the exploration of automated solutions, particularly leveraging artificial intelligence (AI) and deep learning techniques, to improve the efficiency and accuracy of tumor segmentation in MRI-guided RT planning.

Incorporating AI-driven approaches into this workflow could not only reduce the time burden on clinicians but also improve the precision of tumor segmentation, facilitating the adoption of MRI-guided adaptive RT on a broader scale. The potential for real-time, adaptive treatment adjustments is a key aspect of this evolving technology, making MRI-guided RT a promising area for future innovations in cancer care.

Medical image analysis, a cornerstone of computer-aided diagnosis, often grapples with the challenge of limited labeled data, especially for intricate 3D tasks. Self-supervised learning (SSL) has emerged as a promising approach to reduce the dependence on extensive manual annotation by leveraging vast amounts of unlabeled data. However, SSL still demands large unlabeled datasets to effectively learn meaningful feature representations. This need is further compounded by the scarcity of 3D medical data, driven by the high costs of imaging and stringent privacy concerns, resulting in most datasets containing only a few cases [5–7]. Existing segmentation methods, such as U-Nets and 3D CNNs [5], often require large, annotated datasets, which are challenging to obtain in medical imaging due to the time and expertise needed for manual labeling. These methods may also struggle with capturing temporal changes in longitudinal datasets, such as MRI scans collected before and during radiation therapy. Self-supervised learning (SSL) addresses these limitations by leveraging abundant unlabeled data to learn robust spatial and structural features, reducing reliance on labeled datasets and improving generalizability. Vision-LSTM is particularly suitable for this task as it combines spatial encoding with temporal modeling, effectively capturing the progression of tumor changes over time, making it ideal for both pre-RT and mid-RT segmentation tasks.

We introduce a straightforward, yet powerful SSL approach based on the DINOv2 framework [7], aimed at minimizing the need for extensive manual annotation in 3D medical image segmentation. Our method takes advantage of available unlabeled data to create robust pre-trained models that excel in downstream segmentation tasks. In the first stage, we adapt the DINOv2 framework, originally designed for 2D tasks, to handle the complexities of 3D medical imaging using our proposed Vision-LSTM (xLSTM) [8, 9]. This adaptation involves extending the model's architecture and training framework to process volumetric data, enabling it to learn rich, spatially aware representations from unlabeled 3D medical images. This allows the model to capture the intricate structures inherent in 3D medical datasets.

Once pre-training is complete, the learned knowledge is transferred to a specific 3D medical image segmentation task. This is achieved by extracting the encoder from the pre-trained model and fine-tuning it alongside a task-specific decoder tailored for segmentation. The pre-trained encoder serves as a strong foundation, enriched with spatial features learned during the self-supervised phase, significantly enhancing segmentation performance on the target task. To further enhance our approach given the limitations of the labeled HNC dataset, we propose a deep learning-based model for precise segmentation of HNC structures:

1. We present a modified DINOv2 framework that incorporates the 3D Vision-LSTM (xLSTM) model for self-supervised learning in the initial phase, utilizing the limited pre-RT and mid-RT MRI Tumor datasets.
2. For the downstream task, Vision-LSTM (xLSTM) serves as the backbone, with the encoder frozen and the decoder fine-tuned to improve medical image segmentation.
3. We thoroughly evaluate the proposed solution using the pre-RT and mid-RT MRI datasets, comparing its performance against state-of-the-art deep learning models to validate its effectiveness.

This approach not only overcomes the challenges of limited labeled data but also shows promise in advancing accurate 3D medical image segmentation, particularly in the context of HNC diagnosis and treatment.

2 Proposed Method

2.1 Dataset

The dataset comprises 150 T2-weighted (T2w) MRI scans of the head and neck region collected at MD Anderson Cancer Center (MDACC) [10–12]. The images include a mix of fat-suppressed and non-fat-suppressed sequences, with all patients immobilized using a thermoplastic mask. The scans were extracted from an institutional imaging repository (Evercore) and consisted of both pre-radiation therapy (pre-RT) images, taken 1–3 weeks before RT began, and mid-radiation therapy (mid-RT) images, acquired 2–4 weeks into treatment. Each patient's image pairs are consistently either fat-suppressed or non-fat-suppressed. The dataset includes segmentations of primary gross tumor volumes (GTVp) and metastatic lymph nodes (GTVn), with GTVp present in at most one region per patient and a variable number of GTVn. Multiple physician experts (3 to 4 observers) independently annotated the GTVp and GTVn structures on both pre-RT and mid-RT scans. Based on recent research [13], at least three annotators are recommended for reliable segmentation, which were combined using the STAPLE (Simultaneous Truth and Performance Level Estimation) algorithm to create consensus segmentations. The annotators were experienced medical doctors specializing in head and neck cancer, with access to each patient's medical history and prior imaging, such as PET/CT scans. The final segmentations were verified by radiation oncology faculty with over 10 years of experience. In cases of significant disagreement among observers, a single contour from an expert was used. A training dataset comprising 150 cases has been made available for head and neck cancer (HNC) segmentation.

We have proposed methods to address two tasks: pre-RT MRI Tumor Segmentation and mid-RT MRI Tumor Segmentation. The first task focuses on segmenting head and neck cancer (HNC) tumor volumes from MRI scans acquired before the initiation of radiation therapy. The second task involves segmenting HNC tumor volumes from MRI scans obtained during radiation therapy, utilizing a dataset that also includes images from the pre-RT segmentation task.

2.2   Teacher Student SSL Model for 3D Medical Imaging

Our study is implemented using a self-supervised learning (SSL) framework that incorporates an innovative structure with a single online student encoder and a corresponding momentum-based teacher encoder. Both encoders share the same network architecture; however, the teacher encoder's parameters are updated using momentum from the student encoder's parameters. This setup allows the model to progressively improve its learning from unlabeled data as the teacher encoder evolves alongside the student's performance. A key component of SSL is the use of strong data augmentations to generate diverse and informative training samples. For our study, we apply a variety of transformations to the input images, such as flipping, scaling, adding Gaussian noise and blur, and adjusting brightness and contrast. These augmentations produce two distinct views of the same input, which are processed through a Siamese network structure. By comparing these views, the model learns robust feature representations. The entire workflow is illustrated in Fig. 1.

To train the 3D Vision-LSTM (xLSTM) model, we adapted the DINOv2 approach within a self-supervised learning context. This approach enables our model to effectively leverage the unlabeled data, facilitating the learning of rich spatial representations that are essential for downstream tasks like 3D medical image segmentation. This methodology not only optimizes the use of available data but also boosts the segmentation model's performance and resilience.

The momentum teacher encoder's parameters θ_t are updated based on the student encoder's parameters θ_s using a momentum-based approach:

$$\theta_t = m.\theta_t + (1 - m)\theta_t \tag{1}$$

where θ_t are the parameters of the teach encoder, θ_s are the parameters of the student encoder, m is the momentum coefficient typically a value close to 1.

Let x be the original input image. Two different views of the input, x_1 and, x_2 are generated using strong data augmentations:

$$x_1 = Augment(x), x_2 = Augment(x) \tag{2}$$

Both views are then processed through the student encoder f_s and teacher encoder f_t to extract feature representations:

$$h_1 = f_s(x_1; \theta_s), h_2 = f_s(x_2; \theta_s)$$

$$h'_1 = f_s(x_1; \theta_t), h'_2 = f_s(x_2; \theta_t) \tag{3}$$

where h_1 and h_2 are the feature representations from the student encoder, h'_1 and h'_2 are the feature representations from the teacher encoder.

The feature representations h_1, h_2, h'_1, h'_2 are subjected to global average pooling to reduce them into feature vectors:

$$v_1 = GAP(h_1), v_2 = GAP(h_2)$$

$$v'_1 = GAP(h'_1), v'_2 = GAP(h'_2) \tag{4}$$

where v_1, v_2, v'_1 and v'_2 are the resulting feature vectors.

$$z_1 = MLP(v_1), z_2 = MLP(v_2)$$

$$z'_1 = MLP(v'_1), z'_2 = MLP(v'_2) \tag{5}$$

After projection, the teacher's output is centered, sharpened, and passed through a softmax function to produce the supervision signal:

$$q'_1 = Softmax(\frac{Center(z'_1)}{\tau})$$

$$q'_2 = Softmax(\frac{Center(z'_2)}{\tau}) \tag{6}$$

where Center (z) subtracts the mean of the vector to have zero mean. τ is the temperature parameter controlling the sharpness of the distribution. Softmax(z) normalizes the vector into a probability distribution.

The loss function is designed to minimize the divergence between the student's feature vectors and the teacher's processed outputs. A common choice is the cross-entropy loss or mean squared error (MSE) between the student's and teacher's outputs:

$$L = \frac{1}{2}\left(Loss(z_1, q_2') + Loss(z_2, q_1')\right) \tag{7}$$

where the cross-entropy loss is

$$L(z, q') = -\sum_{k=1}^{K} q'[k]\log(Softmax(z)[k]) \tag{8}$$

where this loss function encourages the student encoder to produce feature representations that align closely with the teacher's outputs, thus enabling effective learning from the unlabeled data. The student model is trained to align with the sharpened output of the teacher model.

The 3D Vision-LSTM (xLSTM) model is trained within this SSL framework, where the student encoder's parameters are updated through backpropagation to minimize the loss function L, while the teacher encoder's parameters are adjusted using the momentum mechanism previously described. This integration of self-supervised learning, momentum-based updates for the teacher encoder, and strong data augmentation techniques forms an effective strategy for pre-training the model on unlabeled data. As a result, this approach significantly boosts the model's performance in downstream tasks, including 3D medical image segmentation.

2.3 Model Architecture

The xLSTM-UNet model integrates Vision-LSTM (xLSTM), a refined version of Long Short-Term Memory (LSTM) networks. xLSTM has shown exceptional performance across various fields, such as Natural Language Processing (NLP) and image classification, surpassing models like Vision Transformers and State Space Models (SSMs) such as Mamba [8, 9]. This advanced xLSTM module is embedded within the UNet architecture, which is well-known for its ability to effectively extract local features through convolutional layers. The UNet structure is highly suited for image segmentation due to its encoder-decoder architecture, where the encoder captures hierarchical features using convolutional operations, and the decoder reconstructs and refines these features to generate segmentation maps.

2.4 Self-Supervised Learning (SSL) Approach

Pre-Trained Encoder: In the initial phase of our approach, we employ a self-supervised learning (SSL) technique to pre-train the encoder of the xLSTM-UNet model. This SSL framework utilizes large volumes of unlabeled data to train the encoder, focusing on the Vision-LSTM (xLSTM) model to capture rich, contextual features from the input images. This pre-training stage is essential for developing robust feature representations that generalize well across various medical imaging tasks [14].

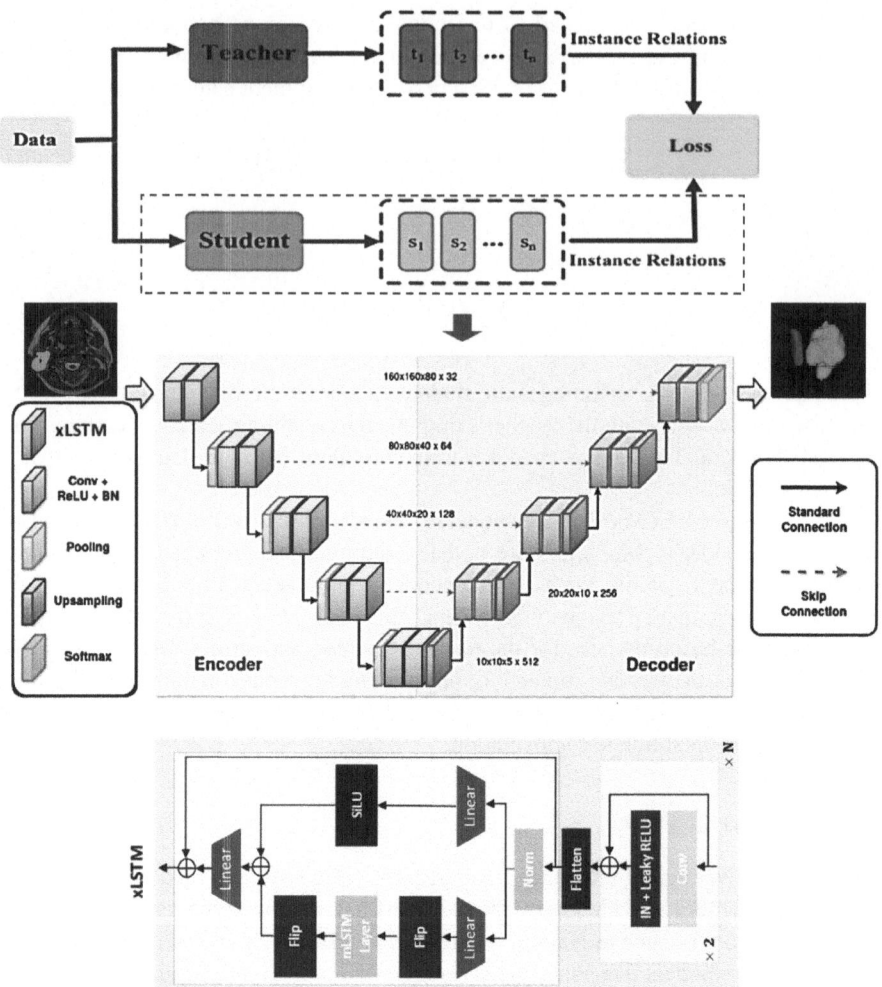

Fig. 1. Proposed SSL model for automated Head and Neck Tumor Segmentation.

Fine-Tuned Decoder: Following the pre-training of the encoder, we fine-tune the decoder in a supervised manner using labeled data. The decoder is designed to convert the high-level feature representations from the encoder into accurate and detailed segmentation maps. Fine-tuning the decoder's parameters optimizes its performance for the specific segmentation task.

Enhanced Long-Range Dependency Capture: By integrating xLSTM—which excels at capturing long-range dependencies—into the UNet architecture, the model efficiently combines local feature extraction with global contextual understanding. This combination improves segmentation accuracy, particularly for complex biomedical images.

Efficient Learning with Limited Labeled Data: The SSL approach enables the effective use of unlabeled data, reducing the need for large labeled datasets. The pre-trained encoder establishes a solid foundation, while the fine-tuned decoder ensures high accuracy for targeted segmentation tasks. The xLSTM-UNet model, with its pre-trained xLSTM encoder and fine-tuned decoder, presents a powerful solution for biomedical image segmentation. This strategy not only capitalizes on advanced deep learning methods but also maximizes the utility of unlabeled data through SSL, resulting in greater accuracy and efficiency for segmentation tasks. To enhance training diversity, strong augmentations such as flipping, scaling, Gaussian noise addition and brightness adjustments are applied to create two distinct views of each input, improving the model's robustness to data variations. In the SSL stage, the model is trained on unlabeled data for 500 epochs to learn rich feature representations, followed by fine-tuning the SSL-pretrained encoder in the segmentation stage with labeled data for 300 epochs. Input images are normalized to zero mean and unit variance, resized to a fixed resolution and processed by the 3D Vision-LSTM model. Segmentation optimization uses a combination of Dice loss and cross-entropy loss to address the class imbalance and improve pixel-wise accuracy. The Adam optimizer, with a low learning rate and scheduling adjustments, ensures stable and efficient convergence during fine-tuning. The PyTorch library is used for training, testing, and optimizing the proposed model.

3   Results

In this paper, we proposed a modified DINOv2 framework, incorporating the 3D Vision-LSTM (xLSTM) model for self-supervised learning, which significantly enhances medical image segmentation tasks in RT datasets. The proposed method leverages two key segmentation datasets: pre-RT segmentation (Task 1) and mid-RT segmentation (Task 2). The innovation lies in employing a self supervised learning approach in the first stage, where the xLSTM model is utilized to process the unlabeled segmentation data, enabling the model to learn meaningful representations without the need for manual labeling. By doing this, we harness a large amount of pre- and mid-RT segmentation data to develop a robust framework for downstream tasks.

For the second stage, the xLSTM model acts as the backbone for the downstream task of medical image segmentation. Here, we freeze the encoder weights learned during the self-supervised phase and fine-tune the decoder to adapt the model to the specific segmentation tasks. This method significantly improves performance by retaining the powerful, generalized features learned during the self-supervised phase, while fine-tuning the decoder allows for the task-specific refinement required for precise segmentation of radiotherapy data.

The performance of our method was rigorously evaluated on two tasks: Task 1 (pre-RT segmentation) and Task 2 (mid-RT segmentation). The results, summarized in Table 1, demonstrate the effectiveness of our approach. For Task 1, the Dice Similarity Coefficient (DSC) achieved was 0.86045 for GTVn (Gross Tumor Volume - nodes), and 0.7690 for GTVp (Gross Tumor Volume - primary), resulting in an Aggregated DSC of 0.8147. For Task 2, which is more challenging due to the mid-RT setting, our model still performed strongly, achieving a DSC of 0.7731 for GTVn and 0.5343 for GTVp, with

Table 1. Test results for task1 and task2 using our proposed method

Proposed method	DSC (GTVn)	DSC (GTVp)	Aggregated DSC
	Task1		
SSL-xSLTM-UNet	0.86045	0.7690	**0.8147**
	Task2		
SSL-xSLTM-UNet	0.7731	0.5343	**0.6537**

Table 2. Validation results for task1 and task2 using our proposed and state-of-the-art methods

Proposed method	DSC (GTVn)	DSC (GTVp)	Aggregated DSC
	Task1		
3D-ResUNet [15]	0.9345	0.7647	0.8496
xSLTM-UNet	0.9333	0.7752	0.8542
SSL-xSLTM-UNet	0.9194	0.8539	0.8867
	Task2		
3D-ResUNet [15]	0.7589	0.0	0.3794
xSLTM-UNet	0.7684	0.0	0.3842
SSL-xSLTM-UNet	0.8028	0.0	0.4014

an Aggregated DSC of 0.6537. Compared to state-of-the-art deep learning models, our proposed framework demonstrates superior segmentation performance, particularly in Task 1, where the model showed excellent segmentation accuracy for both GTVn and GTVp. Task 2, while more challenging, also yielded competitive results, particularly in GTVn segmentation. These results highlight the robustness and generalization ability of our modified DINOv2 framework and xLSTM-based approach in handling complex medical imaging tasks, especially when processing multi-stage radiotherapy datasets. This comprehensive evaluation not only validates the efficiency of the proposed method but also underscores its potential for improving segmentation accuracy in clinical applications, particularly in radiotherapy planning, where accurate segmentation is critical for treatment success. Table 2 presents a comparison of the proposed model with state-of-the-art models, based on results from the leaderboard on two subject validation datasets. The proposed model is evaluated both with and without SSL and is also compared to the base 3D-ResUNet.

In Fig. 2, the results for two segmentation tasks are presented in two rows, each highlighting different aspects of the analysis. The first row focuses on Task 1, which involves segmenting tumor types GTVn (Gross Tumor Volume of lymph nodes) and GTVp (Gross Tumor Volume of the primary tumor). The second column shows the Ground Truth (GT) segmentation masks as a baseline for comparison, while the third column presents the segmentation results produced by the proposed model, which effectively identifies tumor

Image	GT	Pred	GT Volume	Pred Volume

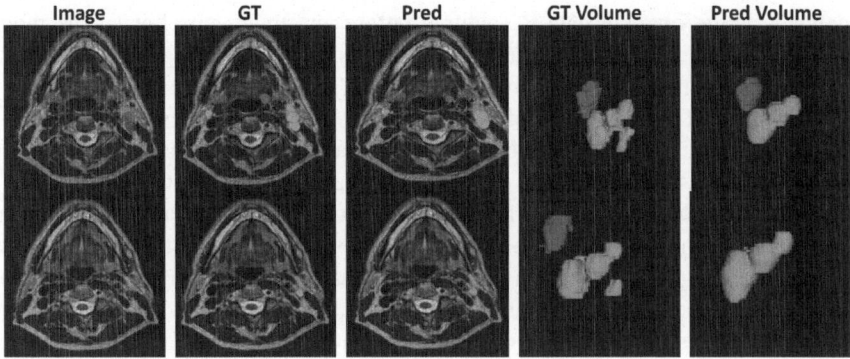

Fig. 2. The first row displays the predictions for Task 1, while the second row presents the predictions for Task 2.

regions in this task. The fourth column displays a 3D visualization of the tumor volume based on the Ground Truth (GT), providing a detailed representation of the actual tumor structure. The fifth column shows the corresponding 3D segmentation volume generated by the proposed model, which aligns well with the GT for Task 1, demonstrating the model's capability in accurately capturing tumor features in 3D.

The second row illustrates the results for Task 2, where the performance of the proposed model is notably less accurate. As seen in the third column, the segmentation masks predicted by the model deviate significantly from the Ground Truth (GT), failing to capture the required structures effectively. Similarly, the 3D segmentation volume generated by the proposed model, shown in the fifth column, does not match the precision of the Ground Truth's 3D representation depicted in the fourth column. These discrepancies highlight the model's limitations in Task 2, especially in capturing complex or nuanced features, both in 2D segmentation masks and 3D volumetric reconstructions. Overall, while the proposed model performs well for Task 1, its underperformance in Task 2 underscores the need for further optimization to handle the specific challenges of this task. The relative underperformance of the proposed model in Task 2 (mid-RT segmentation) compared to Task 1 (pre-RT segmentation) can be attributed to several key factors related to the complexity of mid-RT imaging, the inherent challenges of the dataset, and the model's adaptation capabilities.

1. **Increased Variability in mid-RT Images**

One of the primary reasons for the lower Dice Similarity Coefficient (DSC) in Task 2 is the increased variability in the imaging data during mid-radiotherapy (mid-RT). By the mid-point of radiotherapy, significant physiological changes may occur in the patient due to radiation exposure, including tumor shrinkage, changes in tissue density, and the development of inflammation or edema. These factors lead to less uniform, more complex image features that are harder for the model to segment accurately. For example, as the tumor responds to treatment, its shape, size, and boundaries may become less distinct, making it difficult for the model to differentiate between the Gross Tumor Volume of primary cancer (GTVp) and the surrounding tissue. In Task 1 (pre-RT segmentation), the tumor boundaries are typically more defined, as the imaging data is captured before any treatment effects. The model can rely on more consistent anatomical features, resulting in higher segmentation accuracy and a better DSC for both GTVn and GTVp.

2. **Challenges in Gross Tumor Volume - Primary (GTVp) Segmentation**

The model's performance drop in Task 2 is particularly notable in the segmentation of the Gross Tumor Volume - primary (GTVp), where the DSC drops to 0.5343, compared to 0.7690 in Task 1. This indicates that GTVp segmentation is more challenging in the mid-RT phase. The mid-RT images of the primary tumor (GTVp) are affected by complex factors such as tissue deformation, changes in intensity patterns, and the presence of artifacts from ongoing radiotherapy, all of which contribute to difficulties in distinguishing the tumor from surrounding healthy or affected tissues. These variations may cause the model to misclassify or blur the boundaries between tumor and non-tumor regions, reducing its overall accuracy.

3 **Generalization Limitations of the xLSTM Model**

The xLSTM model, while effective at learning meaningful representations during the self-supervised phase in Task 1, may face challenges when generalizing to more complex scenarios in Task 2. Although the model's encoder was frozen and only the decoder fine-tuned for the segmentation task, the features learned during the pre-RT stage might not be fully transferable to the mid-RT stage. Task 2 involves dynamic changes in the tumor and surrounding tissues that may not have been adequately represented in the pre-RT dataset, making it harder for the model to adjust effectively. The gap between the pre-RT and mid-RT domains means that features learned from the pre-RT data might not align well with the mid-RT variations, leading to reduced performance.

4. **Impact of Limited Labelled Data**

Another factor contributing to the lower performance in Task 2 is the potential limitation of labeled data available for mid-RT segmentation. Unlike pre-RT data, where the images are usually more standardized and available in larger quantities, mid-RT datasets can be scarcer and more heterogeneous, reducing the opportunity for the model to train effectively. The reduced labeled data for fine-tuning during Task 2 could limit the model's ability to generalize well to unseen mid-RT scenarios, especially when dealing with complex tissue changes.

5. **Balancing GTVn and GTVp Performance**

In Task 2, the model performs better on GTVn segmentation (DSC: 0.7731) than GTVp (DSC: 0.5343), suggesting that the anatomical changes and treatment effects might affect the primary tumor (GTVp) more than the lymph nodes (GTVn). Lymph nodes tend to have more consistent anatomical features and may not change as drastically during treatment, allowing the model to retain higher accuracy for their segmentation. In contrast, the primary tumor, being more directly affected by radiation therapy, undergoes more pronounced changes that the model struggles to capture accurately. The lower DSC score in Task 2 compared to Task 1 reflects the greater complexity of segmenting mid-RT images, where treatment-induced anatomical and tissue changes present significant challenges for the model. These challenges, coupled with domain shift between pre- and mid-RT datasets, and potential limitations in labeled data for fine-tuning, result in a decreased ability to segment the primary tumor accurately in Task 2. However, the model still performs reasonably well on

GTVn segmentation, highlighting its strengths in dealing with less variable structures like lymph nodes, while further optimization is needed to improve its robustness in handling the more complex and dynamic changes of the primary tumor during radiotherapy.

4 Conclusion

We have developed a two-stage approach for Pre-RT and mid-RT head and neck cancer (HNC) segmentation, incorporating advanced machine learning techniques. The method begins with a self-supervised learning phase, utilizing the DINOv2 framework and a 3D Vision-LSTM (xLSTM) model, which effectively addresses the challenges of limited data and the absence of annotated samples from the pre-RT and mid-RT HNC dataset. In the second stage, the model is fine-tuned for segmentation by freezing the encoder and optimizing the decoder, resulting in enhanced overall performance. The xLSTM model demonstrates effectiveness in accurately segmenting head and neck tumors, including GTVn (Gross Tumor Volume of lymph nodes) and GTVp (Gross Tumor Volume of the primary tumor), while also enhancing the precision of surface segmentation, as evidenced by the results. This approach offers a robust and scalable solution for automated pre-RT and mid-RT HNC segmentation, leading to more precise diagnoses and personalized treatment planning.

For future directions, we recommend expanding this approach to larger, more diverse datasets to further validate its generalizability and improve its ability to handle a wider range of cases. Integrating additional modalities, such as PET or CT scans, could enhance multimodal learning and provide more comprehensive treatment planning. Moreover, incorporating real-time feedback during treatment could further refine the accuracy of adaptive radiation therapy. By advancing this model's application to other cancer types or anatomical regions, it could become a key tool in precision oncology, contributing to more individualized and effective cancer treatment strategies.

References

1. Pollard, J.M., Wen, Z., Sadagopan, R., Wang, J., Ibbott, G.S.: The future of image-guided radiotherapy will be MR-guided. Br. J. Radiol. **90**(1073), 20160667 (2017)
2. Kiser, K.J., Smith, B.D., Wang, J., Fuller, C.D.: "Après mois, le déluge": preparing for the coming data flood in the MRI-guided radiotherapy era. Front. Oncol. **9**, 983 (2019)
3. Thorwarth, D., Low, D.A.: Technical challenges of real-time adaptive MR-guided radiotherapy. Front. Oncol. **11**, 634507 (2021)
4. Segedin, B., Petric, P.: Uncertainties in target volume delineation in radiotherapy-are they relevant and what can we do about them? Radiol. Oncol. **50**, 254–262 (2016)
5. Mazher, M., et al.: Self-supervised spatial–temporal transformer fusion based federated framework for 4D cardiovascular image segmentation. Inf. Fusion **106**, 102256 (2024)
6. Qayyum, A., et al.: Transforming Heart Chamber Imaging: Self-Supervised Learning for Whole Heart Reconstruction and Segmentation. arXiv preprint arXiv:2406.06643 (2024)
7. Oquab, M., et al.: Dinov2: Learning robust visual features without supervision. arXiv preprintarXiv:2304.07193 (2023)
8. Alkin, B., Beck, M., Pöppel, K., Hochreiter, S., Brandstetter, J.: Vision-LSTM: xLSTM as Generic Vision Backbone. arXiv preprintarXiv:2406.04303 (2024)

9. Chen, T., et al.: xLSTM-UNet can be an Effective 2D&3D Medical Image Segmentation Backbone with Vision-LSTM (ViL) better than its Mamba Counterpart. arXiv preprint arXiv: 2407.01530 (2024)
10. Wahid, K.A., et al.: Evaluation of deep learning-based multiparametric MRI oropharyngeal primary tumor auto-segmentation and investigation of input channel effects: Results from a prospective imaging registry. Clin. Transl. Radiat. Oncol. 32, 6–14 (2022)
11. Cardenas, C.E., et al.: Comprehensive quantitative evaluation of variability in magnetic resonance-guided delineation of oropharyngeal gross tumor volumes and high-risk clinical target volumes: an R-IDEAL stage 0 prospective study. Int. J. Radiat. Oncol. Biol. Phys. 113(2), 426–436 (2022)
12. Hindocha, S., et al.: Artificial intelligence for radiotherapy auto-contouring: current use, perceptions of and barriers to implementation. Clin. Oncol. 35(4), 219–226 (2023)
13. Lin, D., et al.: E pluribus unum: prospective acceptability benchmarking from the contouring collaborative for consensus in radiation oncology crowdsourced initiative for multiobserver segmentation. J. Med. Imaging 10(S1), S11903–S11903 (2023)
14. Eisenmann, M., et al.: Biomedical image analysis competitions: the state of current participation practice. arXiv preprint arXiv:2212.08568 (2022)
15. Qayyum, A., et al.: Hybrid 3D-ResNet deep learning model for automatic segmentation of thoracic organs at risk in CT images. In: 2020 International Conference on Industrial Engineering, Applications and Manufacturing (ICIEAM), pp. 1–5. IEEE (2020)

MRI-Based Head and Neck Tumor Segmentation Using nnU-Net with 15-Fold Cross-Validation Ensemble

Frank N. Mol[1]([⊠]) [ID], Luuk van der Hoek[2] [ID], Baoqiang Ma[2] [ID],
Bharath Chowdhary Nagam[1] [ID], Nanna M. Sijtsema[2] [ID], Lisanne V.
van Dijk[2] [ID], Kerstin Bunte[1] [ID], Rifka Vlijm[1] [ID], and Peter M. A. van Ooijen[2] [ID]

[1] Faculty of Science and Engineering,University of Groningen, Nijenborgh 4, 9747
AG Groningen, The Netherlands
{frank.mol,b.c.nagam,k.bunte,r.vlijm}@rug.nl
[2] University of Groningen, University Medical Center Groningen, 9700 RB
Groningen, The Netherlands
{l.van.der.hoek02,b.ma,n.m.sijtsema,l.v.van.dijk,p.m.a.van.ooijen}@umcg.nl

Abstract. The superior soft tissue differentiation provided by MRI may
enable more accurate tumor segmentation compared to CT and PET,
potentially enhancing adaptive radiotherapy treatment planning. The
Head and Neck Tumor Segmentation for MR-Guided Applications challenge (HNTSMRG-24) comprises two tasks: segmentation of primary
gross tumor volume (GTVp) and metastatic lymph nodes (GTVn) on T2-
weighted MRI volumes obtained at (1) pre-radiotherapy (pre-RT) and (2)
mid-radiotherapy (mid-RT). The training dataset consists of data from
150 patients, including MRI volumes of pre-RT, mid-RT, and pre-RT
registered to the corresponding mid-RT volumes. Each MRI volume is
accompanied by a label mask, generated by merging independent annotations from a minimum of three experts. For both tasks, we propose adopting the nnU-Net V2 framework by the use of a 15-fold cross-validation
ensemble instead of the standard number of 5 folds for increased robustness and variability. For pre-RT segmentation, we augmented the initial
training data (150 pre-RT volumes and masks) with the corresponding
mid-RT data. For mid-RT segmentation, we opted for a three-channel
input, which, in addition to the mid-RT MRI volume, comprises the registered pre-RT MRI volume and the corresponding mask. The mean of
the aggregated Dice Similarity Coefficient for GTVp and GTVn is computed on a blind test set and determines the quality of the proposed
methods. These metrics determine the final ranking of methods for both
tasks separately. The final blind testing (50 patients) of the methods
proposed by our team, *RUG_UMCG*, resulted in an aggregated Dice
Similarity Coefficient of 0.81 (0.77 for GTVp and 0.85 for GTVn) for
Task 1 and 0.70 (0.54 for GTVp and 0.86 for GTVn) for Task 2.

Keywords: Head and Neck Tumor · Segmentation · Deep Learning ·
Radiotherapy · HNTSMR24

© The Author(s) 2025
K. A. Wahid et al. (Eds.): HNTS-MRG 2024, LNCS 15273, pp. 179–190, 2025.
https://doi.org/10.1007/978-3-031-83274-1_13

1   Introduction

Head and neck cancer (HNC) is among the most prevalent cancers globally, with over half a million new cases each year, making it the eighth leading cause of cancer-related deaths [26]. In clinical practice, HNC patients are often treated with (chemo-)radiotherapy, optionally complemented with surgery. Accurate tumor segmentation is essential for precisely targeting tumor regions with the appropriate dose while sparing surrounding healthy tissues. The tumor contours are generally manually delineated by experts on CT scans, using MRI and PET scans as references. However, MRI offers superior soft tissue differentiation compared to CT and PET, and MRI-only data facilitates adaptive RT with MR-linac. The manual delineation is time-consuming and may introduce intra- and inter-observer variability [28]. Automatic tumor segmentation could tackle these challenges by advanced techniques such as deep learning. Similar challenges preceding HNTSMRG-24 predominantly explored automated segmentation of HNC on CT and PET scans, as in the HECKTOR challenge of 2020, 2021, and 2022 [1,2,4].

Previous studies primarily utilized Deep Convolutional Neural Networks (DCNNs) for segmenting head and neck tumors in PET/CT and MRI scans [6,12,13,29]. Despite being introduced almost a decade ago, the U-Net architecture [25] continues to be dominant in the field of medical image segmentation. The (original) 2D variant gained prominence in segmenting the gross tumor volume (GTV) [19]. Further developments have been made in the form of 2.5D (handling multiple slices at once) and 3D (handling complete volumes) [8]. Due to the increasing computation capabilities, the 3D U-Net has widely been adopted for head and neck tumor segmentation, yielding impressive results across multiple studies [7,10,30]. Furthermore, a comparison of the different U-Net approaches resulted in the best segmentation performance for 3D U-Net in MRI segmentation, at the cost of increased computation time [5]. The tuning of hyper-parameters and design choices of such a U-Net have also been automated for medical image segmentation by the nnU-Net framework [14], which recently received a major update [15]. This framework is already widely used in the HNC tumor segmentation for PET and CT modalities [22,31].

Alternatively to the DCNN paradigm, transformer-based models have recently garnered significant attention in the fields of image processing and computer vision, and consequently the field of medical image segmentation. The Swin UNETR [11] has emerged as a popular architecture, demonstrating remarkable success in 3D brain tumor segmentation. Meta introduced the Segment Anything Model (SAM) [16] as a foundation model for zero-shot image segmentation, marking a significant advancement in the field. Despite being pretrained on approximately 11 million natural images, SAM lacks performance in medical applications due to absence of medical images in training. MedSAM [17] was developed to address this issue, by utilizing the SAM model and pretraining it on over a million annotated medical images from more than 30 cancer types. MedSAM has been applied to head and neck tumor segmentation in various recent studies [24], demonstrating its potential in specialized medical imaging tasks.

MRI data offers significant potential for HNC tumor segmentation compared to CT and PET, due to its superior soft tissue contrast provided by multi-sequence settings. Among the commonly employed MRI techniques, including T1-, T2-weighted, and diffusion-weighted imaging (DWI), T2-weighted MRI scans are often favored, in case of single-modality use, for tumor segmentation due to their enhanced sensitivity to water content. This enhanced contrast between tumors and surrounding tissues facilitates more accurate delineation. In radiotherapy, MRI scans typically precede the start of treatment. However, with the emergence adaptive radiotherapy has led to the widespread use of MRI scans during the course of the treatment. The data considered in this study consists of two scans for every patient. A pre-radiotherapy (pre-RT) scan acquired 1–3 weeks before treatment and a mid-radiotherapy (mid-RT) scan acquired 2–4 weeks after the first treatment. These scans are complemented with expert annotations identifying primary gross tumor volumes (GTVp) and metastatic lymph nodes (GTVn) and hence capture changes in these gross tumor volumes for each patient.

This study aims to develop a deep learning model for HNC tumor segmentation based on T2 MRI volumes by comparing a custom Unet, a model trained using nnU-Net V2 and SAM. To increase robustness and variability, the nnU-Net 5-fold cross validation ensemble was adjusted to a 15-fold cross validation ensemble. The performance of the networks was optimized for two tasks: GTVp and GTVn segmentation on either (1) pre-RT or (2) mid-RT MRIs and was assessed by the aggregated Dice Similarity Coefficient with the ground truth segmentations.

2   Materials and Methods

2.1   Data

The training dataset comprises data from 150 patients who were treated for HNC with (chemo)radiation at The University of Texas MD Anderson Cancer Center (MDACC). For each patient, the T2-weighted MRI scans of pre-RT (1–3 weeks before start of RT) and mid-RT (2–4 weeks after start of RT) were provided. Furthermore, registered pre-RT volumes were provided (registered on the corresponding mid-RT MRI volume), for which the SimpleElastix package was used [18]. The data were provided in NIfTI format, with a uniform spacing of $0.5 \times 0.5 \times 2.0$ mm applied across all datasets. Each volume is a cropped MRI containing the region from the top of the clavicles to the bottom of the nasal septum. The sizes of the volumes vary across patients. E.g. the number of slices in the pre-RT training volumes varies between 57 and 162. To ensure uniformity in shape for the training samples, this variability needs to be adjusted.

For each MRI volume, a corresponding label mask was provided, containing a voxel-wise annotation for three classes: background, GTVp, and GTVn (class 0, 1, and 2, respectively). These annotations were based on a combination of manual annotations from at least three experts. Figure 1 shows the pre-RT MRI sample of patient number 30 with corresponding annotation.

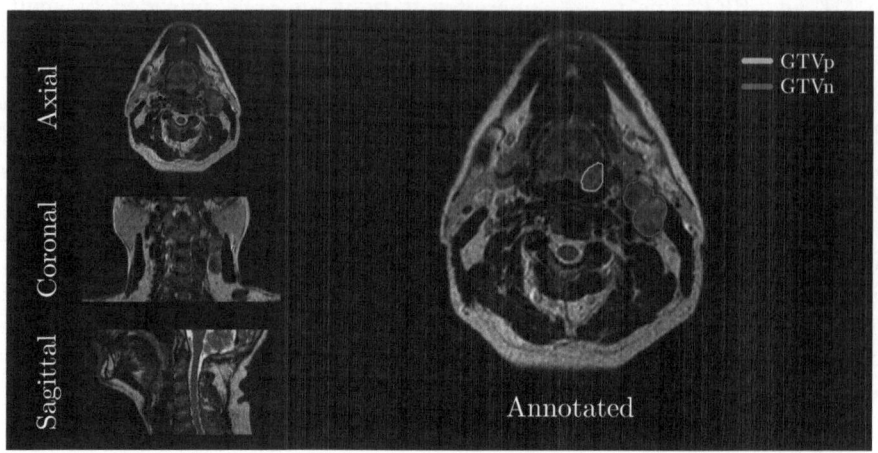

Fig. 1. Example of pre-RT MRI volume with corresponding annotations. The left panel shows the axial, coronal and sagittal view of the center of volume. The right panel displays the annotated labels for GTVn and GTVp in the axial view. The displayed data is from patient number 30 from the training set.

The GTVp volumes in the pre-RT training data have a volume consisting of on average of 22,138 voxels (137 GTVp volumes across 135 patients). In the mid-RT data, the average number of voxels for GTVp significantly dropped to 8,373 (116 GTVp volumes across 114 patients). For GTVn, the average number of voxels per volume had a lower reduction: 14,001 (259 voxels, with 20 patients having no GTVn) for pre-RT to 9,607 for mid-RT (241 voxels, with 21 patients not having GTVn).

During development, the used methods were tested using a Docker image on a preliminary blind test set consisting of data from two patients for both Task 1 and Task 2. Similarly, the final test is conducted blindly on a test set comprising data from 50 patients (not present in the initial training data). For Task 1, a pre-RT MRI volume was provided per patient as a single file as input for testing. For Task 2, five files per patient were provided for input for testing: (1) the mid-RT MRI volume, (2) the corresponding pre-RT MRI volume and (3) its respective label mask, (4) the registered pre-RT MRI volume and (5) its respective label mask.

2.2   Preprocessing

The preprocessing steps were determined by the nnU-Net v2 framework by automatically analyzing the training data, resulting in similar preprocessing for both tasks. The main step of this preprocessing is Z-score normalization of the MRI volumes before training (and for inference). The original voxel size of $0.5 \times 0.5 \times 2.0$ mm^3 was kept, i.e. no resizing of the volumes was performed. Since the data consists of large, variably sized volumes, a crop of the volumes was used during training, for example, crops of shape (48, 224, 192) were used in Task 1. These crops were taken randomly, with oversampling the foreground labels (GTVp and GTVn) using a probability of 0.33. The oversampling was used to overcome the class imbalance in training (e.g. 99.8% of the voxels is background). During training, the data was augmented following the recommended augmentations of the nnU-Net V2 framework [14], e.g. rotations, mirroring, blur. For inference, these volumes are again cropped and fed forward to the network using a sliding-window approach with a step size of 0.5 (the cropped volumes have 50% overlap).

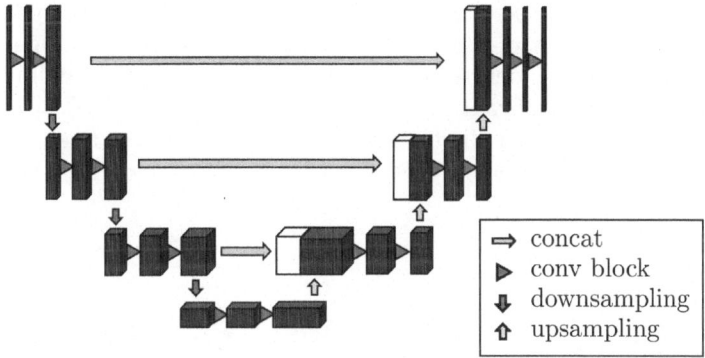

Fig. 2. Schematic of the U-Net architecture, as conceptually proposed by Ronneberger et al. [25]. In our method, we use the volumetric U-Net that processes 3D volumes.

2.3   Architecture

The Deep-Learning architecture used for both tasks was a 3D U-Net [8,25]. In Fig. 2 a schematic of the general U-Net architecture is presented. We have used the nnU-Net V2 framework [14,15] for planning and training the models. We have adopted the *3d_fullres* configuration due to the proven performance of this configuration [5], the framework also provides planning for *2d*, *3d_lowres* and *3d_cascade_fullres*. LeakyReLU activation functions were used with a negative slope set to $\alpha = 0.01$. The loss function that was used is a combination of the Dice Loss and Cross Entropy loss (sum with equal weights). Stochastic Gradient Descent was used as the optimizer, using a learning rate scheduled from the initial 0.01 to 10^{-5} over the used epochs.

2.4   Evaluation

The evaluation of the segmentation quality of the proposed methods is done by computing the aggregated Dice Similarity Coefficient (DSC_{agg}), obtained from

[3]. This score is defined by

$$DSC_{agg} = \frac{2 \times \sum_k TP_k}{2 \times \sum_k TP_k + \sum_k FP_k + \sum_k FN_k} = \frac{2 \sum_k \bar{y}_k y_k}{\sum_k (\bar{y}_k + y_k)} \quad (1)$$

with k as the patient number, TP, FP and FN are the true positives, false positives and false negatives, respectively, and $\bar{y}_k$ is the prediction for the ground truth volume y_k for patient number k. This score is an extended version of the original Dice Similarity Coefficient, which is given by

$$DSC = \frac{2 \times TP_k}{2 \times TP_k + FP_k + FN_k} = \frac{2\bar{y}_k y_k}{(\bar{y}_k + y_k)} \quad (2)$$

for an individual patient k. Typically, to assess the overall performance of a method on an entire test set, the mean DSC is calculated across all patients per type of GTV volume. A disadvantage of this metric is that test samples with small GTV volumes can yield a low individual DSC, as even a small number of incorrect voxels can significantly affect the final result. The DSC_{agg} overcomes this problem by calculating the final metric voxel-wise alternatively to sample-wise, which reduces the large impact of small GTV volumes (Fig. 3).

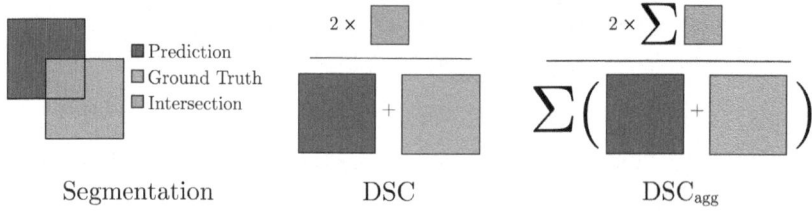

Segmentation   DSC   DSC$_{agg}$

Fig. 3. Visual representation of the Dice Similarity Coefficient (DSC) and its aggregated version (DSC_{agg}). Color marks different volumes, namely the expert annotated ground truth volume in orange, the predicted segmentation in purple, and the intersection volume indicating the true positive is green. (Color figure online)

The DSC_{agg} is computed for both non-background classes GTVp and GTVn, aggregated over all test samples provided by the organizers. The final scoring is determined by the mean of the DSC_{agg} over both types of GTV.

2.5   Experiments

We tested a custom framework on the data for Task 1 based on the MONAI package [21], using the U-Net, 3D U-Net, and Swin UNETR architectures. The resulting training speed, validation results, and tinkering time were less optimal compared to the initial nnU-Net V2 experiments, which resulted in the decision to not further investigate the possibilities for this custom framework.

Furthermore, we also investigated on Task 1 the viability of using SAM [17], the foundation model (FM) pretrained on a large medical dataset. Since SAM needs user input in the form of clicks or bounding boxes, we adopted the Auto-gluon package that uses the SAM model as baseline [9,16,27]. Zero-shot evaluation provided no realistic segmentation, therefore fine-tuning was required. For fine-tuning 3D image mask pairs were converted into 2D slices which served as input for the model. The model was trained for up to 2 h (6 epochs) on the Hàbròk HPC cluster, from which the optimal intermediate model was selected based on the validation Intersection over Union (IoU) score. Since this method is a 2D approach, the individual slices were fed into the network for inference, stacking the output to create the volumetric segmentation. This preliminary test of the fine-tuned SAM approach resulted in a DSC_{agg} of 0.70 on an internal test set. Due to the more promising results obtained by nnU-Net, and the preference for 3D modalities [5], we decided to not further investigate this approach for this challenge.

We adopted the nnU-Net V2 framework and increased the number of folds in the cross-validation ensemble from 5 to 15, to increase the robustness and variability of the ensemble. These benefits result from both an increase in the number of training samples per fold (140 instead of 120) and the use of a greater number of models trained on different subsets (15 instead of 5). The resulting DSC_{agg} scores on the validation sets were promising, leading us to adhere to this regime. For the first task, we explored various numbers of epochs, but this did not prompt us to deviate from the initial 1000 epochs set by the framework. For Task 2, we have investigated the impact of the number of epochs using 15-fold cross-validation and computing the DSC_{agg} aggregated over the validation results over all folds. The number of epochs used in our final method is 1250, based on the improved results compared to the 1000 epochs 15-fold cross-validation. The increase in epochs resulted in a better DSC_{agg} for 11 out of 15 folds for GTVp and 12 out of 15 folds for GTVn, as shown in Fig. 4. The final validation DSC_{agg} computed over all samples increased from 0.73 for 1000 epochs to 0.74 for 1250 epochs. A comparison of training 5-folds and 15-folds resulted in a similar validation DSC_{agg} of 0.74 for both models, both trained for 1250 epochs. The source-code used for this study have been published on Zenodo, see the repository [20].

3   Results

3.1   Pre-RT Segmentation

The resulting method that is used for pre-RT segmentation of GTVp and GTVn (Task 1) is a 15-fold cross-validation ensemble of the nnU-Net V2 $3d_{fullres}$ configuration, in which each fold was trained on 280 training samples. The 280 training samples consisted of 140 pre-RT and 140 mid-RT scans (from the same 140 patients, but treated as separate training inputs) with corresponding label masks. The model was trained over 1000 epochs. The increase of the number of ensembles from 5-fold cross-validation to 15-fold cross-validation ensembles

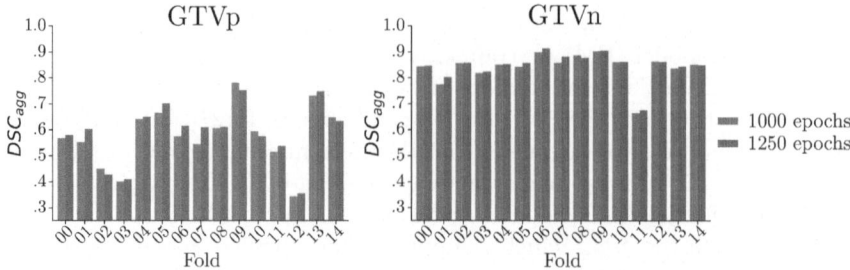

Fig. 4. Comparison of 15-fold cross validation of nnU-Net V2 using 1000 and 1250 epochs for mid-RT segmentation (Task 2). The validation DSC_{agg} is displayed for each fold, displaying an increase in performance by training for 1250 epochs above 1000 epochs. The model trained for 1250 epochs yields increased performance in terms of DSC_{agg} for 11 and 12 out of 15 folds for GTVp and GTVn, respectively. The DSC_{agg} across all validation samples reached 0.74 for 1250 epochs and 0.73 for 1000 epochs.

increases the inference time, i.e. the number of forward passes through a model is increased from 5 to 15 for predicting a single segmentation. Due to the inference time limit of the Docker image on the challenge platform of 20 min (NVIDIA T4 Tensor Core GPU), we decided to disable the test-time augmentation. The reasoning is that increasing variability by increasing the number of folds has a larger impact than the data augmentation, in the form of flips in all dimensions, on the resulting segmentation quality. The preliminary development phase, which consists of 2 patient samples, resulted in a mean DSC_{agg} of 0.89. The final testing (on 50 patients) of the method resulted in a mean DSC_{agg} of 0.77 for GTVp and 0.85 for GTVn. Resulting in a final mean DSC_{agg} of 0.81.

3.2   Mid-RT Segmentation

We applied a similar approach as in Task 1 for GTVp and GTVn segmentation on mid-RT data (Task 2), utilizing a 15-fold cross-validation ensemble of the nnU-Net V2 $3d_{fullres}$ configuration. In contrast to Task 1, the provided data includes more than just a single MRI volume, allowing for the use of multichannel input. The mid-RT MRI volume is complemented by the corresponding pre-RT MRI volume and label mask. As a result, it is not possible to increase the number of initially provided samples when the pre-RT is added to the input. The pre-RT data consisted of MRI volumes with corresponding segmentation masks, along with the registered versions aligned with the mid-RT volumes. We opted for including the registered pre-RT MRI volume (as the second channel) and the corresponding registered annotated mask for GTVp and GTVn (as the third channel) since the aligned data provides local information of the same region. Hence, the training of all folds was performed on 140 3-channels patient samples. Each fold was trained for 1250 epochs, and the resulting models were used for the final ensemble. Again, we disabled test-time augmentation to reduce the inference time such that the inference time per sample falls within the 20-minute time limit.

To infer the model quality, we computed the DSC_{agg} across all validation samples across all 15 folds, which results in a single DSC_{agg} value across the validation test set. This resulted in a DSC_{agg} of 0.62 for GTVp and 0.86 for GTVn, resulting in a final DSC_{agg} of 0.74. The preliminary development phase, which consists of 2 patient samples, resulted in a mean DSC_{agg} of 0.75. The final testing of the method (on data from 50 patients) resulted in a mean DSC_{agg} of 0.54 for GTVp and 0.86 for GTVn. Resulting in a mean DSC_{agg} of 0.70.

4   Discussion

This study investigated the performance of several deep learning methods for segmenting GTVp and GTVn in pre-RT and mid-RT T2-weighted MRI scans of HNC patients. Initial experiments resulted in a final approach using the nnU-Net V2 framework [14,15]. This framework configures a U-Net architecture [25], alongside the hyper-parameters and data processing steps, based on the provided training data. The nnU-Net framework has demonstrated high performance for Head and Neck Cancer segmentation in PET/CT [22,31] and lymph node segmentation in MRI [23]. To increase the variability in the ensemble, and increase the number of training samples per fold, we adapted the framework to work with 15-fold cross-validation ensembles, instead of the original 5-fold cross-validation. For Task 1, the training data was augmented by incorporating the Mid-RT MRI volumes with corresponding label masks. For Task 2, the increased complexity of the three-channel input (Mid-RT MRI, Registered pre-RT MRI and label mask) causes that more epochs were necessary to reach the optimal DSC_{agg} scores in the validation set (1250 instead of 1000). The final trained models yield convincing results in terms of the DSC_{agg} values.

The final blind test on data from 50 patients resulted in a DSC_{agg} of 0.81 for GTVp and GTVn segmentation on pre-RT MRI volumes (Task 1) and a DSC_{agg} of 0.70 for GTVp and GTVn segmentation on mid-RT MRI volumes (Task 2). Task 2 resulted in a lower score compared to Task 1, even though the provided information as input was increased. The metastatic lymph nodes (GTVn) had similar results for both tasks in terms of DSC_{agg}, 0.85 for Task 1 and 0.86 for Task 2. Consequently, the lower result is caused by a worse segmentation of the primary gross tumor volume (GTVp). In Task 1, the GTVp resulted in a DSC_{agg} of 0.77, followed by a lower DSC_{agg} of 0.54 for Task 2. The observed decrease in segmentation performance for GTVp from pre-RT to mid-RT could potentially be related to the large reduction in the size of the volumes (possibly due to effective RT). Furthermore, a decrease in the contrast results in less clear tumor boundaries. Future work aimed at improving segmentation performance for mid-RT GTVp with lower contrasts holds significant potential. Additionally, advancements in the registration process may further improve performance for both labels in mid-RT segmentation.

MRI-based HNC tumor segmentation was less explored in previous studies compared to CT/PET, possibly due to the later inclusion of MRI into clinical radiotherapy flow of HNC. However, with the increasing use of MRI in radiotherapy-such as employing MR-linac for visualization of the head and neck region before each radiotherapy fraction-automatic segmentation of the radiotherapy target volumes (GTVp and GTVn), as well as critical organs in the irradiated volume, is potentially crucial to realize real-time accurate adaptive radiotherapy.

This study has some limitations. Firstly, a low sample size of 150 patients was used for training. To overcome these limitations, future work might benefit from an increase in sample size and utilize federated learning allowing to incorporate multi-center data to enhance the performance and generalization of the study. Secondly, the advantage of MRI over CT/PET for HNC tumor segmentation remains to be proven. In the future, a multi-center and multi-modality (CT/PET/MRI) dataset should be acquired to further evaluate the efficacy of nnU-Net for HNC tumor segmentation.

In conclusion, increasing the number of folds from 5 to 15 for the cross-validation ensemble displayed to have a positive effect on the segmentation performance using the nnU-Net V2 framework for GTVp and GTVn segmentation, in both pre-RT and mid-RT T2-weighted MRI volumes.

Acknowledgments. We thank the Center for Information Technology of the University of Groningen for their support and for providing access to the Hábrók high-performance computing cluster.

Disclosure of Interests. The authors declare no competing interests.

References

1. Andrearczyk, V., et al.: Overview of the HECKTOR challenge at MICCAI 2022: automatic head and neck tumor segmentation and outcome prediction in pet/ct. In: Andrearczyk, V., Oreiller, V., Hatt, M., Depeursinge, A. (eds.) Head and Neck Tumor Segmentation and Outcome Prediction, pp. 1–30. Springer, Cham (2023)
2. Andrearczyk, V., et al.: Overview of the HECKTOR challenge at MICCAI 2021: automatic head and neck tumor segmentation and outcome prediction in PET/CT images. In: 3D Head and Neck Tumor Segmentation in PET/CT Challenge, pp. 1–37. Springer (2021)
3. Andrearczyk, V., Oreiller, V., Jreige, M., Castelli, J., Prior, J.O., Depeursinge, A.: Segmentation and classification of head and neck nodal metastases and primary tumors in PET/CT. In: 2022 44th Annual International Conference of the IEEE Engineering in Medicine & Biology Society (EMBC), pp. 4731–4735 (2022). https://doi.org/10.1109/EMBC48229.2022.9871907
4. Andrearczyk, V., et al.: Overview of the HECKTOR challenge at MICCAI 2020: automatic head and neck tumor segmentation in PET/CT. In: Head and Neck Tumor Segmentation: First Challenge, HECKTOR 2020, Held in Conjunction with MICCAI 2020, Lima, Peru, 4 October 2020, Proceedings 1, pp. 1–21. Springer (2021)

5. Avesta, A., Hossain, S., Lin, M., Aboian, M., Krumholz, H.M., Aneja, S.: Comparing 3D, 2.5D, AND 2D approaches to brain image auto-segmentation. Bioengineering **10**(2) (2023). https://doi.org/10.3390/bioengineering10020181. https://www.mdpi.com/2306-5354/10/2/181

6. Bielak, L., et al.: Convolutional neural networks for head and neck tumor segmentation on 7-channel multiparametric MRI: a leave-one-out analysis. Radiat. Oncol. **15**, 1–9 (2020)

7. Chen, J., Martel, A.L.: Head and neck tumor segmentation with 3D UNet and survival prediction with multiple instance neural network. In: 3D Head and Neck Tumor Segmentation in PET/CT Challenge, pp. 221–229. Springer (2022)

8. Çiçek, Ö., Abdulkadir, A., Lienkamp, S.S., Brox, T., Ronneberger, O.: 3D U-Net: learning dense volumetric segmentation from sparse annotation. In: Ourselin, S., Joskowicz, L., Sabuncu, M.R., Unal, G., Wells, W. (eds.) MICCAI 2016. LNCS, vol. 9901, pp. 424–432. Springer, Cham (2016). https://doi.org/10.1007/978-3-319-46723-8_49

9. Erickson, N., et al.: Autogluon-tabular: robust and accurate automl for structured data. arXiv preprint arXiv:2003.06505 (2020)

10. Ghimire, K., Chen, Q., Feng, X.: Patch-based 3D unet for head and neck tumor segmentation with an ensemble of conventional and dilated convolutions. In: Head and Neck Tumor Segmentation: First Challenge, HECKTOR 2020, Held in Conjunction with MICCAI 2020, Lima, Peru, 4 October 2020, Proceedings 1, pp. 78–84. Springer (2021)

11. Hatamizadeh, A., Nath, V., Tang, Y., Yang, D., Roth, H.R., Xu, D.: Swin UNETR: swin transformers for semantic segmentation of brain tumors in MRI images. In: International MICCAI Brainlesion Workshop, pp. 272–284. Springer (2021)

12. Huang, B., et al.: Fully automated delineation of gross tumor volume for head and neck cancer on PET-CT using deep learning: a dual-center study. Contrast Media Mol. Imaging **2018**(1), 8923028 (2018)

13. Ibragimov, B., Xing, L.: Segmentation of organs-at risks in head and neck CT images using convolutional neural networks. Med. Phys. **44**(2), 547–557 (2017)

14. Isensee, F., et al.: nnU-Net: self-adapting framework for u-net-based medical image segmentation (2018). https://arxiv.org/abs/1809.10486

15. Isensee, F., et al.: nnu-net revisited: a call for rigorous validation in 3D medical image segmentation (2024). https://arxiv.org/abs/2404.09556

16. Kirillov, A., et al.: Segment anything. In: Proceedings of the IEEE/CVF International Conference on Computer Vision (ICCV), pp. 4015–4026 (2023)

17. Ma, J., He, Y., Li, F., Han, L., You, C., Wang, B.: Segment anything in medical images. Nat. Commun. **15**(1), 654 (2024)

18. Marstal, K., Berendsen, F., Staring, M., Klein, S.: Simpleelastix: a user-friendly, multi-lingual library for medical image registration. In: 2016 IEEE Conference on Computer Vision and Pattern Recognition Workshops (CVPRW), pp. 574–582 (2016). https://doi.org/10.1109/CVPRW.2016.78

19. Moe, Y.M., et al.: Deep learning for automatic tumour segmentation in PET/CT images of patients with head and neck cancers. arXiv preprint arXiv:1908.00841 (2019)

20. Mol, F.: HNTSMRG24_RUG_UMCG version: hntsmrg (2024). https://doi.org/10.5281/zenodo.14193311

21. MONAI Consortium: MONAI: Medical open network for AI: version 1.3.2 (2024). https://doi.org/10.5281/zenodo.12542217

22. Murugesan, G.K., et al.: Head and neck primary tumor segmentation using deep neural networks and adaptive ensembling. In: Andrearczyk, V., Oreiller, V., Hatt, M., Depeursinge, A. (eds.) Head and Neck Tumor Segmentation and Outcome Prediction, pp. 224–235. Springer, Cham (2022)

23. Reinders, F.C., et al.: Automatic segmentation for magnetic resonance imaging guided individual elective lymph node irradiation in head and neck cancer patients. Phys. Imaging Radiat. Oncol. **32**, 100655 (2024)

24. Ren, J., Rasmussen, M., Nijkamp, J., Eriksen, J.G., Korreman, S.: Segment anything model for head and neck tumor segmentation with CT, PET and MRI multi-modality images. arXiv preprint arXiv:2402.17454 (2024)

25. Ronneberger, O., Fischer, P., Brox, T.: U-net: convolutional networks for biomedical image segmentation. CoRR abs/1505.04597 (2015). http://arxiv.org/abs/1505.04597

26. Sung, H., et al.: Global cancer statistics 2020: Globocan estimates of incidence and mortality worldwide for 36 cancers in 185 countries. CA: Cancer J. Clin. **71**(3), 209–249 (2021)

27. Tang, Z., et al.: Autogluon-multimodal (automm): supercharging multimodal automl with foundation models. arXiv preprint arXiv:2404.16233 (2024)

28. Valentini, V., Boldrini, L., Damiani, A., Muren, L.P.: Recommendations on how to establish evidence from auto-segmentation software in radiotherapy. Radiother. Oncol. **112**(3), 317–320 (2014). https://doi.org/10.1016/j.radonc.2014.09.014

29. Wahid, K.A., et al.: Evaluation of deep learning-based multiparametric MRI oropharyngeal primary tumor auto-segmentation and investigation of input channel effects: results from a prospective imaging registry. Clin. Transl. Radiat. Oncol. **32**, 6–14 (2022)

30. Xie, J., Peng, Y.: The head and neck tumor segmentation based on 3D U-net. In: 3D Head and Neck Tumor Segmentation in PET/CT Challenge, pp. 92–98. Springer (2021)

31. Xie, J., Peng, Y.: The head and neck tumor segmentation using nnu-net with spatial and channel 'squeeze & excitation' blocks. In: Head and Neck Tumor Segmentation: First Challenge, HECKTOR 2020, Held in Conjunction with MICCAI 2020, Lima, Peru, 4 October 2020, Proceedings 1, pp. 28–36. Springer (2021)

Head and Neck Tumor Segmentation Using Pre-RT MRI Scans and Cascaded DualUNet

Mikko Saukkoriipi[1], Jaakko Sahlsten[1], Joel Jaskari[1], Ahmed Al-Tahmeesschi[3], Laura Ruotsalainen[2], and Kimmo Kaski[1,4(✉)]

[1] Department of Computer Science, Aalto University, Espoo, Finland
kimmo.kaski@aalto.fi
[2] Department of Computer Science, University of Helsinki, Helsinki, Finland
[3] Department of Electronic Engineering, University of York, York, UK
[4] The Alan Turing Institute, London, UK

Abstract. Accurate segmentation of the primary gross tumor volumes and metastatic lymph nodes in head and neck cancer is crucial for radiotherapy but remains challenging due to high interobserver variability, highlighting a need for an effective auto-segmentation tool. Tumor delineation is used throughout radiotherapy for treatment planning, initially for pre-radiotherapy (pre-RT) MRI scans followed-up by mid-radiotherapy (mid-RT) during the treatment. For the pre-RT task, we propose a dual-stage 3D UNet approach using cascaded neural networks for progressive accuracy refinement. The first-stage models produce an initial binary segmentation, which is then refined with an ensemble of second-stage models for a multiclass segmentation. In Head and Neck Tumor Segmentation for MR-Guided Applications (HNTS-MRG) 2024 Task 1, we utilize a dataset consisting of pre-RT and mid-RT T2-weighted MRI scans. The method is trained using 5-fold cross-validation and evaluated as an ensemble of five coarse models and ten refinement models. Our approach (team FinoxyAI) achieves a mean aggregated Dice similarity coefficient of 0.737 on the test set. Moreover, with this metric, our dual-stage approach highlights consistent improvement in segmentation performance across all folds compared to a single-stage segmentation method.

Keywords: Cascaded deep neural networks · Dual-stage refinement · 3D UNet · HNTS-MRG · MRI Head and Neck Tumor Segmentation

1 Introduction

Radiation therapy (RT) remains a cornerstone in the treatment of head and neck cancer (HNC), including oropharyngeal cancer (OPC). RT requires accurate tumor delineation for effective treatment outcomes while minimizing damage to surrounding healthy tissue. This requirement is particularly acute in HNC due

K. A. Wahid et al. (Eds.): HNTS-MRG 2024, LNCS 15273, pp. 191–203, 2025.
https://doi.org/10.1007/978-3-031-83274-1_14

to close proximity of tumors to critical and radiation sensitive structures, such as neural and optic pathways [1]. However, accurately delineating gross primary tumor volumes (GTVp) and metastatic lymph nodes (GTVn) in HNC is challenging, as evidenced by the high interobserver variability among oncologists [1].

In order to address these challenges, various imaging modalities have been explored for their ability to provide distinct tissue and functional contrasts. The primary imaging modalities for HNC detection are computed tomography (CT), positron emission tomography (PET), and magnetic resonance imaging (MRI) [1]. CT has traditionally been preferred for its spatial accuracy, but it suffers from scatter artifacts, especially from dental amalgam, and difficulties to differentiate tumors from adjacent glottic musculature [1]. This leads to high interobserver variability when using CT alone [2]. To improve tumor delineation, PET has been integrated with CT as it provides metabolic information that aids in differentiating tumor cells from healthy tissue due to their elevated metabolic activity. Despite its low spatial resolution and partial volume effects, combining PET with CT or MRI enhances the accuracy of tumor delineation [1].

We note that MRI offers superior spatial resolution and soft tissue contrast, with the added benefit of not exposing patients to radiation, which provide notable advantage over both CT and PET [3]. This capability enhances the differentiation of adjacent soft tissues and allows for precise delineation of tumor margins. This is crucial for mapping deep tissue invasion, detecting small tumors, and characterizing complex tumor structures [4]. The improved contrast resolution also enables more sensitive evaluation of perineural invasion, intracranial extension, vascular infiltration, and bone marrow involvement [4]. However, MRI has also some limitations compared to CT. The procedure is considerably more time-consuming and expensive, which limit its use as a first-line imaging modality. In addition, patients may find it difficult to remain still for the extended duration of the scan [4].

Artificial intelligence (AI) tools have shown potential to assist radiologists in enhancing diagnostic quality while reducing their workload [5]. Although there is substantial research on automatic segmentation of oropharyngeal cancer in cases of multimodal PET-CT [1, 6–9] and PET-MRI [1, 6], the segmentation of HNC from MRI scans alone has received comparatively less attention [10]. Recent studies have explored various approaches to segmenting HNC from MRI scans. Schouten et al. [11] investigated the use of T1-weighted, T1gad, and STIR MRI modalities for HNC segmentation, reporting an average Dice similarity coefficient (DSC) of 0.49. Liedes et al. [12] found that a 2D U-Net struggled to segment HNC using only T1 SPIR, T1 TSE, or T2 TSE MRI images, but observed considerable improvement when these MRI modalities were combined with PET images. Wahid et al. [10] compared various MRI modalities and observed that T2-weighted (T2w) MRI alone achieved an average DSC of 0.72, which was slightly improved by incorporating additional modalities. These findings highlight ongoing efforts to enhance accuracy of HNC segmentation from MRI scans.

Regulatory and privacy constraints require local data processing at hospitals, which may limit the use of powerful cloud computing resources [13]. While real-time tumor delineation is essential in procedures such as MR-Linac, pre-RT setups generally operate without this requirement. Therefore, automatic tumor delineation in this case does not need to take place in real time, thus allowing more processing time compared to many other computer vision tasks.

Previous works have primarily focused on 2D or 3D UNet variants for OPC GTVp and GTVn segmentation. However, the potential benefits of cascaded deep learning have not yet been explored for this task. Cascading involves a sequential architecture in which later stages refine the outputs of earlier ones, often by progressively utilizing more detailed features [14]. This can be achieved either within a single network or through training multiple networks in stages. Cascade networks that segment the region of interest from progressively increasing MRI scan resolution have been studied for brain tumor segmentation [15,16] and organs-at-risk segmentation [17]. There have also been studies that utilize the cascading approach in the full resolution for all the stages, e.g., in bladder cancer segmentation [18] and brain tumor segmentation [19,20] in MRI as well as fetal head and abdomen segmentation in ultrasound images [21].

Inspired by prior work in cascaded deep learning segmentation, we propose a two-stage cascaded ensemble for the segmentation of the OPC GTVp and GTVn using T2w MRI scans. Our approach decomposes the task into two distinct subtasks, both performed by a deep learning model which are trained and evaluated separately. In the first-stage, a model performs binary segmentation to distinguish tumors from healthy tissue, while the second-stage model classifies the tumor regions into the two classes while refining the segmentation boundaries. This sequential approach maintains memory efficiency equivalent to a single model with a cost of additional computational time. Moreover, separating the tasks enables independent evaluation of segmentation and classification performance, providing insights into model-specific errors.

2   Methods

In this section, we describe the dataset and its preprocessing as well as present our DualUNet approach in terms of architecture, model training, and evaluation.

2.1   Datasets

The HNTS-MRG 2024 training dataset comprises both fat-suppressed and non-fat-suppressed T2-weighted MRI scans of 150 head and neck cancer patients imaged at MD Andreson Cancer Center (MDACC). Patient data includes pre-RT scans e.g., imaged 1–3 weeks before radiotherapy and mid-RT scans e.g., imaged 2–4 weeks during radiotherapy. Most of the patients are diagnosed having oropharyngeal cancer and the rest have cancer of unknown primary. Each scan has been segmented for primary gross tumor volumes (GTVp) and metastatic lymph nodes (GTVn) by 3–4 physicians independently and verified by senior radiation oncology faculty. The final ground truth segmentation is provided by simultaneous truth and performance level estimation algorithm (STAPLE) [22].

The scans exhibit considerable variability in both imaging resolution and spatial dimensions. Analysis of the training dataset determined the voxel resolutions to be approximately [0.4–1.0] mm × [0.4–1.0] mm × [1.0–2.5] mm spacing and [512–768] × [480–768] × [64–176] dimension.

2.2   Cascaded Segmentation Framework

DualUnet. The proposed cascaded UNet framework named *DualUNet*, is illustrated in Fig. 1, consists of two individual models, both based on residual UNet [23] due to its reputation for accuracy and robustness in medical imaging tasks. The first-stage UNet uses T2w MRI scans as input and produces a single output channel with sigmoid non-linearity applied separating background and any tumor segmentations. The second-stage UNet functions as a refinement model and it uses the original T2w MRI scans and the output from the first-stage UNet as inputs and produces multiclass outputs for background, GTVp, and GTVn with softmax non-linearity.

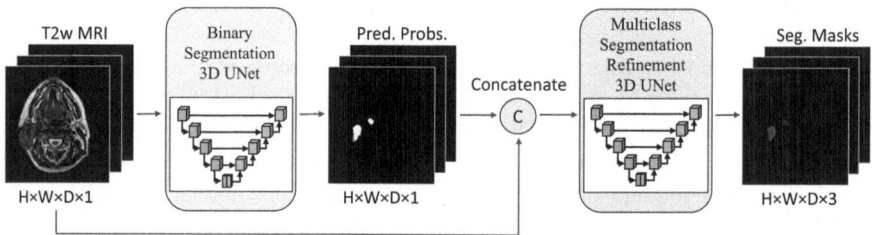

Fig. 1. DualUNet architecture consisting of a binary and a multiclass refinement 3D UNet segmentation models. The input of the refinement UNet is the concatenation of T2-weighted (T2w) MRI and the sigmoid output of the binary UNet.

UNet. Both of the UNet implementations in the framework feature a progressive increase in channel counts for each of the blocks, i.e., 32, 64, 128, 256, and 512, throughout the encoder, enabling detailed feature extraction across multiple scales. A stride pattern of 1, 2, 2, and 2 is employed within each block, starting with a stride of 1 to preserve high-resolution details and increasing to stride 2 in deeper layers, which broadens the receptive field and reduces computational requirements. Each block includes two convolutional layers and a residual connection, the latter of which is used to mitigate the vanishing gradient problem and ensure smooth gradient flow during training. Each convolution layer is followed by Instance Normalization layer to standardize features across each channel and a PReLU non-linearity [24].

The method was implemented with PyTorch 2.2.2 and MONAI 1.3.0 [25], utilizing a 3D UNet model with residual connections [23].

2.3   Training Procedure

Preprocessing. We utilized the HNTS-MRG2024 dataset consisting of 150 head and neck cancer patients with segmented targets. We developed the method using 5-fold cross-validation based on the patients. The training folds included both pre-RT and mid-RT scans of the patients and the validation fold included only pre-RT scans.

In accordance with prior research on tumor segmentation from T2w MRI images [10,26], non-zero voxels were standardized using Z-score normalization (mean = 0, standard deviation = 1) that improves the dynamic range of clinically significant regions and improves numerical stability.

Given the memory constraints of the HNTS-MRG 2024 competition, particularly the Nvidia T4 GPU's 16 GB limit and the variability in volume sizes (up to 768 × 768 × 176 voxels), we implemented cropping during inference and training to avoid exceeding GPU memory. For both the training and testing phases, we used a fixed patch size of 408 × 408 × 64 voxels. In training the first-stage models, we applied the probabilistic crop sampling technique proposed by Myronenko et al. [27], wherein the crop center is selected based on ground truth labels, with a probability distribution of 45% for GTVp, 45% for GTVn, and 10% for background. In contrast, for the second-stage models, random cropping was employed, as the probabilistic approach did not demonstrate improvements in performance. Model inference during the evaluation phase was carried out using a sliding window method, maintaining the same patch size and averaging the results over a 50% overlap.

Data Augmentation. In order to ensure model performance with heterogeneous volumetric medical imaging data, we use a wide range of augmentation methods consisting of spatial and voxel intensity value changes. The spatial augmentations consisted of random mirroring independently on each axis with a probability of 10% as well as random rotation of up to 30 °C and translations within a range of 16 pixels, both with 50% probability. Moreover, the variation in scan resolution needed to be accounted as we used native resolution scans without resampling. During training, we employed a custom resampling augmentation method to cover all resolution ranges identified in the training dataset. The custom resampling method was applied with 50% probability on each axis and with target resolution based on uniform distribution of the training dataset resolution ranges. In terms of intensity augmentations, we used contrast adjustment with gamma range 0.5 to 1.5, intensity shifting with 10% offset, random Gaussian noise with standard deviation of 0.1, and Gaussian blurring, all with 25% probability.

Loss. In our binary segmentation model, we employ a combined Dice loss and binary Cross-Entropy (BCE) loss. For the multiclass refinement model, we utilize a combination of Dice loss and Cross-Entropy (CE) loss. These combined losses have been shown to be an effective strategy for head and neck cancer segmentation from PET-CT images [27,28]. The composite loss functions, with uniform weight of 1, are formulated as follows:

$$\mathcal{L}_{\text{DiceBCE}} = \mathcal{L}_{\text{Dice}} + \mathcal{L}_{\text{BCE}}, \tag{1}$$

$$\mathcal{L}_{\text{DiceCE}} = \mathcal{L}_{\text{Dice}} + \mathcal{L}_{\text{CE}}, \tag{2}$$

where Dice Loss is denoted by $\mathcal{L}_{\text{Dice}}$, Binary Cross-Entropy loss by $\mathcal{L}_{\text{BCE}}$ and Cross entropy by $\mathcal{L}_{\text{CE}}$. The Dice Loss is defined as follows:

$$\mathcal{L}_{\text{Dice}} = 1 - \frac{2\sum_i p_i g_i + \epsilon}{\sum_i p_i + \sum_i g_i + \epsilon}, \tag{3}$$

where p and g represent the model output and ground truth segmentation, respectively, and $\epsilon = 1 \times 10^{-5}$ is a smoothing factor to prevent division by zero. The BCE and CE Losses are defined as:

$$\mathcal{L}_{\text{BCE}} = -\frac{1}{N} \sum_i^N \left[g_i \log(p_i) + (1 - g_i) \log(1 - p_i) \right], \tag{4}$$

$$\mathcal{L}_{\text{CE}} = -\frac{1}{N} \sum_i^N \sum_{c=1}^C g_{i,c} \log(p_{i,c}), \tag{5}$$

where p_i and g_i are the predicted probability and ground truth for pixel i, respectively, and N is the number of pixels. For the multiclass case, C denotes the number of classes, with $p_{i,c}$ and $g_{i,c}$ representing the predicted probability and ground truth for class c at pixel i. These composite loss functions facilitate comprehensive optimization of deep learning segmentation by addressing both the overlap of imbalanced foreground classes and per-pixel classification accuracy [29].

Optimization. Model parameters are optimized using AdamW optimizer with an initial learning rate of 2×10^{-4}, which is decreased to zero at the end of the final epoch using a cosine annealing scheduler. All models are trained for 300 epochs with a mini-batch size of 1 on a single 80 GB NVIDIA A100 machine. Additionally, we implement weight decay regularization set to 1×10^{-5} and use dropout regularization with a probability of 10%.

The first-stage and second-stage models are trained separately. For first-stage model training, we combine the GTVp and GTVn classes, using this as a binary target. For the second-stage model training, the first-stage model generates sigmoid probability volumes, which are then used as an input alongside the original volume. Prior research suggests that such refinement models may become overly reliant on the first-stage segmentation [28]. To mitigate this, we randomly drop the first-stage segmentation input with a probability of 10%.

In addition, the training convergence is improved by leveraging model weight pretraining on the HECKTOR 2022 dataset for OPC GTVp and GTVn segmentation using PET-CT scans [9].

2.4 Model Validation

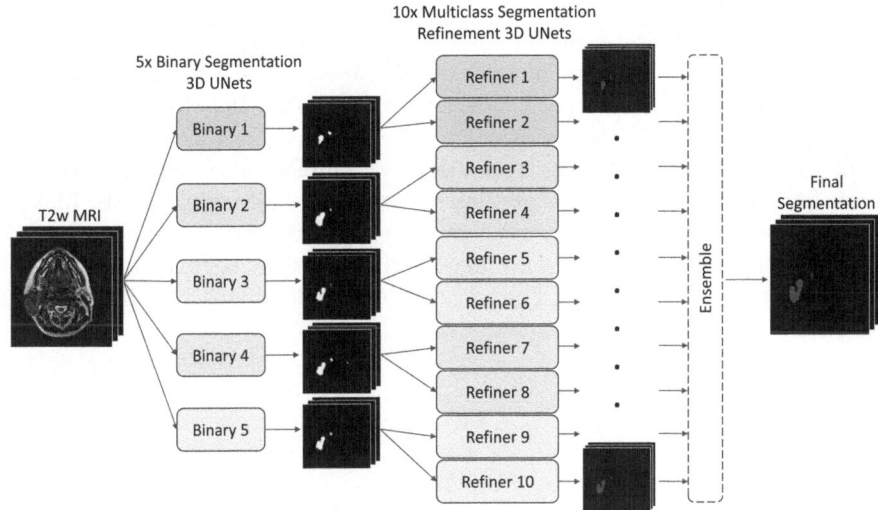

Fig. 2. Test time cascade ensemble with five coarse models and ten refinement models. The five fold used for the five binary and ten refiner model training are colored separately.

We evaluate the segmentation performance using class-wise mean of the Aggregated Dice Similarity Coefficient (DSC$_{\text{agg,mean}}$) which is defined as follows:

$$V_{c,p} = \sum_i [Y_{i,p} = c],$$

$$P_{c,p} = \sum_i [\arg\max_j \hat{Y}_{j,p} = c],$$

$$\text{TP}_{c,p} = \sum_i [Y_{i,p} = c] \cdot [\arg\max_j \hat{Y}_{j,p} = c],$$

$$\text{DSC}_{\text{agg},c} = 2 \frac{\sum_p \text{TP}_{c,p}}{\sum_p (V_{c,p} + P_{c,p})}, \tag{6}$$

$$\text{DSC}_{\text{agg,mean}} = \frac{1}{2}(\text{DSC}_{\text{agg,GTVp}} + \text{DSC}_{\text{agg,GTVn}}), \tag{7}$$

where $V_{c,p}$ and $P_{c,p}$ are the sum of labelled and predicted voxels i and j, respectively, for class c of patient p, $\text{TP}_{c,p}$ is the sum of correctly predicted voxels for class c of patient p, and $\text{DSC}_{\text{agg},c}$ is the aggregated Dice similarity coefficient for class c. Unlike the conventional Dice similarity coefficient in which multiclass segmentation results are averaged with equal weighting on each class, the

DSC_{agg} offers more robust metric, in terms of individual misclassifications, for the overall dataset segmentation performance.

Models were trained and validated using a 5-fold cross-validation approach. The 150 patients in the training dataset were randomly partitioned into five non-overlapping subsets, each containing 30 patients. Each subset served as the validation set, while the remaining 120 patients were used for training. Only pre-RT scans were used during validation to align with the HNTS-MRG 2024 Task 1 focus on pre-RT segmentation.

In the test set performance evaluation, we employ an ensemble consisting of five first-stage binary models and ten second-stage multiclass models. In each fold one first-stage model and two second-stage models were used. This approach results in ten outputs which are averaged and followed by arg max-operation for the final segmentation output. The test time ensemble approach is illustrated in Fig. 2.

3   Results

Table 1. Comparison of UNet and DualUNet performance on the validation folds of the training dataset (N = 150). Results are provided per fold (N = 30) and for the full set. Percentage changes relative to UNet are shown, with increases indicated by green arrows up and decreases by red arrows down.

Metric	Model	Fold 1	Fold 2	Fold 3	Fold 4	Fold 5	Full Set
$DSC_{agg,GTVp}$	UNet	0.706	0.668	0.729	0.620	0.644	0.677
	DualUNet	0.726	0.709	0.736	0.679	0.720	0.716
		↑ 2.8%	↑ 5.8%	↑ 1.0%	↑ 8.7%	↑ 10.6%	↑ 5.4%
$DSC_{agg,GTVn}$	UNet	0.773	0.812	0.685	0.758	0.762	0.758
	DualUNet	0.802	0.805	0.763	0.818	0.793	0.795
		↑ 3.6%	↓ 0.9%	↑ 10.2%	↑ 7.3%	↑ 3.9%	↑ 4.7%
$DSC_{agg,mean}$	UNet	0.740	0.740	0.707	0.689	0.703	0.718
	DualUNet	0.764	0.757	0.750	0.749	0.757	0.755
		↑ 3.1%	↑ 2.2%	↑ 5.7%	↑ 8.0%	↑ 7.1%	↑ 4.9%

On the separate test dataset with (N=50), the DualUNet turned out to have $DSC_{agg,mean}$ value of 0.737. The class specific performances were 0.697 for $DSC_{agg,GTVp}$ and 0.777 for $DSC_{agg,GTVn}$. The 50 test patients were evaluated through the HNTS-MRG 2024 competition official evaluation tool at the Grand-challenge.org platform.

In terms of the overall performance on training dataset, the DualUNet outperformed UNet across all folds with lowest performance on Fold 4 with 0.679 $DSC_{agg,mean}$ and highest performance in Fold 3 with 0.736 $DSC_{agg,mean}$. Specifically, the lowest difference between the methods was with Fold 2 with 2.2% and largest difference with Fold 4 with 8.0% improved performance with DualUNet. The full comparison is shown in Table 1.

When considering only the primary gross tumor volume segmentation on training dataset, the DualUNet outperformed the UNet across all folds with an average increase of 5.4% in $DSC_{agg,GTVp}$ on the full dataset. In terms of the gross tumor volume of nodal disease, DualUNet outperformed UNet in four out of five folds with the average increase of 4.7% in $DSC_{agg,GTVn}$ using the full dataset. Specifically, the largest difference was with Fold 4 in which DualUNet had a value of 0.818 $DSC_{agg,GTVn}$ while UNet had a value of 0.758 $DSC_{agg,GTVn}$. The lowest difference was with Fold 2, where DualUNet had a value of 0.805 $DSC_{agg,GTVn}$ while UNet had a value of 0.812 $DSC_{agg,GTVn}$.

4   Discussion

In this study, we proposed and evaluated a dual-stage 3D UNet architecture to detect and segment primary gross tumor volumes and metastatic lymph nodes in the head and neck area using MRI scans. Our approach employs a cascaded deep neural network with a dual-stage UNet architecture, where the first-stage produces binary segmentations, and the second-stage refines the segmentation and classifies the regions into multiclass segmentations. The performance of the DualUnet method was assessed in the HNTS-MRG 2024 challenge pre-RT segmentation task, yielding a $DSC_{agg,mean}$ of 0.757 with 5-fold cross-validation, a 4.9% improvement in $DSC_{agg,mean}$ over the standard UNet, and a $DSC_{agg,mean}$ of 0.737 on the test set. This study underscores the advantages of cascaded dual-stage deep neural networks for tasks where inference time is less critical. Compared to UNet, DualUNet employs two separate UNets in a serial manner, approximately doubling the inference time.

Interestingly, segmentation performance was generally higher for GTVn than for GTVp. This is in contrast with previous studies using deep learning to segment these tissue types from PET-CT scans, where the performance on GTVp has been generally better [9]. Additional analysis is required to determine whether this is due to difference in scanning modalities, deep learning methods, dataset characteristics, or other factors.

A key strength of our approach is its ability to manage complex segmentation tasks by decomposing them into binary and multiclass subtasks. This separation enables more targeted refinement of the segmentation process and provides clearer insights into the performance of each stage. The performance of the approach on the test set underscores its effectiveness in addressing the variability in segmentation quality typically observed in clinical settings.

Variability in image dimensions and resolution presents two strategies i.e., scaling images to a selected standard resolution such as $1\,mm^3$ [27] or maintaining the native resolution that may need to be accounted for. In this study, we retained the native resolution and employed volume resizing augmentation to enable the trained model to accommodate a range of resolutions identified in the training dataset. In this approach, the proposed resampling augmentation was observed to have a significant impact on model performance. However, comparing these two strategies falls outside the scope of this work and is left for future investigation. As a limitation, the variable resolution strategy may have low performance with rare resolutions that were underrepresented during training. In order to ensure good performance across all resolutions, it is essential to use sufficient resolution augmentation during training and include rare resolutions in the model validation.

Based on our initial tests, the inclusion of mid-RT scans in addition to pre-RT scans in the training data improved the results. Although, the mid-RT scans have distinct properties, we hypothesize that the additional data outweighs the negative effects. However, this analysis is left for future work.

Overall, our dual-stage UNet approach represents a considerable advancement in automated HNC segmentation and may provide a promising tool for clinical practice pending clinical validation. The robust performance of the method and its adaptability to the complexities of clinical data highlight the potential for improving tumor delineation and treatment planning in radiotherapy.

5   Conclusion

In summary, we have demonstrated that our novel dual-stage cascading 3D UNet approach for HNC segmentation results in notable improvements in the segmentation accuracy and explainability in contrast to single stage approach. These findings underscore the potential of this approach to improve tumor delineation and refine treatment planning in radiotherapy.

Acknowledgments. The project was partly supported by Business Finland under project "Medical AI and Immersion" (decision number 10912/31/2022) and Research Council of Finland under Project 345449 (eXplainable AI Technologies for Segmenting 3D Imaging Data). The funders had no role in study design, data collection and analysis, decision to publish, or preparation of the manuscript.

Disclosure of Interests. The authors have no competing interests.

Code Availability. The training and inference code, along with the trained models and Docker configurations utilized by the FinoxyAI team during the competition, are available at **https://version.aalto.fi/gitlab/saukkom3/hnts-mrg2024**.

References

1. Thiagarajan, A., et al.: Target volume delineation in oropharyngeal cancer: impact of PET, MRI, and physical examination. Int. J. Radiat. Oncol. Biol. Phys. **83**(1), 220–227 (2012)
2. Cooper, J.S., et al.: An evaluation of the variability of tumor-shape definition derived by experienced observers from CT images of supraglottic carcinomas (ACRIN protocol 6658). Int. J. Radiat. Oncol. Biol. Phys. **67**(4), 972–975 (2007)
3. Pollard, J.M., Wen, Z., Sadagopan, R., Wang, J., Ibbott, G.S.: The future of image-guided radiotherapy will be MR guided. Br. J. Radiol. **90**(1073), 20160667 (2017)

4. Junn, J.C., Soderlund, K.A., Glastonbury, C.M.: Imaging of head and neck cancer with CT, MRI, and US. Semin. Nucl. Med. **51**(1), 3–12 (2021)
5. Do, H.M., et al.: Augmented radiologist workflow improves report value and saves time: a potential model for implementation of artificial intelligence. Acad. Radiol. **27**(1), 96–105 (2020)
6. Queiroz, M.A., et al.: PET/MRI and PET/CT in follow-up of head and neck cancer patients. Eur. J. Nucl. Med. Mol. Imaging **41**, 1066–1075 (2014)
7. Andrearczyk, V., et al.: Overview of the HECKTOR challenge at MICCAI 2020: automatic head and neck tumor segmentation in PET/CT. In: Head and Neck Tumor Segmentation: First Challenge, HECKTOR 2020, Held in Conjunction with MICCAI 2020, Lima, Peru, 4 October 2020, Proceedings 1, pp. 1–21. Springer (2021)
8. Andrearczyk, V., et al.: Overview of the HECKTOR challenge at MICCAI 2021: automatic head and neck tumor segmentation and outcome prediction in PET/CT images. In: 3D Head and Neck Tumor Segmentation in PET/CT Challenge, pp. 1–37. Springer (2021)
9. Andrearczyk, V., Oreiller, V., Hatt, M., Depeursinge, A.: Head and Neck Tumor Segmentation and Outcome Prediction: Third Challenge, HECKTOR 2022, Held in Conjunction with MICCAI 2022, Singapore, September 22, 2022, Proceedings, vol. 13626. Springer (2023)
10. Wahid, K.A., et al.: Evaluation of deep learning-based multiparametric MRI oropharyngeal primary tumor auto-segmentation and investigation of input channel effects: results from a prospective imaging registry. Clin. Transl. Radiat. Oncol. **32**, 6–14 (2022)
11. Schouten, J.P., et al.: Automatic segmentation of head and neck primary tumors on MRI using a multi-view CNN. Cancer Imaging **22**(1), 8 (2022)
12. Liedes, J., et al.: Automatic segmentation of head and neck cancer from PET-MRI data using deep learning. J. Med. Biol. Eng. **43**(5), 532–540 (2023)
13. Chenthara, S., Ahmed, K., Wang, H., Whittaker, F.: Security and privacy-preserving challenges of e-health solutions in cloud computing. IEEE Access **7**, 74361–74382 (2019)
14. Cai, Z., Vasconcelos, N.: Cascade R-CNN: delving into high quality object detection. In: Proceedings of the IEEE Conference on Computer Vision and Pattern Recognition (CVPR) (2018)
15. Marcinkiewicz, M., Nalepa, J., Lorenzo, P.R., Dudzik, W., Mrukwa, G.: Segmenting brain tumors from MRI using cascaded multi-modal U-nets. In: Crimi, A., Bakas, S., Kuijf, H., Keyvan, F., Reyes, M., van Walsum, T. (eds.) Brainlesion: Glioma, Multiple Sclerosis, Stroke and Traumatic Brain Injuries, pp. 13–24. Springer, Cham (2019)
16. Lyu, C., Shu, H.: A two-stage cascade model with variational autoencoders and attention gates for MRI brain tumor segmentation. In: Crimi, A., Bakas, S. (eds.) Brainlesion: Glioma, Multiple Sclerosis, Stroke and Traumatic Brain Injuries, pp. 435–447. Springer, Cham (2021)
17. Korte, J.C., Hardcastle, N., Ng, S.P., Clark, B., Kron, T., Jackson, P.: Cascaded deep learning-based auto-segmentation for head and neck cancer patients: organs at risk on T2-weighted magnetic resonance imaging. Med. Phys. **48**(12), 7757–7772 (2021)
18. Yu, J., et al.: Cascade path augmentation unet for bladder cancer segmentation in MRI. Med. Phys. **49**(7), 4622–4631 (2022)

19. Li, X., Luo, G., Wang, K.: Multi-step cascaded networks for brain tumor segmentation. In: Crimi, A., Bakas, S. (eds.) Brainlesion: Glioma, Multiple Sclerosis, Stroke and Traumatic Brain Injuries, pp. 163–173. Springer, Cham (2020)

20. Sobhaninia, Z., Rezaei, S., Karimi, N., Emami, A., Samavi, S.: Brain tumor segmentation by cascaded deep neural networks using multiple image scales. In: 2020 28th Iranian Conference on Electrical Engineering (ICEE), pp. 1–4 (2020). https://doi.org/10.1109/ICEE50131.2020.9260876

21. Wu, L., Xin, Y., Li, S., Wang, T., Heng, P.A., Ni, D.: Cascaded fully convolutional networks for automatic prenatal ultrasound image segmentation. In: 2017 IEEE 14th International Symposium on Biomedical Imaging (ISBI 2017), pp. 663–666 (2017). https://doi.org/10.1109/ISBI.2017.7950607

22. Warfield, S.K., Zou, K.H., Wells, W.M.: Simultaneous truth and performance level estimation (STAPLE): an algorithm for the validation of image segmentation. IEEE Trans. Med. Imaging 23(7), 903–921 (2004)

23. Kerfoot, E., Clough, J., Oksuz, I., Lee, J., King, A.P., Schnabel, J.A.: Left-ventricle quantification using residual U-Net. In: Statistical Atlases and Computational Models of the Heart. Atrial Segmentation and LV Quantification Challenges: 9th International Workshop, STACOM 2018, Held in Conjunction with MICCAI 2018, Granada, Spain, 16 September 2018, Revised Selected Papers 9, pp. 371–380. Springer (2019)

24. He, K., Zhang, X., Ren, S., Sun, J.: Delving deep into rectifiers: surpassing human-level performance on imagenet classification. In: Proceedings of the IEEE International Conference on Computer Vision, pp. 1026–1034 (2015)

25. Cardoso, M.J., et al.: MONAI: an open-source framework for deep learning in healthcare (2022). https://arxiv.org/abs/2211.02701

26. Zeineldin, R.A., Karar, M.E., Burgert, O., Mathis-Ullrich, F.: Multimodal CNN networks for brain tumor segmentation in MRI: a BraTS 2022 challenge solution. In: International MICCAI Brainlesion Workshop, pp. 127–137. Springer (2022)

27. Myronenko, A., Siddiquee, M.M.R., Yang, D., He, Y., Xu, D.: Automated head and neck tumor segmentation from 3D PET/CT HECKTOR 2022 challenge report. In: 3D Head and Neck Tumor Segmentation in PET/CT Challenge, pp. 31–37. Springer (2022)

28. Saukkoriipi, M., et al.: Interactive 3D Segmentation for Primary Gross Tumor Volume in Oropharyngeal Cancer (2024). https://arxiv.org/abs/2409.06605

29. Asgari Taghanaki, S., Abhishek, K., Cohen, J.P., Cohen-Adad, J., Hamarneh, G.: Deep semantic segmentation of natural and medical images: a review. Artif. Intell. Rev. 54, 137–178 (2021)

Benchmark of Deep Encoder-Decoder Architectures for Head and Neck Tumor Segmentation in Magnetic Resonance Images: Contribution to the HNTSMRG Challenge

Marek Wodzinski[1,2]([✉]) [iD]

[1] Department of Measurement and Electronics, AGH University of Krakow, Krakow, Poland
wodzinski@agh.edu.pl
[2] Information Systems Institute, University of Applied Sciences Western Switzerland (HES-SO Valais-Wallis), Sierre, Switzerland

Abstract. Radiation therapy is one of the most frequently applied cancer treatments worldwide, especially in the context of head and neck cancer. Today, MRI-guided radiation therapy planning is becoming increasingly popular due to good soft tissue contrast, lack of radiation dose delivered to the patient, and the capability of performing functional imaging. However, MRI-guided radiation therapy requires segmenting of the cancer both before and during radiation therapy. So far, the segmentation was often performed manually by experienced radiologists, however, recent advances in deep learning-based segmentation suggest that it may be possible to perform the segmentation automatically. Nevertheless, the task is arguably more difficult when using MRI compared to e.g. PET-CT because even manual segmentation of head and neck cancer in MRI volumes is challenging and time-consuming. The importance of the problem motivated the researchers to organize the HNTSMRG challenge with the aim of developing the most accurate segmentation methods, both before and during MRI-guided radiation therapy. In this work, we benchmark several different state-of-the-art segmentation architectures to verify whether the recent advances in deep encoder-decoder architectures are impactful for low data regimes and low-contrast tasks like segmenting head and neck cancer in magnetic resonance images. We show that for such cases the traditional residual UNet-based method outperforms (DSC = 0.775/0.701) recent advances such as UNETR (DSC = 0.617/0.657), SwinUNETR (DSC = 0.757/0.700), or SegMamba (DSC = 0.708/0.683). The proposed method (lWM team) achieved a mean aggregated Dice score on the closed test set at the level of 0.771 and 0.707 for the pre- and mid-therapy segmentation tasks, scoring 14th and 6th place, respectively. The results suggest that proper data preparation, objective function, and preprocessing are more influential for the segmentation of head and neck cancer than deep network architecture.

K. A. Wahid et al. (Eds.): HNTS-MRG 2024, LNCS 15273, pp. 204–213, 2025.
https://doi.org/10.1007/978-3-031-83274-1_15

Keywords: Head and Neck Cancer · Deep Learning · Image Segmentation · HNTSMRG · Challenge · Benchmark · MRI

1   Introduction

Radiation therapy is one of the most frequently applied cancer treatments worldwide, especially in the context of head and neck (HN) cancer. Today, radiation therapy (RT) planning using magnetic resonance images (MRI) is becoming increasingly popular, thanks to good soft tissue contrast, lack of radiation dose delivered to the patient, and the availability of functional imaging [15].

However, RT planning requires segmenting the cancer before the therapy (preRT) with potential adaptations requiring updates to the segmentation mask during the therapy (midRT). Manual segmentation is time-consuming and varies strongly between observers, depending on their experience [18]. The task is especially difficult for MR volumes compared to e.g. PET-CT volumes providing immediate tumor metabolic information that usually makes the segmentation easier. The importance of HN tumor segmentation motivated researchers to organize several scientific challenges, e.g. HECKTOR dedicated to the tumor segmentation in PET-CT volumes [1], HaN-Seg combining both CT and MRI modalities in the context of segmenting the organs-at-risk [16,17], SegRap dedicated to segmenting both gross tumor volume and organs-at-risk for radiotherapy planning [13], and now the HNTSMRG challenge dedicated to perform the segmentation using only T2w MR volumes.

The progress in the automatic segmentation of medical volumes is tremendous. Probably the most recognized segmentation framework is nnUNet [11], successfully applied to numerous scientific challenges [5,6]. Nevertheless, research in this domain persists, yielding an increasing number of segmentation architectures and frameworks. Today, the novel contributions are mainly based on transformer architecture, such as UNETR [9], SwinUNETR [8], SegMamba [21], or foundation models like MedSAM [14]. The transformer architecture avoids the inductive bias present in convolutional networks and enables the network to model long-range relations [12], however, at the cost of larger datasets required for training.

All challenges related to HN cancer segmentation (and the majority of segmentation tasks in medical imaging) suffer from a common limitation associated with the amount of available and annotated data [6]. The number of annotated cases for HN cancer is relatively large when compared to other medical datasets (e.g., HECKTOR - 524 training cases, HNTSMRG - 150 training cases, Top-Cow - 90 training cases [22], SEG.A - 56 training cases, SPPIN - 34 training cases), however it is extremely small when compared to the datasets available in computer vision, often exceeding millions of cases.

An open question arises: Are all the advances in medical image segmentation architectures useful in low data regime tasks like the HN cancer segmentation? Do transformer-based networks improve over the baseline based on UNet? We attempt to answer the question by proposing a contribution to the HNTSMRG challenge. The goal of the HNTSMRG challenge is to propose automatic algorithms that correctly segment the primary gross tumor volumes (GTVp) and the metastatic lymph nodes (GTVn). The challenge is divided into two subtasks. The first one (preRT) is dedicated to segmenting the GTVp and GTVn in T2-weighted volumes acquired before the treatment. The second one (midRT) aims to achieve the same goal, however, during the RT, resulting in a considerably more difficult challenge.

Contribution: In this work, we benchmark several encoder-decoder architectures dedicated to the segmentation of 3-D medical volumes. We compare traditional Resiudal UNet (RUNET) to more recent architectures: UNETR, Swin-UNETR, and SegMamba. We show that the use of more recent and advanced transformer-based architectures is not beneficial for the HN tumor segmentation. The proposed segmentation architecture achieved a considerably good score, however, not at the level of the best-performing submissions based on the nnUNet framework and its pre- and post-processing capabilities. Importantly, the goal of the contribution is not to propose any novel segmentation method, but to verify the influence of the currently existing building blocks.

2   Methods

2.1   Dataset

The training dataset consisted of 150 T2-weighted (T2w) sequences of the head and neck region acquired at the MD Anderson Cancer Center. For the preRT task, only the MR volume used for therapy planning was available, while for the midRT task, it was possible to use both the midRT volume and the preRT data registered to the midRT MR volume.

The test set consisted of 50 T2w sequences, sharing similar properties to the training test set. However, the test cases were not released to the participants and the evaluation was performed using the Grand-Challenge (GC) platform.

The ground-truth represented the primary gross tumor volumes (GTVp) and metastatic lymph nodes (GTVn). All the GTVp and GTVn were independently segmented by multiple experts (from 3 to 4 for each volume) and then combined. All observers were medical doctors with at least 2 years of experience in HN tumor segmentation. Finally, the quality of the segmentation was verified by experienced radiologists with more than 10 years of experience and the segmentations were combined using the STAPLE algorithm [19] to provide a single segmentation per patient.

2.2   Segmentation Overview

The proposed method consisted of (i) preprocessing, (ii) splitting the volume into overlapping patches, (iii) inference using encoder-decoder architecture, and (iv) aggregation of results. The pipeline is presented in Fig. 1.

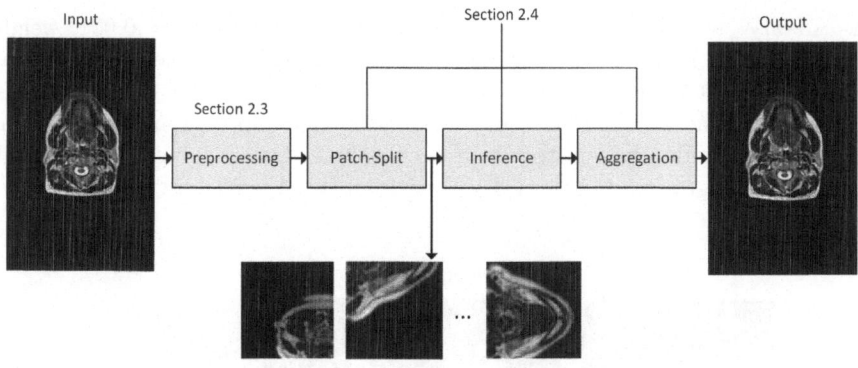

Fig. 1. Visualization of the processing pipeline.

2.3   Preprocessing

The preprocessing was similar for both the preRT and midRT subtasks. It started with resampling all the volumes to isotropic spacing equal to 0.5 mm in each dimension. Then, the pre- and mid-therapy volumes were normalized using the Z-normalization, separately to both volumes.

2.4   Inference

The proposed method was patch-based, which means that the input volume was divided into a given number of overlapping patches, the inference was performed separately for each patch, and then the results were aggregated. The patch-based approach was used for two reasons: (i) the limited amount of VRAM that made it impossible to process the volumes directly, (ii) it can be considered as a natural form of augmentation enforcing the network to be able to be more resistant to false positives.

The patch size for all experiments was equal to 128^3 voxels, allowing the network to use a relatively large region of interest while maintaining computational efficiency. The overlap between patches was set to 32^3 voxels. The output for each class was calculated by applying the argmax operator to the channel dimension after aggregating the activations from each patch. Due to the heterogeneity of the segmentation masks and the evaluation metric based on the aggregated Dice score, no additional post-processing was applied.

During internal evaluation, only a single model was used for each fold. During the final evaluation using the GC platform, activation maps were aggregated from models trained using each training fold.

2.5   Training

The training was implemented using the Lightning [7] and MONAI [3] libraries. The objective function was a weighted combination of SoftDice and Focal losses.

The AdamW was used as the optimizer (initial learning rate = 0.001, weight decay = 0.005), and the learning rate was automatically reduced on a plateau by a factor of 0.9 if the validation loss had not improved for more than 10 epochs. The training was accelerated using automatic mixed precision. The training patches were randomly selected from each volume by cropping to each available class (background, GTVn, GTVp) with an equal probability.

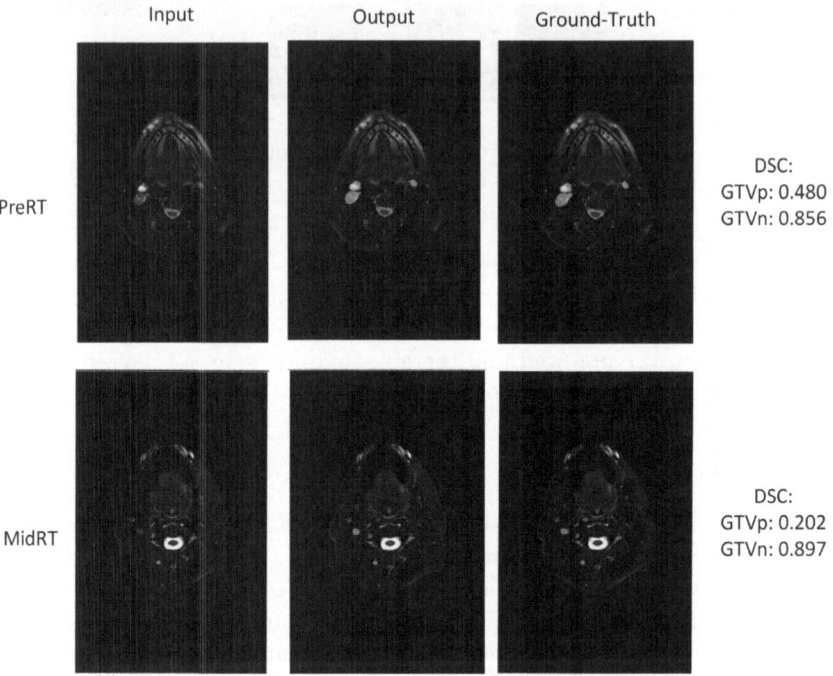

Fig. 2. Exemplary visualization of the results for the preRT and midRT segmentation tasks (GTVp in green, GTVn in yellow). (Color figure online)

2.6 Experimental Setup

Two separate networks were trained for Task 1 and Task 2: (i) a single-channel preRT network with the preRT volume as input for Task 1 and (ii) a three-channel midRT network using the concatenated midRT volume, the registered preRT volume, and the associated registered preRT ground truth, for Task 2. All experiments were trained until convergence. The MR volumes (both midRT and preRT) were augmented by random axis flipping, Gaussian and Rician noise, and Gaussian smoothing. The batch size was set to 16, resulting in batches containing 4 random patches from 4 randomly selected volumes. Network architectures, augmentation methods, objective functions, and training utilities were taken from the MONAI library [3]. All experiments were performed using 5-fold cross-validation where each fold consisted of 120 training and 30 validation cases. Any claim about a statistical improvement was supported by a Wilcoxon signed-rank test with a p-value below 0.05.

The training was performed using the PLGRID Helios supercomputing infrastructure with nodes containing $4\times$ NVIDIA GH200 accelerators ($4\times96\,$GB VRAM) with automatic mixed-precision enabled. All volumes were initially transferred to the RAM, allowing one to perform fast and efficient training. The training and inference scripts are available in the associated repository [20].

3   Results

We follow the convention of the challenge organizers and use the aggregated Dice score (DSCAgg) to compare the architectures. The results using an internal 5-fold cross-validation are presented in Table 1. Based on the internal evaluation, we decided to use the RUNet as the final architecture and evaluated it on the closed test set. The final results for both the preRT and midRT subtasks are presented in Table 2. The results of our experiments performed well in comparison to prior contributions (DSCAgg at the best-performing submission level in the HECKTOR Challenge dedicated to HN cancer segmentation in PET-CT [1]), however, considerably worse than other methods based on nnUNet [10]. Figure 2 presents exemplary segmentation results with strong performance.

Table 1. Quantitative results for the internal evaluation using 5-fold cross-validation split based on aggregating results from 5 folds in terms of the aggregated Dice score (DSCAgg)

Model	GTVn ↑	GTVp ↑	Mean GTV ↑
Before Radiation Therapy (preRT)			
RUNet	**0.809**	**0.741**	**0.775**
UNETR	0.642	0.592	0.617
SwinUNETR	0.789	0.724	0.757
SegMamba	0.743	0.672	0.708
During Radiation Therapy (midRT)			
RUNet	**0.787**	0.614	**0.701**
UNETR	0.751	0.562	0.657
SwinUNETR	0.785	**0.616**	0.700
SegMamba	0.772	0.594	0.683

Table 2. Quantitative results on the closed test set using the Grand-Challenge evaluation platform in terms of the aggregated Dice score (DSCAgg) for both the subtasks (before and during the radiation therapy).

Task	GTVn ↑	GTVp ↑	Mean GTV ↑
preRT	0.826	0.717	0.771
midRT	0.836	0.579	0.707

4 Discussion

The results confirm that the complexity of network architecture is not beneficial for the HN tumor segmentation, at least in low data regimes. The inductive bias in CNNs may be beneficial for tasks suffering from low amounts of data. The traditional RUNet outperforms more recent architectures like UNETR, Swin-UNETR, or SegMamba. The most probable reason behind such an observation is connected with a relatively low amount of training data, limiting the expressiveness of transformer-based architectures. Therefore, in the final submission, we decided to use the RUNet, especially when it turned out to be impossible to combine RUNet and SwinUNETR models for external validation without reducing the number of parameters or volume resolution, due to the limitations of the GC hardware (T4 GPU with 16 GB of available VRAM). The models could not be loaded and process the volume simultaneously, and processing them independently with late aggregation would increase the inference time beyond the allowed limits.

Interestingly, the differences between the RUNet and transformer-based architectures are lower for midRT tasks. In terms of the GTVp, the SwinUNETR acquired even slightly better results (however, without statistically significant differences, p-value ¿ 0.05). The reason behind that is probably connected with the fact that we decided to use the three-channel input for the midRT tasks (the midRT volume, the registered preRT volume, and the registered preRT ground truth). By design, this helps the network to perform the initial detection and helps with the training convergence. Nevertheless, such an approach has also disadvantages. The network is unable to segment tumors that are not available (e.g. metastases) or were incorrectly segmented before the radiation therapy. However, such cases are outside the scope of the challenge. Due to the metrics used for the evaluation, we decided to use the three-channel input instead of using only the midRT volume. The decision is connected with the DSCAgg used for evaluation. The metric calculates the volumetric overlap using the entire test set. As a result, accurate segmentation of large segmentation masks is preferred over sensitivity to small lesions that barely impact the final metric. The correct detection of small changes in midRT volume is less influential than the accurate segmentation of the large tumors already visible in the associated preRT volume. For such cases, it would be more beneficial to use the average or median of individual Dice scores instead.

Importantly, we used only data provided by the challenge organizers. Probably external data used for pretraining the transformer-based architectures using self-supervised techniques (e.g. masked autoencoding) could improve the results. However, it would require to either use large-scale MRI datasets or to use the already existing foundational models dedicated to MR volumes [4]. The self-supervised pretraining to each task separately would be suboptimal, it would require performing large-scale and computationally expensive initialization for each task separately. Therefore, a universal foundation model is necessary to make the approach useful. Current advances in deep learning suggest that this may be the most influential research direction to improve segmentation results [2].

To conclude, the paper presents a benchmark of several encoder-decoder architectures and suggests that in low data regimes, the more recent transformer-based architectures are not improving the segmentation results. Probably large-scale self-supervised pretraining could change the outcomes, however, to make it useful and scalable, such an approach would require proposing a universal foundation model dedicated to MR volumes. Therefore, the next research step would be related to developing or fine-tuning a large-scale foundation model dedicated to medical image segmentation.

Acknowledgements. We gratefully acknowledge Polish HPC infrastructure PLGrid support within computational grant no. PLG/2024/017079. The research was partially supported by the program "Excellence Initiative - Research University" for AGH University.

Disclosure of Interests. The authors have no competing interests to declare that are relevant to the content of this article.

References

1. Andrearczyk, V., et al.: Automatic head and neck tumor segmentation and outcome prediction relying on FDG-PET/CT images: findings from the second edition of the HECKTOR challenge. Med. Image Anal. **90**, 102972 (2023)
2. Azad, B., et al.: Foundational models in medical imaging: a comprehensive survey and future vision. arXiv preprint arXiv:2310.18689 (2023)
3. Consortium, M.: MONAI: medical open network for AI (2024). https://doi.org/10.5281/zenodo.12542217
4. Cox, J., et al.: BrainSegFounder: towards 3D foundation models for neuroimage segmentation. Med. Image Anal. **97**, 103301 (2024)
5. Eisenmann, M., et al.: Why is the winner the best? In: Proceedings of the IEEE/CVF Conference on Computer Vision and Pattern Recognition, pp. 19955–19966 (2023)
6. Eisenmann, M., et al.: Biomedical image analysis competitions: the state of current participation practice. arXiv preprint arXiv:2212.08568 (2022)
7. Falcon, W.: The PyTorch Lightning team: PyTorch Lightning (2019)

8. Hatamizadeh, A., Nath, V., Tang, Y., Yang, D., Roth, H.R., Xu, D.: Swin unetr: swin transformers for semantic segmentation of brain tumors in mri images. In: International MICCAI Brainlesion Workshop, pp. 272–284. Springer, Heidelberg (2021). https://doi.org/10.1007/978-3-031-08999-2_22

9. Hatamizadeh, A., et al.: Unetr: transformers for 3d medical image segmentation. In: Proceedings of the IEEE/CVF Winter Conference on Applications of Computer Vision, pp. 574–584 (2022)

10. HNTSMRG-Organizers: HNTSMRG Leaderboard (2024). https://hntsmrg24.grand-challenge.org/evaluation/final-test-task-2-mid-rt-segmentation/leaderboard/

11. Isensee, F., Jaeger, P.F., Kohl, S.A., Petersen, J., Maier-Hein, K.H.: nnU-Net: a self-configuring method for deep learning-based biomedical image segmentation. Nat. Methods **18**(2), 203–211 (2021)

12. Khan, S., Naseer, M., Hayat, M., Zamir, S.W., Khan, F.S., Shah, M.: Transformers in vision: a survey. ACM Comput. Surv. (CSUR) **54**(10s), 1–41 (2022)

13. Luo, X., et al.: Segrap2023: a benchmark of organs-at-risk and gross tumor volume segmentation for radiotherapy planning of nasopharyngeal carcinoma. arXiv preprint arXiv:2312.09576 (2023)

14. Ma, J., He, Y., Li, F., Han, L., You, C., Wang, B.: Segment anything in medical images. Nat. Commun. **15**(1), 654 (2024)

15. McDonald, B.A., Dal Bello, R., Fuller, C.D., Balermpas, P.: The use of MR-guided radiation therapy for head and neck cancer and recommended reporting guidance. In: Seminars in Radiation Oncology, vol. 34, pp. 69–83. Elsevier (2024)

16. Podobnik, G., et al.: HaN-Seg: the head and neck organ-at-risk CT and MR segmentation challenge. Radiother. Oncol. **198**, 110410 (2024)

17. Podobnik, G., Strojan, P., Peterlin, P., Ibragimov, B., Vrtovec, T.: HaN-Seg: the head and neck organ-at-risk CT and MR segmentation dataset. Med. Phys. **50**(3), 1917–1927 (2023)

18. Preim, B., Botha, C.P.: Visual computing for medicine: theory, algorithms, and applications. Newnes (2013)

19. Warfield, S.K., Zou, K.H., Wells, W.M.: Simultaneous truth and performance level estimation (STAPLE): an algorithm for the validation of image segmentation. IEEE Trans. Med. Imaging **23**(7), 903–921 (2004)

20. Wodzinski, M.: HNTSMRG Code Repository. https://github.com/MWod/HNTSMRG_2024 (2024)

21. Xing, Z., Ye, T., Yang, Y., Liu, G., Zhu, L.: Segmamba: long-range sequential modeling mamba for 3d medical image segmentation. arXiv preprint arXiv:2401.13560 (2024)

22. Yang, K., et al.: Benchmarking the cow with the topcow challenge: topology-aware anatomical segmentation of the circle of willis for cta and mra. ArXiv (2023)

Ensemble of LinkNet Networks for Head and Neck Tumor Segmentation

Maria Baldeon-Calisto[✉]

Departamento de Ingeniería Industrial y Instituto de Innovación en Productividad y Logística CATENA-USFQ, Universidad San Francisco de Quito USFQ, Quito, Ecuador
mbaldeonc@usfq.edu.ec

Abstract. The segmentation of head and neck cancer (HNC) tumors is a critical step in radiotherapy treatment planning. The development of automatic segmentation algorithms has the potential to streamline the radiation oncology process. In this work, we develop an ensemble of LinkNet networks for HNC tumor segmentation as part of the HNTS-MRG 2024 Grand Challenge. A single LinkNet network, pretrained on the Imagenet dataset, was trained for 200 epochs on the HNC dataset provided by the challenge. Eight good performing weights from the internal validation set were selected to create an ensemble of 2D networks. Specifically, each selected weight was used to generate a LinkNet architecture, resulting in eight networks whose predictions were averaged to produce the final predicted segmentation. Our experiments demonstrate that the ensemble network performs better than each individual architecture, leveraging the benefits of ensemble learning without the computational cost of training each network from scratch. In the challenge's test set, the LinkNet Ensemble (team ECU) achieved an aggregated Dice score of 64.60% and 49.53% for metastatic lymph nodes and primary gross tumor segmentation, respectively, and a mean score of 57.06%.

Keywords: Head and Neck Cancer · Tumor Segmentation · Convolutional Neural Networks · Deep Learning

1 Introduction

Head and neck cancers (HNC) comprise a diverse group of malignancies that affect the upper aerodigestive tract. Among these, squamous cell carcinoma is the most prevalent type [1], with most cases occurring in the oral cavity, oropharynx, and larynx. Globally, HNC accounts for approximately 5% of all cancers, with an annual incidence of 600,000 new cases and 300,000 related deaths [2]. Major risk factors include alcohol consumption, infections by human papilloma virus and Epstein-Barr virus, and tobacco use. Despite advances in treatment, the prognosis for patients remains poor, with a 5-year survival rate of approximately 45.7% [3]. Existing treatment modalities, such as surgery, radiotherapy, and targeted therapies, face significant challenges, including late diagnosis, metastasis,

K. A. Wahid et al. (Eds.): HNTS-MRG 2024, LNCS 15273, pp. 214–221, 2025.
https://doi.org/10.1007/978-3-031-83274-1_16

and therapy resistance. Moreover, most therapies demonstrate higher efficacy in early-stage disease.

Accurate segmentation of medical images is critical in radiotherapy treatment planning, as it allows for precise delineation of tumors and surrounding healthy tissues. This is essential for delivering targeted radiation doses while minimizing exposure to nearby organs. However, manual segmentation, which requires radiation oncologists to manually outline regions of interest on imaging scans, is not only time-consuming and labor-intensive but also subject to inter-observer variability. Hence, there is a pressing need for the development of automated segmentation methods, which can significantly reduce time and cost involved while improving consistency.

Recent advances in deep learning have shown promising results in the automatic segmentation of head and neck organs-at-risk (OARs). In [4], a fully convolutional neural network combined with a shape representation model is proposed. The network achieved high levels of accuracy across multiple OARs, surpassing the performance of atlas-based methods. Similarly, Nikolov et al. [5] presented a 3D U-Net architecture that provided segmentation results comparable to expert-level performance in the delineation of various head and neck OARs. Jinzhong et al. [6] proposed a multichannel Gaussian mixture model algorithm that integrates data from CT, PET, and MRI scans for tumor volume delineation. The model also showed to have a strong concordance with physician-defined gross tumor volumes. Despite these advancements, automatic segmentation techniques are not yet sufficiently robust to fully replace physician-drawn volumes, particularly in the delineation of gross tumor volumes.

In this work, we propose an ensemble of LinkNet models for the segmentation of metastatic lymph nodes and primary gross tumor in T2-weighted magnetic resonance images. A LinkNet architecture [7] was trained on a head and neck tumor dataset provided by the HNTS-MRG 2024 Grand Challenge. Based on internal validation, eight well-performing model weights were selected to create an ensemble of LinkNet networks. The final predicted segmentation was generated by averaging the predictions from these eight networks. Our experiments show that the ensemble approach surpasses individual network performance in segmenting metastatic lymph nodes (GTV_n) and primary gross tumors (GTV_p). On the challenge test set, the LinkNet ensemble achieved an aggregated Dice score of 64.60% for GTV_n segmentation and 49.53% for GTV_p segmentation.

2   Methods

2.1   Imaging Data

The dataset used in this work was sourced from the HNTS-MRG 2024 Grand Challenge. It is comprised of 150 training and 50 test cases, all from patients with histologically confirmed HNC who received radiotherapy at the University of Texas MD Anderson Cancer Center. The dataset includes T2-weighted magnetic resonance anatomical sequences of the head and neck region with a mix of fat-suppressed and non-fat suppressed sequences. The images have been annotated by 3 to 4 experts that delineate the GTV_p and GTV_n and the concesus segmentation are made available to the participants.

2.2 Image Preprocessing

The preprocessing pipeline was designed to standardize image resolution and intensity values across the dataset. First, all images were resampled to an isotropic resolution of $0.5 \times 0.5 \times 2$ mm using the B-Spline interpolation method for the images and nearest neighbor interpolation for the ground truth segmentation masks. Next, pixel intensities were normalized by truncating values that fell outside three standard deviations from the mean intensity. Any pixel intensities beyond this threshold were clipped to the respective 3-standard deviation limits. Lastly, intensities were rescaled to a range between 0 and 1 by subtracting the minimum intensity value and dividing by the range.

The dataset, consisting of 150 cases, was split into training and validation sets with an 80-20 ratio. This resulted in 120 cases for training, totaling 8,808 slices, and 30 cases for validation, consisting of 2,186 slices.

2.3 Model Architecture

We employed an ensemble of LinkNet architectures [7], as this model has demonstrated strong performance in various segmentation tasks and offers computational efficiency [8,9]. LinkNet is a fully-convolutional network featuring a down-sampling and up-sampling path, as ilustrated in Fig. 1a). The down-sampling path consists of an initial block followed by 4 encoder blocks, while the up-sampling path comprises 4 decoder blocks and a final block. The initial block applies a convolutional layer with a 7×7 kernel and a stride of 2, followed by spatial max-pooling with a 3×3 kernel and a stride of 2. Convolutional layers have a ReLU activation function, with batch normalization applied between layers. Each encoder block contains two residual blocks, as presented in Fig. 1b). The decoder blocks in the up-sampling path are composed of three convolutional layers, depicted in Fig. 1c). The final block includes an up-sampling full convolutional layer with a 3×3 kernel size, a convolutional layer of the same kernel size, and a final full convolutional layer. The LinkNet model was imported from pytorch's segmentation models library. As LinkNet is a 2D architecture, the input to the model were the 2D slices from the 3D pre-RT images.

2.4 Training Configuration

The LinkNet model was initialized with pre-trained weights from the Imagenet dataset and fine-tuned in the HNC dataset for 200 epochs, stopping once the validation Dice score showed no further improvement. We utilized an Adam optimizer with a learning rate of 2×10^{-3} and a batch size of 20 slices. Moreover, the model was trained with a dice loss function to optimize performance. During training, data augmentation on-the-fly was applied to increase the size and

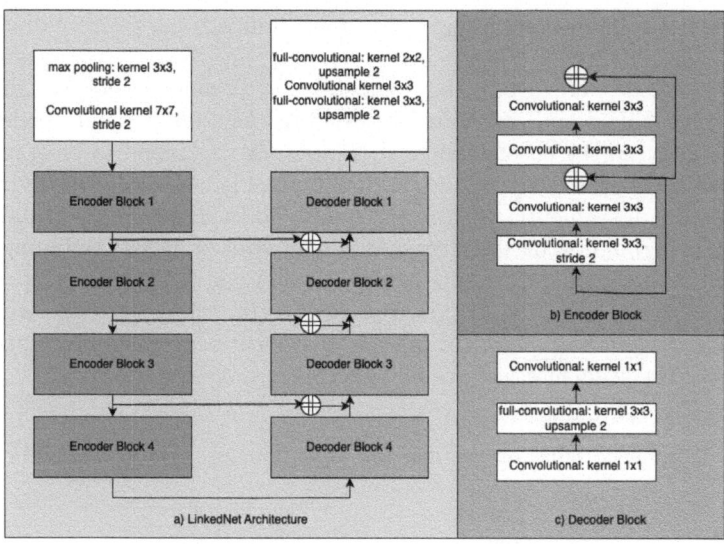

Fig. 1. LinkNet architecture [7] comprised of an up-sampling and down-sampling path.

diversity of the dataset using the albumentations library. Specifically, horizontal and vertical flips were applied with a probability of 0.5, along with rotations and transpositions. The implementation use Python 3.8.18, Pytorch 2.4.0 and executed on a Nvidia Tesla V100 GPU.

To construct the ensemble, we selected eight weights that exhibited both high validation and training Dice scores. To ensure diversity among the selected models, a minimum difference of at least eight epochs between them was enforced. Furthermore, we verified that the models had different performance in the validation cases and hence made distinct mistakes. This approach proved effective, as the ensemble performed better than individual models, demonstrating the benefits of combining multiple networks with complementary strengths.

3   Results and Discussion

The performance of the eight ensemble members, referred to as LinkNet, and the eight-network ensemble, called Ensemble_8, on the internal validation set in terms of mean Dice score, is presented in Table 1. Additionally, we evaluated a ten-network ensemble by selecting 10 weights from the training process, denoted as Ensemble_10. We use the conventional Dice score, that assigns a value of 0 for a false positive prediction, and a dice of 1 for an true negative prediction. As shown, Ensemble_8 and Ensemble_10 exhibit nearly identical performance, both achieving higher segmentation accuracy than any individual ensemble member. Furthermore, the results suggests that increasing the number of networks beyond eight does not substantially improve accuracy, indicating a point of diminishing returns.

Paired t-tests were conducted to assess the significance of the difference between Ensemble_8 and each LinkNet network. At a 95% confidence level, the difference between the ensemble and each network was found to be statistically significant, allowing us to conclude that the ensemble network achieves the highest mean score in segmenting metastatic lymph nodes and primary gross tumors. This finding highlights the advantages of ensemble learning, as combining predictions from multiple models results in a more robust and accurate segmentation, demonstrating the potential of this approach in clinical applications where precise tumor delineation is critical for effective treatment planning. Qualitative examples of the segmentation can be seen in Fig. 2.

Table 1. Ablation Studies on the internal validation set. Mean dice values are presented.

Model	Dice GTV_p	Dice GTV_n
LinkNet_1	0.769	0.825
LinkNet_2	0.789	0.818
LinkNet_3	0.781	0.824
LinkNet_4	0.798	0.807
LinkNet_5	0.791	0.818
LinkNet_6	0.795	0.809
LinkNet_7	0.800	0.812
LinkNet_8	0.794	0.811
Ensemble_8	0.822	0.845
Ensemble_10	0.821	0.846

For external validation, we submited our model to the challenge's website through a docker container. The LinkNet ensemble network achieves a mean aggregated Dice score of 64.60% and 49.53% for GTV_n and GTV_p segmentation, respectively.

The limitations of our work include the use of a 2D network for the segmentation of 3D images, a choice driven by time and computational constraints. Additionally, as the ensemble was created from a single LinkNet network, there is a degree of correlation between the performance of the ensemble members. Incorporating models initialized with different weights or using distinct architectures could potentially improve performance. However, our statistical results demonstrate that integrating the networks through an ensemble does improve segmentation accuracy for both types of tumors. This suggests that our approach could be a viable strategy in situations where computational resources and time are limited. Additionally, while our study focused exclusively on Task 1 of the challenge (pre-RT segmentation), extending our method to mid-RT segmentation

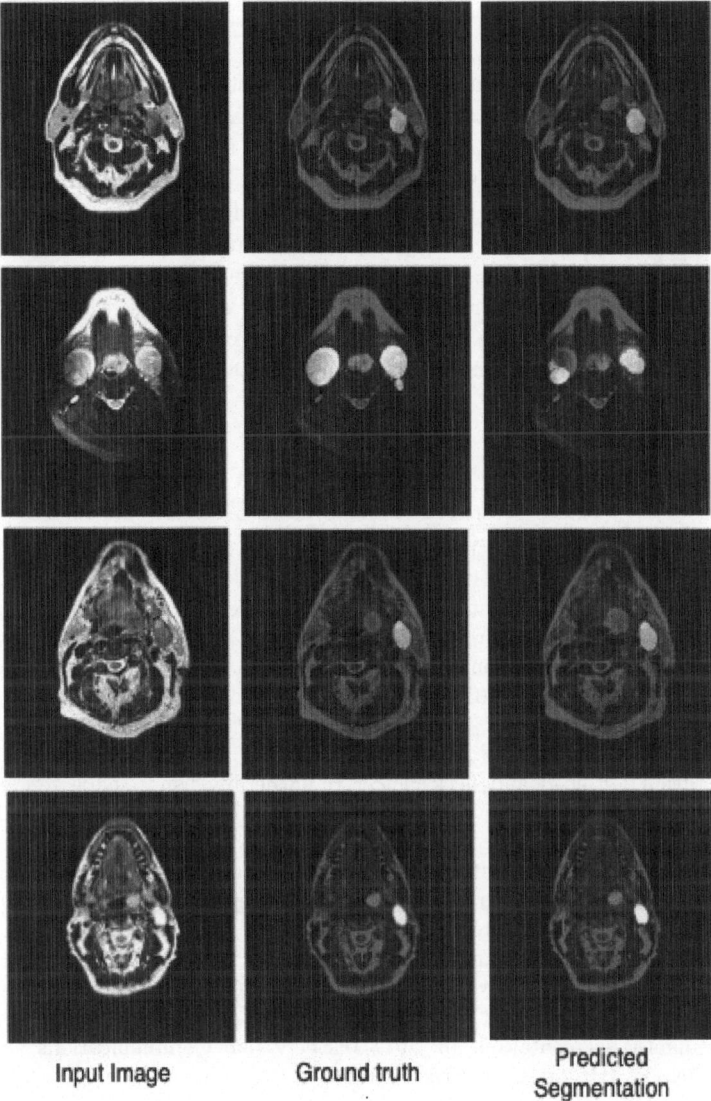

Input Image   Ground truth   Predicted
                                               Segmentation

Fig. 2. Examples of segmentations generated by the ensemble of LinkNet networks. The red region is the primary gross tumor and the yellow region the metastatic lymph nodes. (Color figure online)

prediction (Task 2) could be a valuable direction for future research, potentially demonstrating the model's adaptability to dynamic treatment scenarios.

4   Conclusions

In this work, we developed an ensemble of LinkNet networks for head and neck cancer (HNC) tumor segmentation. By leveraging a single LinkNet network pretrained on the ImageNet dataset, we trained the model for 200 epochs on the HNC dataset provided by the HNTS-MRG 2024 Grand Challenge. From the internal validation set, we selected eight high-performing weights to create an ensemble, resulting in improved segmentation accuracy compared to individual networks. This ensemble approach offers the advantages of ensemble learning without the added computational cost of training each network independently. In the challenge's test set, the LinkNet ensemble achieved aggregated Dice scores of 64.60% and 49.53% for metastatic lymph nodes and primary gross tumor segmentation, respectively.

References

1. Mody, M.D., Rocco, J.W., Yom, S.S., Haddad, R.I., Saba, N.F.: Head and neck cancer. The Lancet **398**(10318), 2289–2299 (2021)
2. Konings, H., et al.: A literature review of the potential diagnostic biomarkers of head and neck neoplasms. Front. Oncol. **10**, 1020 (2020)
3. Gore, M.R.: Survival in sinonasal and middle ear malignancies: a population-based study using the seer 1973–2015 database. BMC Ear Nose Throat Disord. **18**, 1–11 (2018)
4. Tong, N., Gou, S., Yang, S., Ruan, D., Sheng, K.: Fully automatic multi-organ segmentation for head and neck cancer radiotherapy using shape representation model constrained fully convolutional neural networks. Med. Phys. **45**(10), 4558–4567 (2018)
5. Nikolov, S., et al.: Deep learning to achieve clinically applicable segmentation of head and neck anatomy for radiotherapy. arXiv preprint arXiv:1809.04430 (2018)
6. Yang, J., Beadle, B.M., Garden, A.S., Schwartz, D.L., Aristophanous, M.: A multimodality segmentation framework for automatic target delineation in head and neck radiotherapy. Med. Phys. **42**(9), 5310–5320 (2015)
7. Chaurasia, A., Culurciello, E.: Linknet: exploiting encoder representations for efficient semantic segmentation. In: 2017 IEEE Visual Communications and Image Processing (VCIP), pp. 1–4. IEEE (2017)
8. Araújo, R.L., Araújo, F.H.D., Silva, R.R.E.: Automatic segmentation of melanoma skin cancer using transfer learning and fine-tuning. Multimedia Syst. **28**(4), 1239–1250 (2022)
9. Granizo, S., et al.: A comparative analysis of vision transformers and convolutional neural networks in cardiac image segmentation. In: 2024 12th International Symposium on Digital Forensics and Security (ISDFS), pp. 1–7. IEEE (2024)

Enhancing nnUNetv2 Training with Autoencoder Architecture for Improved Medical Image Segmentation

Yichen An[1], Zhimin Wang[1], Eric Ma[1], Hao Jiang[1(✉)], and Weiguo Lu[2]

[1] NeuralRad LLC, Madison, WI, USA
hao.jiang@neuralrad.com
[2] Department of Radiation Oncology, UT Southwestern Medical Center, Dallas, TX, USA

Abstract. Auto-segmentation of gross tumor volumes (GTVs) in head and neck cancer (HNC) using MRI-guided radiotherapy (RT) images presents a significant challenge that can greatly enhance clinical workflows in radiation oncology. In this study, we developed a novel deep learning model based on the nnUNetv2 framework, augmented with an autoencoder architecture. Our model introduces the original training images as an additional input channel and incorporates an MSE loss function to improve segmentation accuracy. The model was trained on a dataset of 150 HNC patients, with a private evaluation of 50 test patients as part of the HNTS-MRG 2024 challenge. The aggregated Dice similarity coefficient (DSCagg) for metastatic lymph nodes (GTVn) reached 0.8516, while the primary tumor (GTVp) scored 0.7318, with an average DSCagg of 0.7917 across both structures. By introducing an autoencoder output channel and combining dice loss with mean squared error (MSE) loss, the enhanced nnUNet architecture effectively learned additional image features to enhance segmentation accuracy. These findings suggest that deep learning models like our modified nnUNetv2 framework can significantly improve auto-segmentation accuracy in MRI-guided RT for HNC, contributing to more precise and efficient clinical workflows.

Keywords: MRI-guided radiotherapy · nnUNetv2 · Autoencoder · Deep learning · Head and neck cancer · Tumor segmentation · Dice similarity coefficient · Medical image segmentation

1 Introduction

Radiation therapy (RT) is a cornerstone in the treatment of head and neck cancer (HNC), and recent advances in MRI-guided RT have significantly improved treatment precision. The superior soft tissue contrast of MRI allows for better tumor delineation compared to conventional CT imaging, leading to enhanced

© The Author(s) 2025
K. A. Wahid et al. (Eds.): HNTS-MRG 2024, LNCS 15273, pp. 222–229, 2025.
https://doi.org/10.1007/978-3-031-83274-1_17

treatment planning [1]. However, manual tumor segmentation is both time-consuming and error-prone due to the complex anatomy of HNC tumors [2]. This challenge has led to the increased use of artificial intelligence (AI) in automatic tumor segmentation. Deep learning, specifically through the nnU-Net framework [3], has shown tremendous potential in addressing the segmentation challenges in RT planning. The HNTS-MRG 2024 challenge, focused on head and neck tumor segmentation, encourages participants to leverage MRI data to develop robust segmentation algorithms for pre- and mid-radiotherapy images. This challenge offers a unique opportunity to evaluate how incorporating multi-time point data can improve segmentation outcomes. In this work, we present a novel approach utilizing a nnUNet-based model with an autoencoder architecture to address these segmentation challenges. By introducing additional input channels, such as training images, and incorporating mean squared error (MSE) loss, we aim to enhance segmentation accuracy. Our model was trained over multiple folds and demonstrated improved performance over the baseline nnUNet model, achieving promising results after 1000 epochs and seven stages of training. This paper details our approach and evaluates its effectiveness in the HNTS-MRG 2024 challenge.

2   Related Works

The nnU-Netv2 framework [3] has become a leading solution for medical image segmentation due to its robust, self-adapting architecture. It automatically configures its hyperparameters based on the dataset, making it highly versatile across different segmentation tasks. nnU-Net's core design includes both 2D and 3D U-Net models, offering flexibility for a range of medical imaging modalities. A key feature is its dynamic patch size adjustment and depth scaling, allowing the architecture to adapt to specific tasks without manual intervention. Previously, our group developed a nnUNet-based platform for automatic delineation of Head & Neck GTV, which ranked No. 3 in the HECKTOR 2022 challenge [4]. Recent developments, such as the **Residual Encoder Presets** [5] introduced in nnUNetv2, have shown improvements in segmentation performance by incorporating deeper encoder stages (up to seven) to extract more detailed feature representations. These presets integrate residual connections, which enhance the model's ability to retain learned information across layers, addressing the vanishing gradient problem. By using **7-stages** in our task, we leverage these architectural improvements to achieve more accurate segmentation results, particularly in multi-channel settings, such as incorporating auto-encoder predictions along with traditional segmentation targets.

This evolution of the nnU-Net framework demonstrates its continuous adaptability and high performance, making it a fitting choice for the HNTS-MRG 2024 challenge. In this challenge, we apply these advances to segment head and neck tumors in MRI images, targeting both pre- and mid-radiotherapy scans to optimize radiation therapy workflows.

3 Methods

The model training was conducted on an Intel i9-14900 CPU and an Nvidia GeForce RTX 4090 GPU with 24 GB memory. The training process spanned 1000 epochs, with each epoch consisting of 250 iterations and 50 validation iterations. Stochastic Gradient Descent (SGD) was employed as the optimizer, initialized with a learning rate of 1e-2, momentum set to 0.99, and a weight decay of 3e-5. Deep supervision was applied during training, along with an oversampling strategy featuring a foreground sampling percentage of 33%. Additionally, five-fold cross-validation was utilized to ensure robustness and generalizability, optimizing the model's ability to handle complex computations and large datasets effectively across all epochs.

The image processing pipeline is built upon nnUNetv2's framework with both standard and custom modifications. The dataset was preprocessed using fingerprint information, including voxel spacing, intensity normalization, and size standardization, ensuring consistency across inputs (in Fig. 1). All images were cropped to their non-zero regions during preprocessing, resampled to isotropic spacing, and normalized to a fixed intensity range. Additionally, a comprehensive voxel connectivity analysis was conducted to identify and remove isolated noise artifacts erroneously labeled as tumor regions. These refinements enhanced the dataset's quality for more accurate downstream analysis.

A critical modification introduced was the inclusion of original input data as an additional input channel for training. Specifically, the original training images parsed from nnUNetv2 were stored and passed into the training process without concatenation. This stored image data was processed separately during autoencoder training to improve segmentation accuracy. Instead of being merged with other channels, the original image was preserved as its own channel and treated as an auxiliary target in the autoencoder structure. For instance, a tensor originally shaped as [3, 67, 512, 512] (representing three segmentation targets) was transformed into [4, 67, 512, 512], where the additional channel corresponds to the original image data. This transformation enables the model to jointly process segmentation labels and input image data for enhanced feature learning. During the data loading phase, the pipeline was extended to allow storage and utilization of the parsed image data as an additional custom input channel. Moreover, the target format was modified to accept 32-bit floating-point (float32) data rather than the default 16-bit integer (int16) format, ensuring compatibility with our extended training pipeline. This adjustment was critical for integrating the mean squared error (MSE) loss function, which requires continuous floating-point data for accurate gradient computation. The enhanced pipeline enabled the model to optimize for both segmentation and image reconstruction in a unified framework.

Specifically, we assert that the predicted output and target shapes match, applying the MSE loss as:

$$\text{MSE} = \frac{1}{N} \sum_{i=1}^{N} (x_i - y_i)^2 \qquad (1)$$

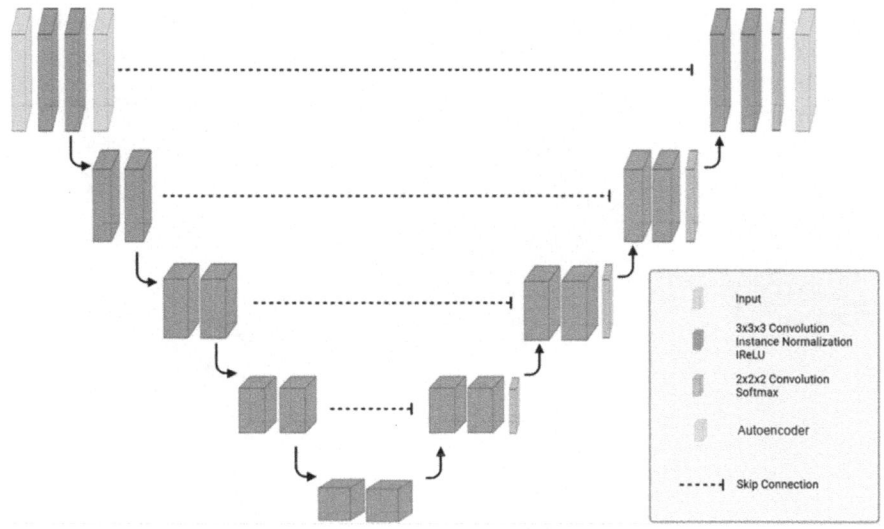

Fig. 1. Autoencoder layer for our nnUNetv2

$$\text{Dice Loss} = 1 - \frac{2 \times \sum(P \times G)}{\sum P + \sum G} \tag{2}$$

To enhance the model's performance, we introduced a compound loss function, combining Dice loss for segmentation and MSE for autoencoder reconstruction (DC and MSE loss). This function separates the final image layer for MSE loss calculation, while the remaining layers compute the segmentation loss via Dice. This dual-objective approach ensures both segmentation accuracy and image reconstruction quality in training.

$$L = 1.2 \times MSE + 1.0 \times Dice \tag{3}$$

During the inference phase, we configured the autoencoder with an additional output channel dedicated to predicting the original input images. This added output channel allows the model to generate both segmentation results and auto-encoder predictions simultaneously. As shown in Fig. 2, the structure of the network ensures that the auto-encoder prediction is handled as a separate task, which enables better alignment between the predicted auto-encoder image and the segmented regions, supporting improved overall performance in tasks requiring auto-encoder images and segmentation outputs.

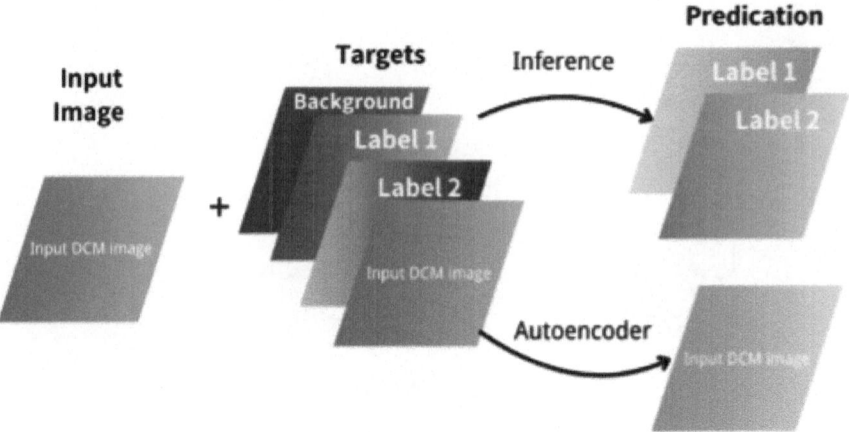

Fig. 2. Input/Output layers

4   Results

The learning curve indicates stable model performance over 1000 epochs, as shown in Fig. 3. Both training and validation losses decrease rapidly during the initial training stages and plateau after approximately 600 epochs. This pattern suggests minimal overfitting and highlights the model's ability to generalize well across the dataset. Simultaneously, the pseudo-Dice score shows a steady improvement, stabilizing after around 600 epochs, further demonstrating the robustness of the model's learning process.

In our experiment, the input image was reconstructed alongside the segmentation task. As depicted in Fig. 4, the reconstructed image closely resembles the original input, demonstrating the autoencoder's capacity to capture image features effectively. This indicates that the autoencoder positively contributes to feature extraction, ultimately enhancing the learning process. Additionally, Fig. 5 highlights the modified nnUNetv2's strong predictive capabilities, with the model accurately identifying the target labels.

The final results from the private evaluation phase of the HNTS-MRG 2024 challenge, involving 50 test patients, confirm the model's promising performance. Using the aggregated Dice Similarity Coefficient (DSCagg) metric, the metastatic lymph nodes (GTVn) achieved a DSC of 0.8516, while the primary tumor (GTVp) scored 0.7318. The mean DSCagg across both structures was 0.7917. These findings underscore the model's robust performance in segmenting lymph nodes while revealing areas for improvement in primary tumor segmentation. Future iterations of the model will focus on addressing this discrepancy to achieve more balanced performance across all target structures.

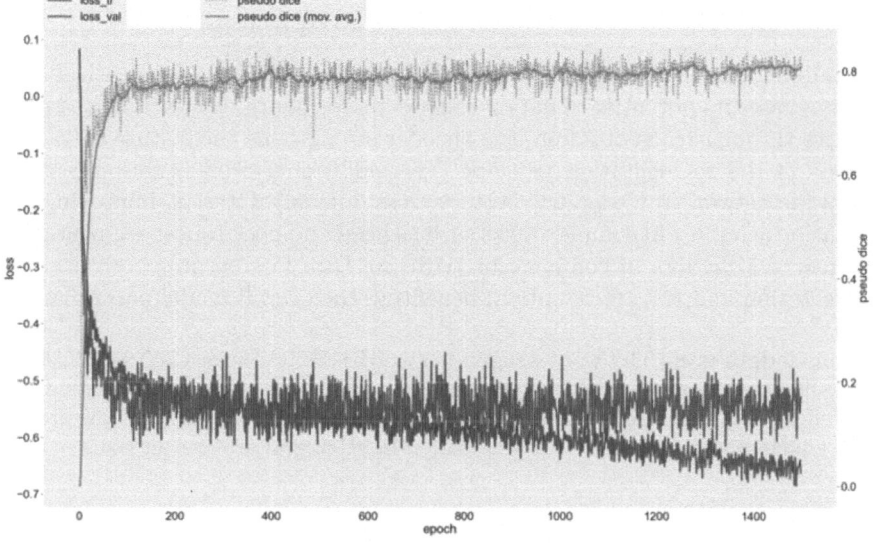

Fig. 3. Learning curve

Fig. 4. Original Input Image vs Auto-encoder Output

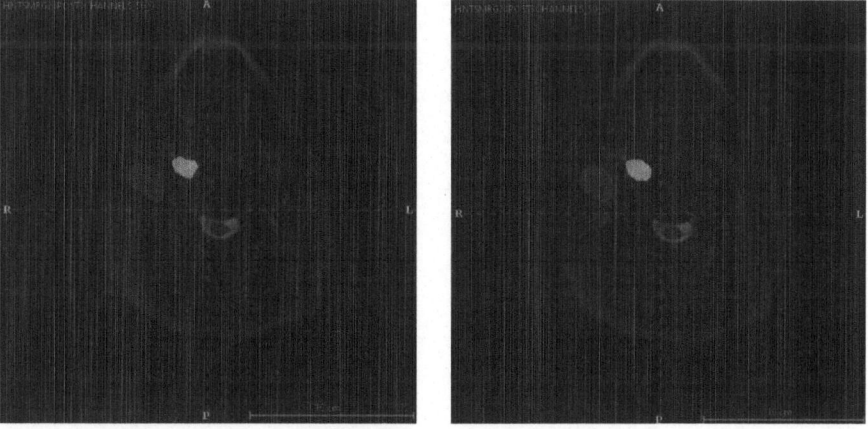

Fig. 5. Ground Truth Label vs Prediction with nnUNetv2 7-stages

5 Discussion

The autoencoder's ability to replicate the input image in a nearly identical manner showcases its potential to improve model performance. By learning the input features through reconstruction, the encoder strengthens the feature extraction process in the segmentation network. This additional representation learning allows the network to generalize better across different datasets, improving segmentation accuracy in complex tasks such as head and neck cancer segmentation. The use of MSE loss in combination with the Dice loss ensures both accurate segmentation and feature retention, benefiting the overall model performance.

Acknowledgments. This work is supported by NIH SBIR Contract 75N91023C00032. We would like to express our sincere gratitude to Michael Tang, Robert Qu, Justin Chen, and Jeff Qian for their invaluable support and contributions throughout this project. Their guidance and expertise have been crucial in helping us achieve our goals.

References

1. Zwanenburg, A., Leger, S., Vallières, M., Löck, S.: Image biomarker standardisation initiative. Radiother. Oncol. **150**, 20–22 (2020). https://doi.org/10.1016/j.radonc. 2020.04.032
2. Menze, B.H., Jakab, A., Bauer, S., et al.: The multimodal brain tumor image segmentation benchmark (BRATS). IEEE Trans. Med. Imaging **34**(10), 1993–2024 (2015). https://doi.org/10.1109/TMI.2014.2377694
3. Isensee, F., Jaeger, P.F., Kohl, S.A.A., Petersen, J., Maier-Hein, K.H.: nnU-Net: a self-configuring method for deep learning-based biomedical image segmentation. Nat. Methods **18**(2), 203–211 (2021). https://doi.org/10.1038/s41592-020-01008-z
4. Jiang, H., Haimerl, J., Gu, X., Lu, W.: A general web-based platform for automatic delineation of head and neck gross tumor volumes in PET/CT images. In: Head and Neck Tumor Segmentation and Outcome Prediction: Third Challenge, HECKTOR 2022, Held in Conjunction with MICCAI 2022, Singapore, 22 September 2022, Proceedings, vol. 13666, pp. 47–53. Springer, Cham (2023). https://doi.org/10.1007/978-3-031-27420-6_4
5. Tang, M., Qu, R., Chen, J., Qian, J.: Auto-segmentation of head and neck tumors using MRI: HNTS-MRG 2024 challenge. arXiv (2024). https://arxiv.org/abs/2404.09556

Improving the U-Net Configuration for Automated Delineation of Head and Neck Cancer on MRI

Andrei Iantsen[✉][iD]

Moscow, Russia
andrei.iantsen@gmail.com

Abstract. Tumor volume segmentation on MRI is a challenging and time-consuming process that is performed manually in typical clinical settings. This work presents an approach to automated delineation of head and neck tumors on MRI scans, developed in the context of the MICCAI Head and Neck Tumor Segmentation for MR-Guided Applications (HNTS-MRG) 2024 Challenge. Rather than designing a new, task-specific convolutional neural network, the focus of this research was to propose improvements to the configuration commonly used in medical segmentation tasks, relying solely on the traditional U-Net architecture. The empirical results presented in this article suggest the superiority of patch-wise normalization used for both training and sliding window inference. They also indicate that the performance of segmentation models can be enhanced by applying a scheduled data augmentation policy during training. Finally, it is shown that a small improvement in quality can be achieved by using Gaussian weighting to combine predictions for individual patches during sliding window inference. The model with the best configuration obtained an aggregated Dice Similarity Coefficient (DSCagg) of 0.749 in Task 1 and 0.710 in Task 2 on five cross-validation folds. The ensemble of five models (one best model per validation fold) showed consistent results on a private test set of 50 patients with an DSCagg of 0.752 in Task 1 and 0.718 in Task 2 (team name: andrei.iantsen). The source code and model weights are freely available at www.github.com/iantsen/hntsmrg.

Keywords: MRI segmentation · Radiation therapy · U-Net · Patch-wise normalization · Scheduled augmentation · Gaussian weighting

1 Introduction

Radiation therapy (RT) plays a crucial role in oncology with more than 40% of patients worldwide undergoing RT at least once as part of cancer treatment [10]. Modern linear accelerators can deliver radiation beams to tissues with

---

A. Iantsen—Independent Researcher.

K. A. Wahid et al. (Eds.): HNTS-MRG 2024, LNCS 15273, pp. 230–240, 2025.
https://doi.org/10.1007/978-3-031-83274-1_18

submillimeter precision and further advances in RT necessitate the integration of increasingly accurate imaging systems for tumor targeting. Computed tomography (CT) is widely used for cancer staging and RT planning. However, due to the limited contrast between soft tissues on CT, magnetic resonance imaging (MRI) is often applied instead of or in addition to CT to better distinguish the tumor from surrounding normal tissues in anatomical areas, such as the brain, nasopharynx and pelvis. Moreover, emerging MR-guided linear accelerators can monitor the target volume and organs at risk in real-time during dose delivery and adjust the treatment plan daily. Despite the good contrast and high spatial resolution, tumor volume segmentation on MRI scans is a challenging and time-consuming process that is performed manually in typical clinical settings. Consequently, the resulting tumor contours are subject to significant intra- and inter-observer variability, which can lead to deleterious consequences in downstream applications (e.g., skewed dose distributions during RT planning; low repeatability/reproducibility of image-based biomarkers in radiomics [1,11]). Hence, fully automated methods for MRI segmentation are of particular interest from a clinical perspective.

Due to the rapid advances in deep learning and computing technologies over the last decade, data-driven models based on convolutional neural networks (CNNs) have achieved impressive results in a wide range of computer vision tasks, including image segmentation. In the medical imaging domain, U-Net has remained a workhorse since its introduction in 2015 [7]. Furthermore, despite a variety of alternative, task-specific models reported in the literature, the vast majority of them actually constitute some variants of U-Net, often with only cosmetic changes. Finally, as shown in the nnU-Net framework [2,3], other components of the overall configuration (e.g., data pre- and post-processing methods, augmentation techniques, training procedures, etc.) often have a greater impact on performance than the choice of architecture per se.

This paper presents an approach to automated delineation of head and neck cancer on MRI, developed in the context of the MICCAI Head and Neck Tumor Segmentation for MR-Guided Applications (HNTS-MRG) 2024 Challenge. The main goal of this research was to propose some improvements to the configuration commonly used in medical segmentation tasks, relying only on the traditional U-Net architecture without significant changes.

2   HNTS-MRG 2024 Challenge

2.1   Data Description

For the purpose of the challenge, the University of Texas MD Anderson Cancer Center provided a dataset of 150 patients with histologically proven head and neck cancer [12]. Two T2-weighted (T2w) MRI scans were available for each patient: a pre-RT scan acquired 1–3 weeks before RT and a mid-RT scan after 2–4 weeks of RT. Manual delineation of primary gross tumor volume (GTVp) and metastatic lymph nodes (GTVn) for each patient was performed independently by 3 to 4 clinical experts, whose results were then combined using the STAPLE algorithm [15]. The resulting segmentation with three target classes (background, GTVp, GTVn) served as the ground truth.

2.2 Segmentation Tasks

Two segmentation tasks were proposed in the HNTS-MRG 2024 Challenge. In **Task 1**, it was required to build an automated solution for segmenting GTVp and GTVn volumes only on pre-RT scans. While in **Task 2**, the goal was to delineate the target volumes on mid-RT scans, optionally using the corresponding pre-RT scans and ground truth annotations for them as input data.

2.3 Evaluation Metric

The Dice Similarity Coefficient (DSC) is a widely used metric for evaluating performance in segmentation tasks. For a binary ground truth y and a binary prediction $\hat{y}$, the DSC is calculated as

$$\text{DSC}(y, \hat{y}) = 2\frac{\sum_i y_i \hat{y}_i}{\sum_i y_i + \sum_i \hat{y}_i}, \tag{1}$$

where y_i and $\hat{y}_i$ are the true and predicted labels for the ith element (voxel), respectively. If the ground truth has no elements of the target class (i.e., $y_i = 0$ for any i), this metric is not informative since DSC $= 0$ for any prediction $\hat{y}$. Accordingly, the aggregated Dice Similarity Coefficient (DSC_{agg}) was used for evaluation in the challenge. For a set of $N_{\mathcal{S}}$ pairs of binary ground truth and predicted masks, $\mathcal{S} = \{(y^{(n)}, \hat{y}^{(n)})\}_{n=1}^{N_{\mathcal{S}}}$, the DSC_{agg} is defined as

$$\text{DSC}_{\text{agg}}(\mathcal{S}) = 2\frac{\sum_{n,i} y_i^{(n)} \hat{y}_i^{(n)}}{\sum_{n,i} y_i^{(n)} + \sum_{n,i} \hat{y}_i^{(n)}}. \tag{2}$$

The average DSC_{agg} for the GTVp and GTVn classes on a test set of 50 patients was used to evaluate performance in both tasks presented in the challenge.

3 Methods

3.1 Network Architecture

The network used for both tasks followed the design principles of the traditional U-Net [7] (see Fig. 1). It was built using convolutional blocks, each consisting of a 3D convolutional layer, instance normalization, and ReLU nonlinearity. In the encoder, the number of feature maps (i.e., channels) was doubled after downsampling, which was preformed with a $2 \times 2 \times 2$ max pooling. Upsampling in the decoder was carried out with a $1 \times 1 \times 1$ convolutional block to halve the number of feature maps, followed by nearest-neighbor interpolation to double their

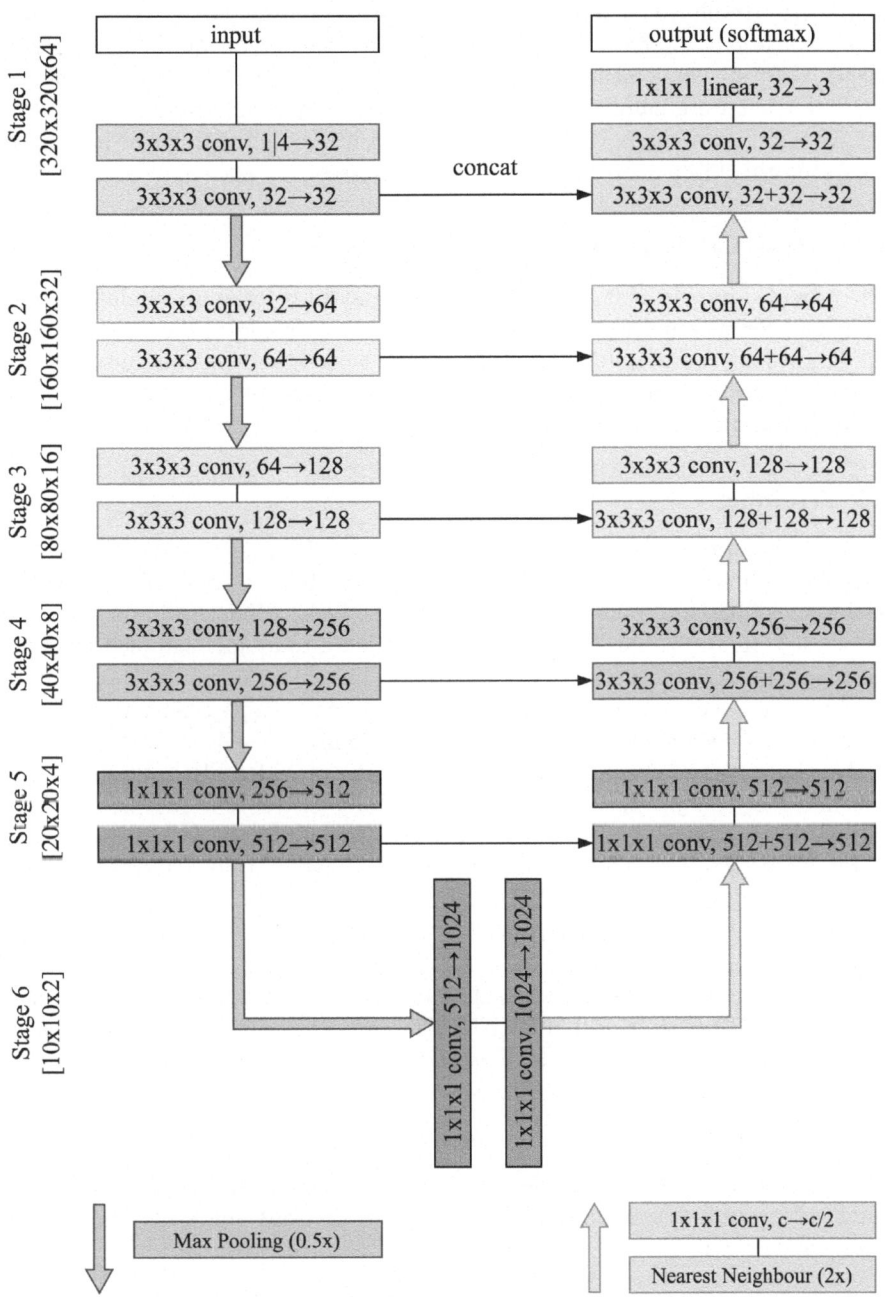

Fig. 1. The network architecture used in both task. It is the traditional U-Net, in which convolutional blocks in Stages 5–6 are implemented with $1 \times 1 \times 1$ kernels to reduce the number of model parameters. Spacial sizes of feature maps are provided in square brackets.

spatial size (resolution). Feature maps from five resolution stages in the encoder were transferred to the decoder via skip-connections. Convolutional blocks in Stages 1–4 were implemented with $3 \times 3 \times 3$ kernels, whereas smaller kernels of size $1 \times 1 \times 1$ were employed in Stages 5–6 to substantially decrease the number of model parameters (86M to 14M). The softmax activation was applied to generate probability scores for three output classes.

3.2   Cross-Validation Folds

The provided dataset was divided into five equal folds for cross-validation, with all images for each individual patient placed in only one fold. The results on the validation folds were used to compare different configurations and evaluate the generalization performance of the model (i.e., the expected performance on new data examples).

3.3   Training

Before being processed by the model, all the MRI scans and segmentation masks were first resampled to a voxel size of $0.5 \times 0.5 \times 2$ mm using linear and nearest-neighbor interpolation methods, respectively. Model training was done on patches of size $320 \times 320 \times 64$ voxels that were randomly sampled from the entire images. The position of each patch was chosen such that 90% of training patches contained some voxels of a target class (i.e., GTVp or GTVn), while the remaining 10% were extracted completely randomly.

For Task 1, the network was trained on all the provided MRI scans (i.e., pre-RT, mid-RT, and pre-RT registered to mid-RT) for 100K iterations (batches) with a batch size of 2. The model for Task 2 was trained for 50K iterations and had four input channels: a mid-RT scan, a registered pre-RT scan, and two binary masks for GTVp and GTVn on the registered pre-RT. Performance on validation examples was evaluated after every 5K training iterations. For both tasks, Adam optimizer [4] was used with a learning rate decreasing from 10^{-3} to 10^{-5} following the cosine decay schedule [5]. The model training was performed on a single GPU with 16 GB of VRAM using mixed precision [6] to significantly reduce the required memory and shorten the execution time.

3.4   Loss Function

During training, the loss function was the Dice Loss computed on the entire batch of ground truth masks and model predictions for each class, $\mathcal{B} = \{(y^{(n)}, \hat{p}^{(n)})\}_{n=1}^{N_\mathcal{B}}$:

$$
L_{Dice}(\mathcal{B}) = 1 - 2 \frac{\sum\limits_{n,i} y_i^{(n)} \hat{p}_i^{(n)}}{\sum\limits_{n,i} y_i^{(n)} + \sum\limits_{n,i} \hat{p}_i^{(n)}} . \tag{3}
$$

The second term in Eq. 3 is the smooth approximation of the DSC_{agg} function for one class in batch $\mathcal{B}$. Because data examples can have only a subset of classes in their ground truth masks (e.g., one or both tumor volumes may be completely missing in some patients, or the tumor may not be in the patch due to sampling), the average loss is calculated only for classes present in the training batch, including the background.

3.5 Sliding Window Inference

Inference on entire MRI scans was performed relying on a sliding window approach: predictions were obtained for consecutive patches of size $320 \times 320 \times 64$ using a stride of $80 \times 80 \times 16$ voxels. Note that predictions on overlapping voxels can be combined in different ways. The default option is to average them by assigning equal weights to all voxels. However, it is known that the accuracy of patch-based predictions decreases towards the patch edges, which can lead to different artifacts on the combined output mask [14]. Alternative methods are based on weighting voxels according to their positions in the patch, so that voxels closer to the patch center have higher weights. For both tasks, predictions for individual patches were combined using the *Gaussian weights* ranging from 1 at the patch center to 0.1 at the edges. This approach slightly improved the equality of sliding window predictions compared to equal weighting (see Sect. 4). After inference, the model predictions were converted into class labels by applying the argmax function.

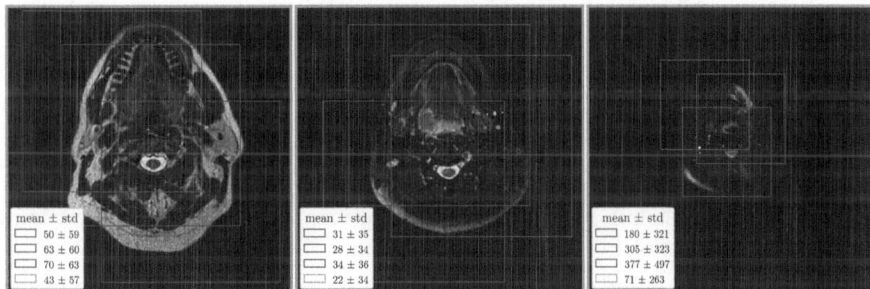

Fig. 2. The mean and standard deviation computed for different patches (first three contours), each of size $320 \times 320 \times 64$, and entire images (last contours) from the training set after resampling to the same voxel size. An axial slice is shown for each example.

3.6 Intensity Normalization

In contrast to CT, the intensity scale in MRI is not standardized and therefore intensity normalization is particularly important, especially when working with images from different MRI scanners [13]. Furthermore, input normalization is generally required to improve the convergence of optimization methods based on gradient descent. Z-score normalization, where all intensities are first shifted by the mean and then scaled by the standard deviation, works well in practice. However, this type of normalization is often applied with the mean and standard deviation computed on the entire image (*image-wise normalization*), even when training is carried out on image patches (see nnU-Net [2]). As a result, the intensity distribution in each patch (i.e., its mean and variance) depends on the patch location, which can significantly hinder the convergence of optimization methods. Similarly, the performance of models trained with *patch-wise normalization* (i.e., normalization is done *after* patch extraction) is affected by covariate shift [8,9] when inference is performed on normalized images in a sliding window manner. Figure 2 shows the differences in means and standard deviations between different patches and between images from the training set. To mitigate the effect of covariate shift, only patch-wise normalization was applied to the model inputs in both tasks (except for the binary masks in Task 2). At inference, the normalization was integrated directly into the sliding window approach.

3.7   Data Augmentation

Data augmentation was used to increase the size and diversity of training examples. Transformations such as image mirroring, rotations, contrast adjustment, imitation of MRI bias field and motion artifacts, as well as distortions by additive Gaussian noise were applied. Each transformation was performed independently with the same probability, which was linearly increasing from 0.05 to 0.25 over the course of training with adjustments made every 1K batches. This *scheduled augmentation policy* was aimed to generate more data examples towards the end of training when overfitting was more likely to occur. The comparison between the scheduled data augmentation and the augmentation with a constant probability of 0.15 is provided in Sect. 4.

3.8   Model Ensembling

The configuration with the best performance on the validation folds was used for the final submission in the challenge. Predictions were obtained by averaging the softmax outputs of five models (one best model per fold) and applying the argmax function to obtain the class labels. The output was then resampled using nearest-neighbor interpolation to restore the original resolution (voxel size) of the input. The source code and model weights to reproduce the submitted results are freely available at www.github.com/iantsen/hntsmrg.

4   Results and Discussion

The main focus of this research, largely inspired by the nnU-Net framework, was to improve some components of the configuration commonly used in medical segmentation tasks without making significant changes to the architecture of the traditional U-Net. The following configuration was used as a baseline for comparison: (1) the intensities of MRI scans were normalized image-wise (i.e., before patch extraction), (2) data augmentation was applied with a probability of 0.15, which remained constant during training, and (3) all voxels in the patch had equal weights during sliding window inference.

Table 1. Average values of DSC$_{agg}$ for the GTVp and GTVn classes on *five validation folds*. "Baseline configuration" refers to training with the image-wise normalization of input intensities and the augmentation policy with a constant probability of 15%, as well as with equal weights for all voxels in the patch during sliding window inference.

Configuration	Task 1			Task 2		
	GTVp	GTVn	Average	GTVp	GTVn	Average
Baseline	0.672	0.705	0.689	0.581	0.815	0.698
+ patch-wise normalization	0.682	0.768	0.725	0.600	0.824	0.712
+ scheduled augmentation	0.691	0.800	0.745	0.598	0.825	0.711
+ Gaussian weighting	0.695	0.804	0.749	0.598	0.823	0.710

Table 2. Average values of DSC$_{agg}$ for the GTVp and GTVn classes on the *test set* with 50 patients. Predictions for each task were obtained using the ensemble of five models.

Task 1			Task 2		
GTVp	GTVn	Average	GTVp	GTVn	Average
0.709	0.794	0.752	0.592	0.845	0.718

The empirical results for different configurations on five validation folds are summarized in Table 1. The use of patch-wise normalization for both training and inference improved performance in both tasks compared to the baseline: the average DSC$_{agg}$ increased from 0.689 to 0.725 in Task 1, and from 0.698 to 0.712 in Task 2. This improvement can be attributed to a reduction in covariate shift, as this type of normalization guarantees that all inputs have zero mean and unit variance. Training the model on a progressively increasing number of examples created with data augmentation techniques produced better results in Task 1 (the metric changed from 0.725 to 0.745), but had no significant impact in Task 2 (0.712 vs. 0.711). Similarly, using the Gaussian weights to combine predictions for individual patches during sliding window inference led to slightly higher results in Task 1 (0.745 vs. 0.749), but not in Task 2 (0.711 vs. 0.710). As for the results for each target class, the configuration with all three modifications achieved more accurate predictions for GTVn than for GTVp in Task 1 (0.804 and 0.695, respectively) and in Task 2 (0.823 and 0.598, respectively). Surprisingly, the model showed significantly worse results for GTVp in Task 2, although registered pre-RT masks were provided as inputs, compared to Task 1 where segmentation was performed only on pre-RT scans. The configuration with all three modifications was selected for the final submission in both tasks.

In addition to the aggregated metrics in Table 1, Fig. 3 shows the results in terms of DSC, Precision, and Recall for individual data examples (i.e., patients) from the validation folds. A visual comparison of the predicted and ground truth masks for three different patients is provided in Fig. 4.

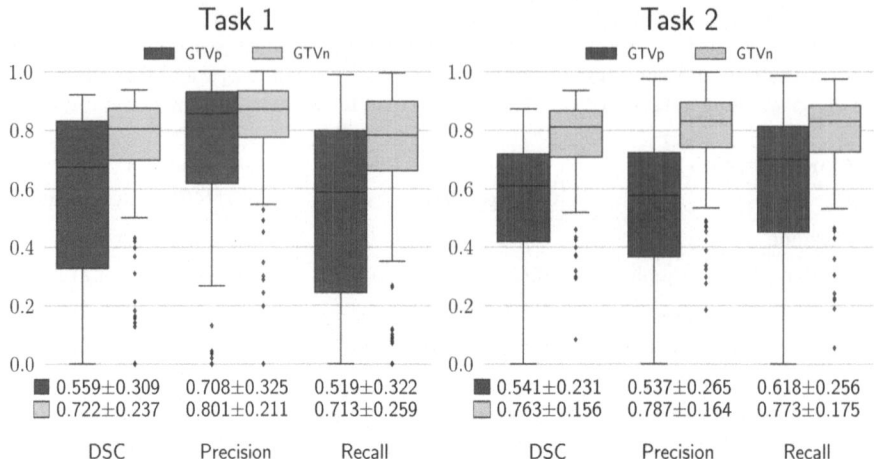

Fig. 3. The distribution of results for patients from five validation folds. All metrics were calculated using only examples with non-zero ground truth masks.

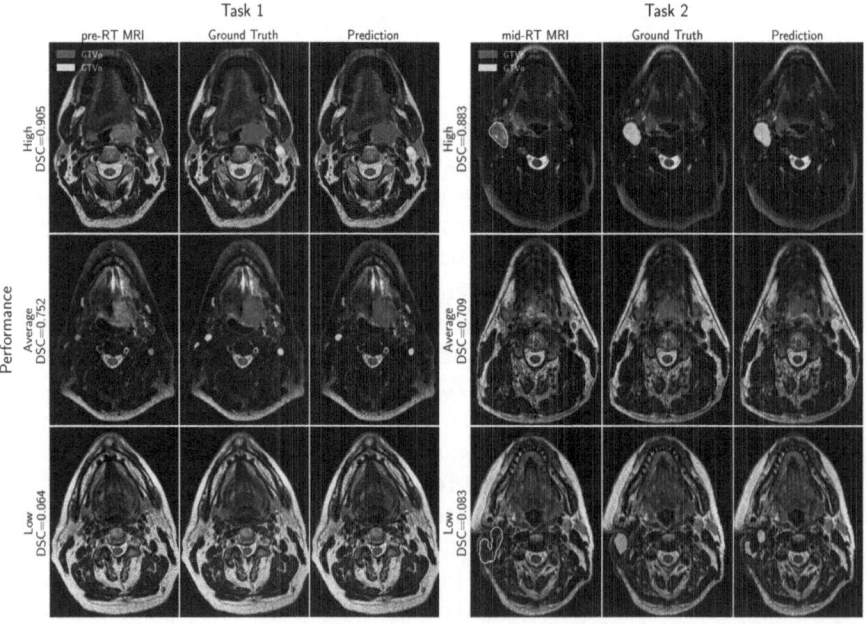

Fig. 4. Examples for visual comparison of model predictions and ground truth mask for both tasks (filled contours). "DSC" is the average DSC for the GTVp and GTVn classes. Contours of both target classes from registered pre-RT scans are drawn on corresponding mid-RT scans (Task 2, unfilled contours). Examples of "average performance" were selected based on the average DSC_{agg} on validation folds in both tasks.

The final results on the test set of 50 patients are shown in Table 2. The ensemble of five models achieved an average DSC_{agg} of 0.752 in Task 1 (0.709 and 0.794 for the GTVp and GTVn classes, respectively) and 0.718 in Task 2 (0.592 and 0.845).

5   Conclusion

This paper presents a number of modifications to a configuration commonly used with convolutional neural networks for medical image segmentation. The empirical results were obtained using the traditional U-Net architecture to address two segmentation tasks proposed in the context of the MICCAI Head and Neck Tumor Segmentation for MR-Guided Applications (HNTS-MRG) 2024 challenge. First, it was shown that patch-wise normalization (i.e., normalization applied after patch extraction) used for both training and sliding window inference improved performance by reducing covariance shift. Second, the scheduled data augmentation policy, where each transformation was applied with a probability linearly increasing towards the end of training, produced better results compared to augmentation with a fixed probability. Finally, using Gaussian weighting to combine predictions for individual patches during sliding window inference resulted in slightly more accurate predictions than those obtained with equal weighting.

References

1. Desseroit, M.C., et al.: Reliability of PET/CT shape and heterogeneity features in functional and morphologic components of non-small cell lung cancer tumors: a repeatability analysis in a prospective multicenter cohort. J. Nucl. Med. **58**(3), 406–411 (2016). https://doi.org/10.2967/jnumed.116.180919
2. Isensee, F., Jaeger, P.F., Kohl, S.A.A., Petersen, J., Maier-Hein, K.H.: nnU-Net: a self-configuring method for deep learning-based biomedical image segmentation. Nat. Methods **18**(2), 203–211 (2020). https://doi.org/10.1038/s41592-020-01008-z
3. Isensee, F., et al.: nnU-Net revisited: a call for rigorous validation in 3d medical image segmentation. In: Medical Image Computing and Computer Assisted Intervention - MICCAI 2024, pp. 488–498. Springer, Cham (2024). https://doi.org/10.1007/978-3-031-72114-4_47
4. Kingma, D.P., Ba, J.: Adam: a method for stochastic optimization (2017). https://arxiv.org/abs/1412.6980
5. Loshchilov, I., Hutter, F.: SGDR: stochastic gradient descent with warm restarts (2017). https://arxiv.org/abs/1608.03983
6. Micikevicius, P., et al.: Mixed precision training (2018). https://arxiv.org/abs/1710.03740
7. Ronneberger, O., Fischer, P., Brox, T.: U-Net: convolutional networks for biomedical image segmentation (2015). https://arxiv.org/abs/1505.04597
8. Shimodaira, H.: Improving predictive inference under covariate shift by weighting the log-likelihood function. J. Stat. Plan. Inference **90**(2), 227–244 (2000). https://doi.org/10.1016/s0378-3758(00)00115-4

9. Sugiyama, M., Krauledat, M., Müller, K.R.: Covariate shift adaptation by importance weighted cross validation. J. Mach. Learn. Res. **8**(35), 985–1005 (2007). http://jmlr.org/papers/v8/sugiyama07a.html

10. Thompson, M.K., et al.: Practice-changing radiation therapy trials for the treatment of cancer: where are we 150 years after the birth of Marie Curie? Br. J. Cancer **119**(4), 389–407 (2018). https://doi.org/10.1038/s41416-018-0201-z

11. Traverso, A., Wee, L., Dekker, A., Gillies, R.: Repeatability and reproducibility of radiomic features: a systematic review. Int. J. Radiat. Oncol.* Biol.* Phys. **102**(4), 1143-1158 (2018). https://doi.org/10.1016/j.ijrobp.2018.05.053

12. Wahid, K., Dede, C., Naser, M., Fuller, C.: Training dataset for HNTSMRG 2024 challenge (2024). https://doi.org/10.5281/ZENODO.11199559

13. Wahid, K.A., et al.: Intensity standardization methods in magnetic resonance imaging of head and neck cancer. Phys. Imaging Radiat. Oncol. **20**, 88–93 (2021). https://doi.org/10.1016/j.phro.2021.11.001

14. Wang, S., et al.: A deep learning-based stripe self-correction method for stitched microscopic images. Nat. Commun. **14**(1) (2023). https://doi.org/10.1038/s41467-023-41165-1

15. Warfield, S., Zou, K., Wells, W.: Simultaneous truth and performance level estimation (STAPLE): an algorithm for the validation of image segmentation. IEEE Trans. Med. Imaging **23**(7), 903–921 (2004). https://doi.org/10.1109/tmi.2004.828354

Head and Neck Gross Tumor Volume Automatic Segmentation Using PocketNet

Awj Twam[1]([✉]) [ID], Adrian Celaya[1] [ID], Evan Lim[2] [ID], Khaled Elsayes[3] [ID], David Fuentes[1] [ID], and Tucker Netherton[4] [ID]

[1] Department of Imaging Physics, The University of Texas MD Anderson Cancer Center, Houston, TX, USA
atwam@mdanderson.org
[2] Institute of Data Science in Oncology, The University of Texas MD Anderson Cancer Center, Houston, TX, USA
[3] Department of Abdominal Imaging, Division of Diagnostic Imaging, The University of Texas MD Anderson Cancer Center, Houston, TX, USA
[4] Department of Radiation Physics, The University of Texas MD Anderson Cancer Center, Houston, TX, USA

Abstract. Head and neck cancer (HNC) represents a significant global health burden, often requiring complex treatment strategies, including surgery, chemotherapy, and radiation therapy. Accurate delineation of tumor volumes is critical for effective treatment, particularly in MR-guided interventions, where soft tissue contrast enhances visualization of tumor boundaries. Manual segmentation of gross tumor volumes (GTV) is labor intensive, time-consuming and prone to variability, motivating the development of automated segmentation techniques. Convolutional neural networks (CNNs) have emerged as powerful tools in this task, offering significant improvements in speed and consistency. In this study, we participated as Team Pocket in Task 1 of the HNTS-MRG 2024 Grand Challenge, which focuses on the segmentation of gross tumor volumes of the primary tumor (GTVp) and the nodal tumor (GTVn) in pre-radiotherapy MR images for HNC. We evaluated the application of PocketNet, a lightweight CNN architecture, for this task. Results for the final test phase of the challenge show that PocketNet achieved an aggregated Dice Sorensen Coefficient (DSCagg) of 0.808 for GTVn and 0.732 for GTVp, with an overall mean performance of 0.77. These findings demonstrate the potential of PocketNet as an efficient and accurate solution for automated tumor segmentation in MR-guided HNC treatment workflows, with opportunities for further optimization to enhance performance.

Keywords: Head and Neck Cancer · Gross Tumor Volume · Automated Segmentation · Convolutional Neural Networks

1 Introduction

Head and neck cancer (HNC) is among the most prevalent cancers worldwide, with over 500,000 new cases diagnosed annually [1]. Treatment strategies for HNC include radiation therapy (RT) as a critical component of a multimodal approach. Essential to RT

The original version of the chapter has been revised. An Acknowledgement section has been added. A correction to this chapter can be found at
https://doi.org/10.1007/978-3-031-83274-1_22

K. A. Wahid et al. (Eds.): HNTS-MRG 2024, LNCS 15273, pp. 241–249, 2025.
https://doi.org/10.1007/978-3-031-83274-1_19

treatment planning is the routine contouring of gross tumor volumes (GTVs) and organs at risk. Accurate delineation of tumor volumes is critical for effective treatment, particularly in MR-guided interventions, where soft tissue contrast enhances visualization of tumor boundaries [2]. However, manual contouring of these structures is both time and labor intensive and associated with known interobserver variability that can impact treatment outcomes [3, 4]. The development of automated tumor segmentation techniques using deep learning offers the potential to enhance the accuracy, reproducibility, and efficiency of tumor delineation.

Historically, traditional methods for tumor segmentation relied on atlas-based approaches or semi-automated tools that required manual adjustments and were limited by anatomical variability in HNC [5]. With the advent of deep learning, convolutional neural networks (CNNs) have emerged as the standard for automated medical image segmentation. State-of-the-art CNNs, such as U-Net and its variants, have demonstrated exceptional performance in segmenting tumors and organs from medical images [6]. However, these architectures often require extensive computational resources, limiting their accessibility and scalability in clinical settings. To address these limitations, we explored the application of PocketNet, a lightweight CNN architecture designed to reduce computational overhead while maintaining segmentation accuracy [7].

The Grand Challenge is a prominent platform designed to host competitions in medical imaging analysis, promoting advancements in machine learning. The HNTS-MRG24 challenge targeted the segmentation of GTV in HNC using MRI. For this study, we participated in Task 1 of the HNTS-MRG24 challenge, which focuses on the autosegmentation of GTVs from pre-radiotherapy (pre-RT) MR images and evaluated the performance of PocketNet for segmenting GTV in HNC patients.

2 Methods

2.1 Imaging Data

The data set used in this study was provided by the HNTS-MRG 2024 Challenge, which consisted of MR images of 150 pre-RT cases for Task 1 segmentation. These images included manually contoured masks of primary tumor (GTVp) and nodal tumor (GTVn), serving as the ground truth for training and validation.

2.2 Image Processing

The preprocessing pipeline from the Medical Imaging Segmentation Toolkit (MIST) was used to prepare the data. This pipeline processes NIfTI files and outputs reoriented, resampled, windowed, and normalized NumPy arrays. Pre-RT images were prepared using a patch size of $256 \times 256 \times 64$ and a pixel spacing of $0.5 \times 0.5 \times 1.2$ mm. The patch size was derived from the median resampled image size by selecting the nearest power of two less than or equal to each dimension, constrained by a maximum patch size. Initial target spacing is anisotropic with the maximum and minimum spacing along the lowest resolution axis [8]. All images were reoriented to right-anterior-inferior (RAI) to ensure uniformity Table 1.

Table 1. Preprocessing of data

Dimension	Value
Patch Size	$256 \times 256 \times 64$
Pixel Spacing	$0.5 \times 0.5 \times 1.2$ mm

2.3 Model Architecture

PocketNet is a lightweight, deep learning model based on the U-Net architecture. Unlike traditional CNNs, which double the number of feature maps at lower resolutions, Pocket-Net maintains a constant number of feature maps across all resolution levels. With the use of PocketNet, this results in faster training and inference times while simultaneously lowering memory usage and requirements [7, 8].

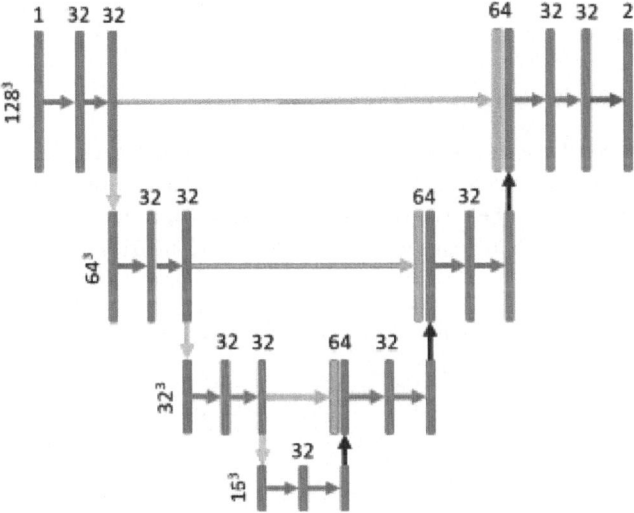

Fig. 1. PocketNet architecture

Similar to a U-Net structure, our network is constructed from a composition of convolution and down sampling operations that extract features along a contracting path. An expanding path consists of convolution and up-sampling operations with 'long skip' connections to integrate features from the corresponding down sampling operations. Four resolution levels are used. At a given resolution, the feature-maps of all preceding layers are used as inputs, and its own feature maps are used as inputs into all subsequent layers. The key difference in our architecture is that the number of layers is not heuristically increased at each resolution. Each convolutional operation uses a 3×3 kernel size and is followed by a batch normalization and a ReLU activation function Fig. 1.

2.4 Model Implementation

The model was trained using an Nvidia Quadro RTX 8000 graphical processing unit (GPU) with a batch size equal to the number of patients per fold. A five-fold cross validation was conducted to ensure robust and generalizable results. For each fold, the network was trained for 1,000 epochs. During training, network weights were determined using the Adam optimizer. Additionally, automatic mixed precision was enabled to optimize memory usage and speed. The loss function combined Dice with categorical cross-entropy to ensure effective segmentation of both GTVp and GTVn. A constant learning rate of 0.0003 was used by default. Postprocessing included selecting the largest connected component from the predicted mask to reduce noise. To assess the model's performance, two primary outcome metrics were used: the Dice Similarity Coefficient (DSC), which measures the overlap between predicted and ground truth segmentations, and the 95th percentile Hausdorff distance (Haus95), which provides insight into the boundary accuracy of segmentations.

3 Results

3.1 Quantitative Evaluation

The performance of the PocketNet model for segmentation gross tumor volumes for HNC was evaluated using two key metrics: DSC and Haus95 for both primary tumor and nodal tumor. Table 2 below provides a summary of the statistics from training on 150 pre-RT MRIs, which includes the conventional DSC metrics.

Table 2. Summary statistics of Dice and Haus95 metrics

	GTVp Dice	GTVn Dice	GTVp Haus95	GTVn Haus95
Mean	0.549	0.641	130.139	173.818
Std	0.314	0.300	212.402	238.689
25th Percentile	0.416	0.546	9.124	12.283
Median	0.648	0.763	37.184	82.717
75th Percentile	0.815	0.867	145.986	185.642

The PocketNet model achieved the following segmentation performance for the final test phase (50 patients): 81% for GTVn and 73% for GTVp with an overall mean of 77%.

Figure 2 presents the conventional DSC and Haus95 metrics in the form of box and whisker plots. PocketNet, when trained via five-fold cross validation, achieved a mean DSC of 55% for GTVp and 64% for GTVn with a standard deviation (Std) of 0.314 for GTVp and 0.299 for GTVn. The distribution of DSC and Haus95 values can be visualized through the box and whisker plots. GTVp showed a wider distribution with whiskers extending lower compared to GTVn, which reflected a greater variability in performance on tumors. For GTVp, the median DSC is 0.648 and 0.763 for GTVn. GTVn displayed one outlier below the lower whisker, suggesting a base where the model struggled, likely due to atypical nodal morphology, smaller tumor size, and/or poor boundary contrast. The overall consistency for GTVn, with a median DSC of 0.763, underscores the model's general robustness for nodal tumor segmentation. The mean Haus95 for GTVp was 130.139 mm and 173.818 mm for GTVn with a Std of 212.402 for GTVp and 238.689 for GTVn. For GTVp, 50% of the error values lie between 9.12 and 145.99 mm, while for GTVn, the range is between 12.28 mm and 185.64 mm. For both GTVp and GTVn, the larger upper quartile range reflected the difficulty the model faced in some cases, particularly when tumors are harder to delineate.

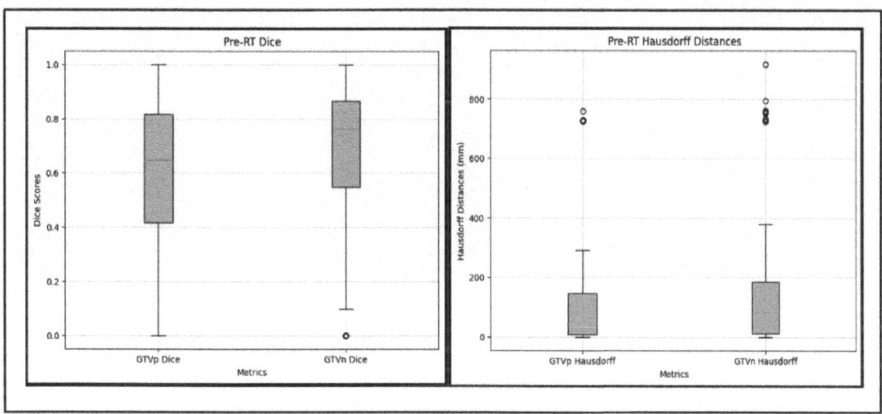

Fig. 2. Box and whisker plots of conventional DSC and Haus95 metrics for Pre-RT

Figures 3 and 4 illustrate example cases of segmentation. Figure 3 presents two cases of segmentation with good agreement. Case 1 had DSC scores of 0.88 for GTVp and 0.91 for GTVn, while Case 2 below it had DSC values of 0.90 for GTVp and 0.896 for GTVn. In contrast to Fig. 3, Fig. 4 presents two cases of segmentation with adequate to poor agreement. Case 3 in Fig. 4 had a DSC value of 0.51 for GTVp and 0.78 for GTVn and Case 4 below it had a DSC of 0.75 for GTVp and 0.58 for GTVn.

3.2 Qualitative Evaluation

To further understand the performance of the automated segmentation model, we evaluated the results by comparing the DSC coefficients between the ground truth and the automated contours, identifying key factors that may have contributed to discrepancies in segmentation accuracy. The variation in Dice values across the two cases in Fig. 4 is primarily due to the model's overestimation or underestimation of the tumor volumes relative to the ground truth. In Case 3, the model underestimated the GTVp volume resulting in a lower DSC of 0.51. In Case 4, the GTVp segmentation model showed a higher degree of overlap (DSC of 0.75), but autosegmentation of GTVn overestimated portions of the nodal tumor. Tumor shapes in HNC can vary significantly. The lower Dice scores, in particular the GTVp in Case 3 of Fig. 4, may be due to the model's difficulty

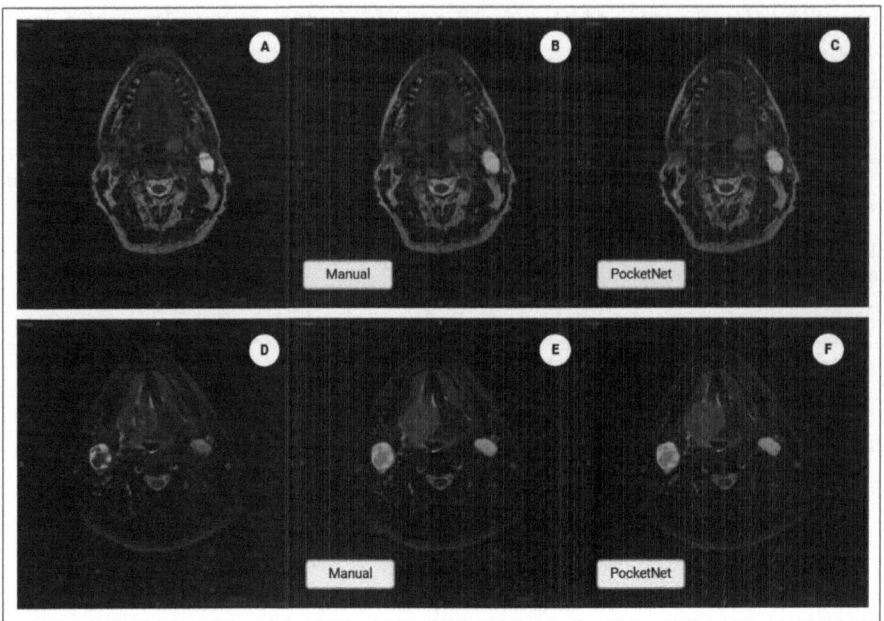

Fig. 3. Manual and automated segmentation in two patients. A) Axial MRI of Case 1. B) Case 1 manual segmentation of GTVp (red) and GTVn (green). C) Case 1 Automated segmentation of GTVp (red) and GTVn (green). D) Axial MRI of Case 2. E) Case 2 manual segmentation of GTVp (red) and GTVn (green). F) Case 2 Automated segmentation of GTVp (red) and GTVn (green). The manual and automated tumor contours demonstrate good agreement (GTVp DSC = 0.88, 0.90 and GTVn DSC = 0.91,0.896).\CFO{}

in capturing accurately the full extent of the tumor, especially in areas where the tumor had fewer clear boundaries. This challenge is compounded by the DSC inherent bias toward volume, as smaller regions of interest contribute less to the overlap metric.

4   Discussion

PocketNet demonstrated strong performance in Task 1 (pre-RT) of the HNTS-MRG 2024 challenge, achieving mean DSCagg scores of 0.808 for GTVn and 0.732 for GTVp in the final phase of testing. These results are comparable to those from similar segmentation challenges. For instance, in the HEKTOR challenge at MICCAI 2022, the winning model achieved DSCagg values of 0.801 for GTVp and 0.775 for GTVn [9]. PocketNet's comparable performance highlights its potential as an efficient segmentation tool with reduced computational requirements, though opportunities for further improvement remain.

A recent study on CT-only GTV segmentation for palliative head and neck radiotherapy highlights the complexities of automated contouring in anatomically challenging regions. Using the nnU-Net architecture, the study evaluated five approaches across two datasets of non-contrast and contrast-enhanced CT scans. Median DSC ranged from 0.6

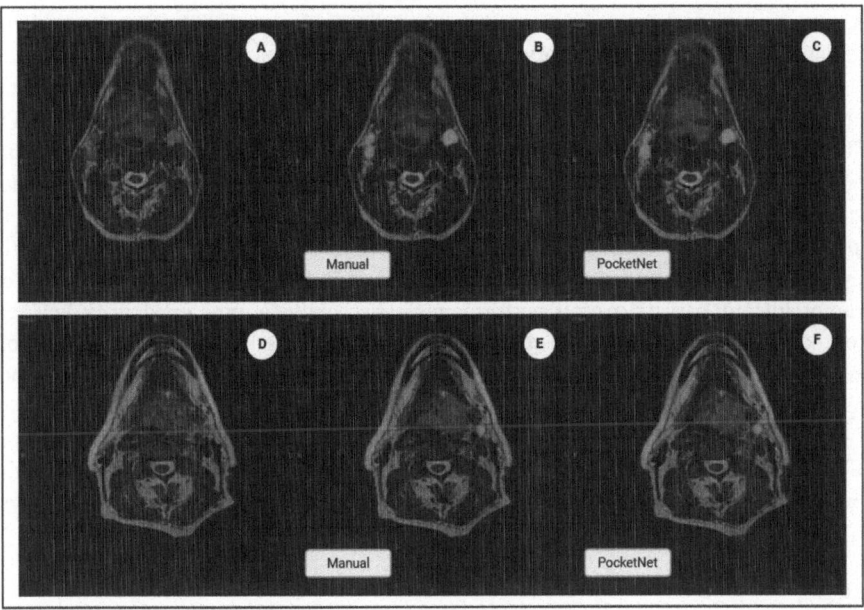

Fig. 4. Visual results of segmentation in two patients. A) Axial MRI of Case 3. B) Case 3 manual segmentation of GTVp (red) and GTVn (green). C) Case 3 automated seg-mentation of GTVp (red) and GTVn (green). D) Axial MRI of Case 4. E) Case 4 manual segmentation of GTVp (red) and GTVn (green). F) Case 4 automated segmentation of GTVp (red) and GTVn (green). The segmentations demonstrate poor agreement of gross tumor volumes with a DSC of 0.51 for GTVp and 0.78 for GTVn for Case 3 and 0.75 DSC for GTVp and 0.58 for GTVn for Case 4.\CFO{}

to 0.7, with Haus95 between 14.7 mm and 19.7 mm. While the performance was competitive, the study emphasized the challenges of delineating nodal involvement and poor tissue soft contrast in CT-only workflows [10]. The CT-only segmentation results align with the challenges addressed by PocketNet, reinforcing the importance of developing models that pereform well in single-modality imaging.

The advantages of multi-modal imaging are well-documented in segmentation tasks. For example, a study using 3D CNNs to automate the delineation of GTV for HNC compared single-modality (CT) and multi-modality (CT + PET and CT + MRI) data. The results showed that multi-modality CNNs performed better, achieving a mean DSC of 69% for GTVp and 79% for GTVn [11]. These results, while slightly lower for GTVp compared to PocketNet, underscore the advantages of incorporating multi-modal imaging, particularly for anatomically complex regions like the head and neck. A separate study conducted by Ren et al. demonstrated that combining CT, PET, and MRI for head and neck tumor segmentation using a 3D Residual U-Net achieved a Dice coefficient of up to 0.74 when utilizing ensemble bi-modal combinations, with PET inclusion shown to be particularly impactful [12]. Similarly, Wahid et al. found that incorporating multiple parametric MRI (mpMRI) channels (e.g., T2 + T1) significantly improved segmentation performance for oropharyngeal primary tumors, achieving a Dice coefficient of 0.73 and also demonstrated non-inferior results compared to ground truth segmentations in a Turing test with clinical experts, further validating the potential of multi-modality imaging in enhancing segmentation accuracy [13].

While this study was focused solely on MR imaging for pre-RT segmentation, the methodology could be extended to mid-RT segmentation tasks (Task 2 of the HNTS-MRG 2024 challenge). Mid-RT segmentation prediction is critical for adaptive radiotherapy, where tumor shrinkage or progression during treatment necessitates updated contours. While we did not evaluate PocketNet on Task 2, future work could explore its application in this context. Additionally, we are conducting an ablation study using the pre-RT data to better understand the contribution of individual components of Pocket-Net's architecture to its overall performance. This analysis aims to identify key features that most significantly impact segmentation accuracy, guiding future model refinements. By optimizing the model's architecture and fine-tuning its parameters, we hope to enhance segmentation accuracy.

Overall, PocketNet represents a promising solution for automated segmentation of GTVp and GTVn in MR images of HNC patients. Its lightweight design, combined with competitive segmentation accuracy, positions it as a robust tool for clinical workflows.

5 Conclusion

Manual contouring of GTV and organs-at-risk can lead to interobserver variability and impact treatment outcomes for HNC patients. Our proposed deep learning segmentation model, PocketNet, with its compact architecture and optimizations for light GPU memory usage and fast inference times, addresses key challenges in automating the segmentation of anatomical regions of interest for HNC radiation treatment planning. These results demonstrate that PocketNet achieved an average DSCagg of 0.808 for GTVn and 0.732 for GTVp for the final test phase of Task 1 in the HNTS-MRG 2024 challenge. Future work employing this model is anticipated to further refine and validate this approach for automated segmentation.

Acknowledgement. This research was partially supported by the Tumor Measurement Initiative through the MD Anderson Strategic Research Initiative Development (STRIDE) and QIAC Partnership in Research (QPR).

Disclosure of Interests. The authors have no competing interests to declare that are relevant to the content of this article.

References

1. Starzyńska, A., Sobocki, B.K., Alterio, D.: Current challenges in head and neck cancer management. Cancers (Basel). **14**(2), 358 (2022). https://doi.org/10.3390/cancers14020358. PMID:35053520;PMCID:PMC8773596
2. Zukauskaite, R., et al.: Delineation uncertainties of tumour volumes on MRI of head and neck cancer patients. Clin. Trans. Radiat. Oncol. **6**(36), 121–126 (2022). https://doi.org/10.1016/j.ctro.2022.08.005.PMID:36017132;PMCID:PMC9395751

3. van der Veen, J., Gulyban, A., Willems, S., Maes, F., Nuyts, S.: Interobserver variability in organ at risk delineation in head and neck cancer. Radiat. Oncol. **16**(1), 120 (2021). https://doi.org/10.1186/s13014-020-01677-2.PMID:34183040;PMCID:PMC8240214

4. Guzene, L., et al.: Assessing interobserver variability in the delineation of structures in radiation oncology: a systematic review. Int. J. Radiat. Oncol. Biol. Phys. **115**(5), 1047–1060 (2023). https://doi.org/10.1016/j.ijrobp.2022.11.021. Epub 2022 Nov 22 PMID: 36423741

5. Franzese, C., et al.: Enhancing radiotherapy workflow for head and neck cancer with artificial intelligence: a systematic review. J. Pers. Med. **13**(6), 946 (2023). https://doi.org/10.3390/jpm13060946.PMID:37373935;PMCID:PMC10301548

6. Sarvamangala, D.R., Kulkarni, R.V.: Convolutional neural networks in medical image understanding: a survey. Evol. Intell. **15**(1), 1–22 (2022). https://doi.org/10.1007/s12065-020-00540-3. Epub 2021 Jan 3. PMID: 33425040; PMCID: PMC7778711

7. Celaya, A., et al.: PocketNet: A Smaller Neural Network for 3D Medical Image Segmentation (2021). ArXiv, abs/2104.10745

8. Celaya, A., et al.: MIST: A Simple and Scalable End-To-End 3D Medical Imaging Segmentation Framework [v2] (2024). arXiv. https://arxiv.org/abs/2407.15476

9. Andrearczyk, V., et al.: Overview of the HECKTOR challenge at MICCAI 2022: automatic head and neck tumor segmentation and outcome prediction in PET/CT. Head Neck Tumor. Chall. 1–30 (2022). https://doi.org/10.1007/978-3-031-27420-6_1. Epub 2023 Mar 18. PMID: 37195050; PMCID: PMC10171217

10. Gay, S.S., Cardenas, C.E., Nguyen, C., et al.: Fully-automated, CT-only GTV contouring for palliative head and neck radiotherapy. Sci. Rep. **13**, 21797 (2023). https://doi.org/10.1038/s41598-023-48944-2

11. Bollen, H., Willems, S., Wegge, M., Maes, F., Nuyts, S: Benefits of automated gross tumor volume segmentation in head and neck cancer using multi-modality information. Radiother. Oncol. **182**, 109574 (2023) ISSN 0167-8140, https://doi.org/10.1016/j.radonc.2023.109574

12. Ren, J., Eriksen, J.G., Nijkamp, J., Korreman, S.S.: Comparing different CT, PET and MRI multi-modality image combinations for deep learning-based head and neck tumor segmentation. Acta Oncol. **60**(11), 1399–1406 (2021). https://doi.org/10.1080/0284186X.2021.1949034. Epub 2021 Jul 15 PMID: 34264157

13. Wahid, K.A., et al.: Evaluation of deep learning-based multiparametric MRI oropharyngeal primary tumor auto-segmentation and investigation of input channel effects: results from a prospective imaging registry. Clin. Trans. Radiat. Oncol. **16**(32), 6–14 (2021). https://doi.org/10.1016/j.ctro.2021.10.003.PMID:34765748;PMCID:PMC8570930

Application of 3D nnU-Net with Residual Encoder in the 2024 MICCAI Head and Neck Tumor Segmentation Challenge

Kaiyuan Ji[1] (ID), Zhihan Wu[2] (ID), Jing Han[2], Jun Jia[3], Guangtao Zhai[3(✉)] (ID), and Jiannan Liu[2(✉)]

[1] School of Communication and Electronic Engineering, East China Normal University, Shanghai, China

[2] Department of Oral and Maxillofacial Head and Neck Oncology, Shanghai Ninth People's Hospital, Shanghai Jiao Tong University School of Medicine, Shanghai, China
laurence_ljn@163.com

[3] School of Electronic Information and Electrical Engineering, Shanghai Jiao Tong University, Shanghai, China
zhaiguangtao@sjtu.edu.cn

Abstract. This article explores the potential of deep learning technologies for the automated identification and delineation of primary tumor volumes (GTVp) and metastatic lymph nodes (GTVn) in radiation therapy planning, specifically using MRI data. Utilizing the high-quality dataset provided by the 2024 MICCAI Head and Neck Tumor Segmentation Challenge, this study employs the 3DnnU-Net model for automatic tumor segmentation. Our experiments revealed that the model performs poorly with high background ratios, which prompted a retraining with selected data of specific background ratios to improve segmentation performance . The results demonstrate that the model performs well on data with low background ratios, but optimization is still needed for high background ratios. Additionally, the model shows better performance in segmenting GTVn compared to GTVp, with DSCagg scores of 0.6381 and 0.8064 for Task 1 and Task 2, respectively, during the final test phase. Future work will focus on optimizing the model and adjusting the network architecture, aiming to enhance the segmentation of GTVp while maintaining the effectiveness of GTVn segmentation to increase accuracy and reliability in clinical applications.

Keywords: Artificial Intelligence · Deep Learning · Medical Image · Tumor Segmentation

---

Team name: Sjtu&Ninth People's Hospital.
K. Ji and Z. Wu—Contribute equally to this work.

K. A. Wahid et al. (Eds.): HNTS-MRG 2024, LNCS 15273, pp. 250–258, 2025.
https://doi.org/10.1007/978-3-031-83274-1_20

1 Introduction

In the field of oncology, the assessment and treatment of primary gross tumor volumes (GTVp) and metastatic mymph dodes (GTVn) hold significant importance in clinical decision-making and research [1]. Traditionally, in radiation therapy planning, physicians were required to manually identify and delineate tumors and affected lymph nodes, a process that was not only time-consuming but also susceptible to subjective biases from individual physicians [2, 3]. In recent years, with the advancement of artificial intelligence technologies, deep learning has offered substantial potential in the field of medical image segmentation [4, 5]. By training models to recognize GTVp and GTVn in medical images, the accuracy of segmentation has been further enhanced [6]. These models learn a variety of visual features of tumors, including shape, size, and contrast with surrounding tissues, through extensive training data [7, 8]. Once trained, these models are capable of automatically processing new images, swiftly identifying tumor regions and affected lymph nodes, significantly accelerating the workflow for treatment planning [9, 10].

The annual Medical Image Computing and Computer Assisted Intervention Society (MICCAI) HEAD AND NECK TUMOR SEGMENTATION FOR MR-GUIDED: APPLICATIONS (HNTS-MRG) 2024 Challenge provided a series of high-quality, annotated T2-weighted datasets of patients who underwent radiotherapy (RT), including pre-RT (1–3 weeks before the start of RT) and mid-RT (2–4 weeks intra-RT) [11]. This offered a pathway for the systematic evaluation of automated segmentation methods for GTVp and GTVn. The study primarily utilized a residual encoder version of 3DnnU-Net to perform tumor segmentation for both the pre and mid-RT tasks.

2 Methods

T2w MRI imaging data from head and neck cancer patients who underwent radiotherapy (Sect. 2.1) was used to develope deep learning models for auto-segmentation of tumors and multiple lymph nodes. The ground truth manual segmentations and the classified imaging data (Sect. 2.2) were uesd to train our models (Sect. 2.3).

2.1 Imaging Data

This study utilized the training dataset from the MICCAI 2024 Head and Neck Tumor Segmentation for MR-Guided Applications (HNTSMRG) 2024 Challenge [11]. The dataset consists of 150 patients with histologically proven head and neck cancer who underwent radiotherapy (RT) at The University of Texas MD Anderson Cancer Center. For each patient, the dataset includes a pre-RT T2w MRI scan (1–3 weeks before start of RT) and a mid-RT T2w MRI scan (2–4 weeks intra-RT). Images consist of a mix of fat-suppressed and non-fat-suppressed MRI sequences.

All imaging data in the training set (from 150 patients) were manually segmented by multiple physician experts, and the segmentations were combined using the STAPLE (Simultaneous Truth and Performance Level Estimation) algorithm to generate the final ground truth segmentation labels. Both training and testing data were provided in Neuroimaging Informatics Technology Initiative (NIfTI) format.

2.2 Data Processing

All images were cropped from the top of the clavicles to the bottom of the nasal septum (oropharynx region to shoulders) by HNTSMRG2024 Challenge organizers to ensure consistency in the field of view and to remove identifiable facial structures. We divided the dataset into two groups based on the proportion of background present in the MRI images: 1) a group with 70%-90% background, containing 60 cases; and 2) a group with less than 70% background, containing 90 cases. This classification was made to explore the potential impact of different background ratios on model performance. To further enhance the smoothness of the segmentation mask boundaries, we employed a smoothing algorithm using 3D Slicer software [12]. This preprocessing step was applied to the manually annotated training data to refine the segmentation masks by reducing isolated segmented regions and smoothing the boundaries. The improved segmentation contributes to the robustness and accuracy of the model during training.

In addition, we applied various data augmentation techniques to further improve the generalization capability of the model. These techniques included random rotations, scaling, and mirroring (horizontal and vertical) of images and their corresponding labels. We also introduced Gaussian noise and Gaussian blur to enhance the model's robustness to noise and varying image quality. Adjustments to brightness and contrast were applied randomly to simulate diverse lighting conditions. To mimic low-resolution images, we utilized average pooling for downsampling, simulating a lower resolution. Gamma correction with random gamma values was used to perform nonlinear grayscale mapping. These augmentations were applied dynamically during training using nnU-Net, which randomly combines them as a foundational and general data augmentation strategy to enrich the training dataset and boost model performance.

2.3 Model Architecture

The nnU-Net residual architecture for 3D medical imaging processes data with a detailed and structured approach. Initially, it normalizes the images using Z-Score normalization and resamples the data and segmentation masks to achieve uniform voxel spacing and dimensions. It handles a patch size of $48 \times 192 \times 192$ voxels, with spacing configured to $1.2 \, \text{mm} \times 0.5 \, \text{mm} \times 0.5 \, \text{mm}$. The architecture employs 3D convolution operations with $3 \times 3 \times 3$ kernels and utilizes strides such as $[1, 2]$ and $[2]$ to progressively downsample feature maps, simultaneously increasing the depth and number of features, which start at 32 and rise to 320 in deeper stages. It incorporates 1 to 6 residual blocks per stage, featuring instance normalization and LeakyReLU activation to maintain non-linearity and computational efficiency. Overall, this complex setup with six stages is designed to capture a comprehensive range of features from medical images, facilitating accurate segmentation tasks through a deep learning network. For more detailed information and specifics, refer to the nnU-Net's residual encoder-generated plans.json file. For more detailed information, refer to the plans.json file generated after using the nnU-Net residual architecture code.

2.4 Model Training Approach

In the training phase, we initially conducted mixed training with 450 samples containing pre, mid, and pre-registered images, completing a total of 1600 iterations. The learning rate was initialized at 0.01 and gradually decreased to 0 by the end of training. We used an SGD optimizer with a weight decay of 3×10^{-5}, and the batch size was set to 2. Additionally, we employed 5-fold cross-validation to ensure robust evaluation. Dice and cross-entropy were used as loss functions to guide the optimization process.

During the subsequent testing phase, we observed that the model performed poorly on certain parts of the dataset. Analysis revealed that this issue was primarily caused by a high proportion of background (labeled as 0) in the dataset. Notably, in these imaging data, the proportion of background ranged from approximately 4% to 90%.

To more effectively segment data with a higher background ratio, we selected samples where the background constituted about 70% to 90%, forming a separate dataset for training. Similarly, we included pre-registered data with a background ratio of 50% to 80% in the training. Both of these additional training phases underwent 500 iterations. For mid images, we adopted a similar strategy, selecting samples with a background ratio of 70% to 90% for independent training, which also involved 500 iterations. Importantly, during these subsequent training phases, all hyperparameters, including the learning rate schedule, optimizer (SGD with a weight decay of 3×10^{-5}), batch size (2), and loss functions (Dice and cross-entropy), were kept consistent. The only difference was the number of iterations, which was adjusted based on the dataset.

The associated training curves are shown in Fig. 1. The visualization metric used here was the Dice coefficient, rather than the DSCagg (aggregated Dice Similarity Coefficient) used by the official competition. The Dice coefficient served only as a display, with final results based on DSCagg. Through stratified training, we discovered that segmentation accuracy improved when the model was specifically trained on images with higher background ratios, suggesting that adapting the training process to the characteristics of the dataset, such as background proportions influenced by the field of view and underlying tumor volumes, can enhance model performance. Additionally, considering the inherent variations in imaging quality, such as resolution indicated by background ratio, we found that training dedicated models for these specific data distributions yielded better results. A simple cropping operation, while effective in reducing unnecessary elements, was evaluated but ultimately deemed less critical than focused training, as the trained model tends to ignore large zero-value areas and concentrate on regions with pixel activity. For images with a high background ratio, after analysis by the model, train the model using preprocessing parameters different from those used for images with a low background ratio, so that it focuses solely on their internal distributions, thereby enhancing the segmentation capabilities of the model to a certain extent.

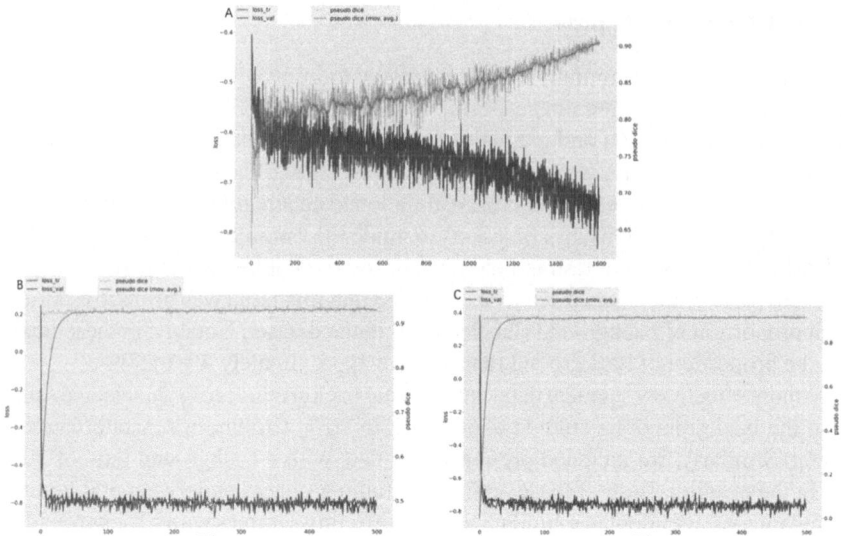

Fig. 1. Training curves for three models (A) Training curve for the mixed model (B) Training curve for the Pre-RT segmentation model with a high background ratio (C) Training curve for the Mid-RT segmentation model.

3 Results

Figure 2 shows the segmentation example results from our study. It is evident that for imaging data with a lower background proportion, the segmentation performance is quite satisfactory, effectively extracting the target areas with high accuracy and robustness. This indicates that our model performs well in scenarios with minimal distractions, accomplishing the segmentation tasks effectively. However, for imaging data with a higher background proportion, the results are less than ideal, possibly due to interference from background information, which complicates the model's ability to distinguish between target and background [13]. We believe further optimization of the model is necessary to enhance its performance in complex background situations. Additionally, Table 1 presents the results from the test set after our competition submission. It is observed that, compared to the segmentation of GTVp, our model demonstrates superior performance in handling the segmentation tasks for GTVn. Specifically, the mean DSCagg values, which were used in the actual challenge ranking, are 0.8380 for GTVn (combining 0.8453 from one test subset and 0.8307 from another) and 0.6065 for GTVp (combining 0.7674 from one test subset and 0.4456 from another). These results underline our model's enhanced capability in accurately segmenting metastatic lymph nodes compared to primary tumor volumes.

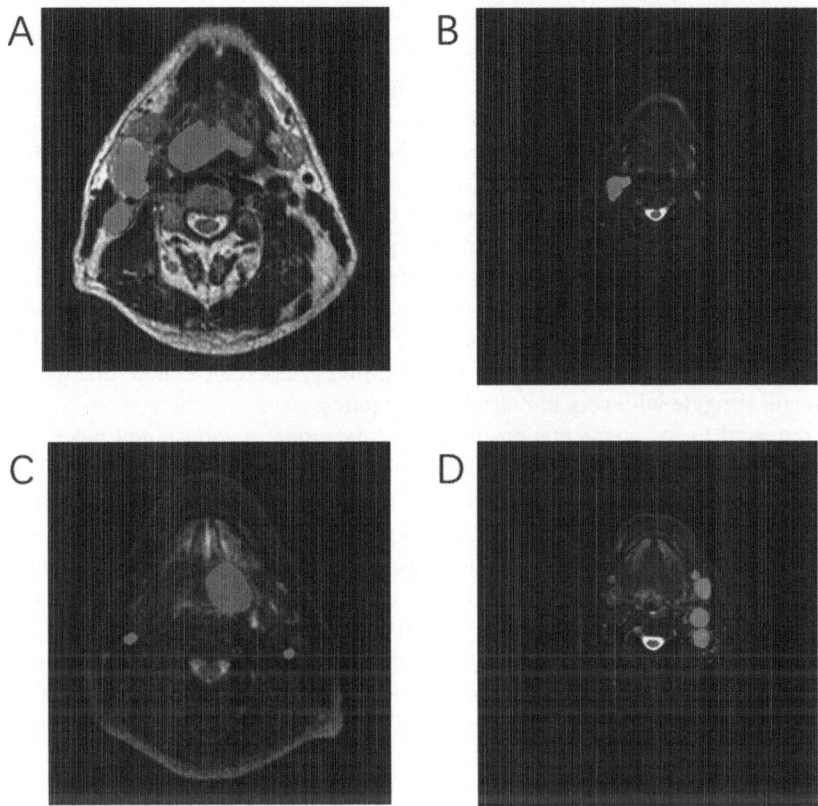

Fig. 2. Head tumor segmentation results for Pre-RT and Mid-RT segmentation. Green and yellow represent the annotation results for GTVp and GTVn, respectively, while red and blue represent the segmentation results for GTVp and GTVn, respectively. (A) Pre-RT tumor segmentation with low background proportion. (B) Pre-RT tumor segmentation with high background proportion. (C) Mid-RT tumor segmentation with low background proportion. (D) Mid-RT tumor segmentation with high background proportion. (Color figure online)

Table 1. Test set results for ensemble models. Metrics are reported from the HNTS-MRG 2024 submission portal.

Task	Tumor	DSCagg
Pre-RT segmentation	GTVn	0.8453
	GTVp	0.7674
Mid-RT segmentation	GTVn	0.8307
	GTVp	0.4456

4 Discussion

In this study, we explored the impact of varying background ratios on the performance of segmentation models. By training with a mix of 450 samples, we initially established a baseline model. However, during the testing phase, we observed that the model performed poorly on images with higher background ratios. This phenomenon highlighted the importance of dataset distribution balance in model training [14, 15].

To address this issue, we implemented targeted measures by selecting samples with higher background ratios from the dataset to construct a new training set [16]. By concentrating the training on these samples, we aimed to enhance the model's ability to segment complex backgrounds. This strategy somewhat mitigated the impact of background noise on segmentation effectiveness, yet it is important to note that the model might still struggle with very high background ratios.

Compared to the segmentation of GTVp, our model demonstrated better performance in segmenting GTVn. This could be due to the more distinct features of GTVn in images, or because the model learned the characteristics of GTVn more thoroughly. This difference indicates that, although the model is generally stable, there is still room for optimization when processing different types of tumor volumes. Understanding these subtle differences will help us further adjust and optimize the algorithm to enhance the model's applicability and accuracy in various clinical scenarios [17, 18].

In future research, we plan to continue optimizing the model structure and explore using more data augmentation techniques and regularization methods to improve the model's generalization ability in complex backgrounds [19]. Additionally, we hope to introduce more representative samples to better adapt the model to varying imaging data distributions, ultimately achieving more ideal segmentation results. Through ongoing experiments and improvements, we look forward to making significant progress in the field of medical image segmentation and providing more reliable technological support for clinical applications [20, 21].

Our study still faces limitations, as we have not improved the model to enhance the performance of GTVp. Additionally, we have not fully utilized the information from the other four inputs in the Mid-RT segmentation task.

5 Conclusion

This study employed 3D nnU-Net with a residual encoder, focusing on the complex task of head and neck tumor segmentation. We utilized meticulous data preprocessing and a multi-model training strategy to cater to various clinical segmentation needs. Specifically, according to Table 2, we achieved an excellent performance in the Pre-RT segmentation task with a DSCagg mean value of 0.8064, while the Mid-RT segmentation task yielded a mean DSCagg of 0.6381, indicating room for improvement.

Moreover, an analysis of the specific GTVp label performance in these two tasks, and their mean values, clearly highlighted the deficiencies in the model's segmentation capabilities for this label. Future efforts were planned to focus on optimizing algorithms and adjusting the network structure, particularly improvements in GTVp segmentation, to enhance model accuracy and reliability in complex clinical applications. These improvements were not only expected to increase the precision of tumor monitoring before and after treatment but were also anticipated to provide more accurate treatment planning support in clinical settings.

Acknowledgements. During the completion of this research, we would like to express our gratitude to East China Normal University, Shanghai Jiao Tong University and Shanghai Ninth People's Hospital for providing the materials and support. This research was supported by the National Natural Science Foundation of China (NSFC, grant No. 62322114); and the Fundamental Research Funds for the Central Universities, China (No. YG2023LC06). The funders participated in the study design, data collection, data analysis, data interpretation, manuscript preparation, and the decision to submit the article for publication.

References

1. Richmon, J.D., et al.: Does current training in radiation oncology prepare radiation oncologists to optimally manage patients with head and neck cancer? Am. J. Clin. Oncol. **46**, 281–283 (2023). https://doi.org/10.1097/COC.0000000000001019

2. Cardenas, C.E., et al.: Comprehensive quantitative evaluation of variability in magnetic resonance-guided delineation of oropharyngeal gross tumor volumes and high-risk clinical target volumes: an R-IDEAL stage 0 prospective study. Int. J. Radiat. Oncol. **113**, 426–436 (2022). https://doi.org/10.1016/j.ijrobp.2022.01.050

3. Sherer, M.V., et al.: Metrics to evaluate the performance of auto-segmentation for radiation treatment planning: a critical review. Radiother. Oncol. **160**, 185–191 (2021). https://doi.org/10.1016/j.radonc.2021.05.003

4. Tajbakhsh, N., Jeyaseelan, L., Li, Q., Chiang, J.N., Wu, Z., Ding, X.: Embracing imperfect datasets: a review of deep learning solutions for medical image segmentation. Med. Image Anal. **63**, 101693 (2020). https://doi.org/10.1016/j.media.2020.101693

5. Isensee, F., Jaeger, P.F., Kohl, S.A.A., Petersen, J., Maier-Hein, K.H.: NnU-Net: a self-configuring method for deep learning-based biomedical image segmentation. Nat. Methods **18**, 203–211 (2021). https://doi.org/10.1038/s41592-020-01008-z

6. Li, Y., et al.: NPCNet: jointly segment primary nasopharyngeal carcinoma tumors and metastatic lymph nodes in MR images. IEEE Trans. Med. Imaging **41**, 1639–1650 (2022). https://doi.org/10.1109/TMI.2022.3144274

7. Zhang, R., et al.: Deep learning for the automatic detection and segmentation of parotid gland tumors on MRI. Oral Oncol. **152**, 106796 (2024). https://doi.org/10.1016/j.oraloncology.2024.106796

8. Choi, Y., Bang, J., Kim, S.-Y., Seo, M., Jang, J.: Deep learning–based multimodal segmentation of oropharyngeal squamous cell carcinoma on CT and MRI using self-configuring nnU-Net. Eur. Radiol. **34**, 5389–5400 (2024). https://doi.org/10.1007/s00330-024-10585-y

9. Silva, H.E.C.D., et al.: The use of artificial intelligence tools in cancer detection compared to the traditional diagnostic imaging methods: an overview of the systematic reviews. PLoS ONE **18**, e0292063 (2023). https://doi.org/10.1371/journal.pone.0292063

10. Hindocha, S., et al.: Artificial intelligence for radiotherapy auto-contouring: current use, perceptions of and barriers to implementation. Clin. Oncol. **35**, 219–226 (2023). https://doi.org/10.1016/j.clon.2023.01.014

11. Wahid, K., Dede, C., Naser, M., Fuller, C.: Training Dataset for HNTSMRG 2024 Challenge (2024). https://doi.org/10.5281/ZENODO.11199559

12. Bahkali, I.M., Semwal, S.K.: Multiple ways for medical data visualization using 3D slicer. In: 2020 International Conference on Computational Science and Computational Intelligence (CSCI), pp. 793–800. IEEE, Las Vegas (2020). https://doi.org/10.1109/CSCI51800.2020.00149

13. Aniraj, A., Dantas, C.F., Ienco, D., Marcos, D.: Masking strategies for background bias removal in computer vision models. In: 2023 IEEE/CVF International Conference on Computer Vision Workshops (ICCVW), pp. 4399–4407. IEEE, Paris (2023). https://doi.org/10.1109/ICCVW60793.2023.00474

14. Cao, K., Wei, C., Gaidon, A., Arechiga, N., Ma, T.: learning imbalanced datasets with label-distribution-aware margin loss (2019)

15. Buda, M., Maki, A., Mazurowski, M.A.: A systematic study of the class imbalance problem in convolutional neural networks. Neural Netw. **106**, 249–259 (2018). https://doi.org/10.1016/j.neunet.2018.07.011

16. Liu, Y., Yu, X., Huang, J.X., An, A.: Combining integrated sampling with SVM ensembles for learning from imbalanced datasets. Inf. Process. Manag. **47**, 617–631 (2011). https://doi.org/10.1016/j.ipm.2010.11.007

17. Surucu, M., et al.: Verification of a machine learning algorithm that predict volume reduction in primary and nodal tumor volumes in head and neck cancer during treatment. Int. J. Radiat. Oncol. **99**, E375 (2017). https://doi.org/10.1016/j.ijrobp.2017.06.1496

18. Tanaka, S., et al.: A deep learning-based radiomics approach to predict head and neck tumor regression for adaptive radiotherapy. Sci. Rep. **12**, 8899 (2022). https://doi.org/10.1038/s41598-022-12170-z

19. Wang, Z., Xie, X., Yang, J., Shi, G.: Soft focal loss: evaluating sample quality for dense object detection. Neurocomputing **480**, 271–280 (2022). https://doi.org/10.1016/j.neucom.2021.12.102

20. Luo, X., et al.: SegRap2023: A benchmark of organs-at-risk and gross tumor volume segmentation for radiotherapy planning of nasopharyngeal carcinoma (2023). https://arxiv.org/abs/2312.09576. https://doi.org/10.48550/ARXIV.2312.09576

21. Oreiller, V., et al.: Head and neck tumor segmentation in PET/CT: the HECKTOR challenge. Med. Image Anal. **77**, 102336 (2022). https://doi.org/10.1016/j.media.2021.102336

Ensemble Deep Learning Models for Automated Segmentation of Tumor and Lymph Node Volumes in Head and Neck Cancer Using Pre- and Mid-Treatment MRI: Application of Auto3DSeg and SegResNet

Dominic LaBella[✉] [iD]

Department of Radiation Oncology, Duke University Medical Center, Durham, NC 27710, USA
dominic.labella@duke.edu

Abstract. Automated segmentation of gross tumor volumes (GTVp) and lymph nodes (GTVn) in head and neck cancer using MRI presents a critical challenge with significant potential to enhance radiation oncology workflows. In this study, we developed a deep learning pipeline based on the SegResNet architecture, integrated into the Auto3DSeg framework, to achieve fully-automated segmentation on pre-treatment (pre-RT) and mid-treatment (mid-RT) MRI scans as part of the DLaBella29 team submission to the HNTS-MRG 2024 challenge. For Task 1, we used an ensemble of six SegResNet models with predictions fused via weighted majority voting. The models were pre-trained on both pre-RT and mid-RT image-mask pairs, then fine-tuned on pre-RT data, without any pre-processing. For Task 2, an ensemble of five SegResNet models was employed, with predictions fused using majority voting. Pre-processing for Task 2 involved setting all voxels more than 1 cm from the registered pre-RT masks to background (value 0), followed by applying a bounding box to the image. Post-processing for both tasks included removing tumor predictions smaller than 175–200 mm^3 and node predictions under 50–60 mm^3. Our models achieved testing DSCagg scores of 0.72 and 0.82 for GTVn and GTVp in Task 1 (pre-RT MRI) and testing DSCagg scores of 0.81 and 0.49 for GTVn and GTVp in Task 2 (mid-RT MRI). This study underscores the feasibility and promise of deep learning-based auto-segmentation for improving clinical workflows in radiation oncology, particularly in adaptive radiotherapy. Future efforts will focus on refining mid-RT segmentation performance and further investigating the clinical implications of automated tumor delineation.

Keywords: Artificial Intelligence · Automated Segmentation · Deep learning · Auto3DSeg · MONAI · Head and neck tumors · Radiation Oncology

1 Introduction

The 2024 Head and Neck Tumor Segmentation (HNTS-MRG 24) challenge aims to advance the precision of oropharyngeal cancer (OPC) treatment by improving gross tumor volume (GTV) delineation through novel imaging techniques, particularly focusing on MRI T2 sequences in both pre-radiotherapy and mid-radiotherapy settings. OPC,

© The Author(s) 2025
K. A. Wahid et al. (Eds.): HNTS-MRG 2024, LNCS 15273, pp. 259–273, 2025.
https://doi.org/10.1007/978-3-031-83274-1_21

a subtype of head and neck squamous cell carcinoma, continues to affect a significant population worldwide [1]. Accurate tumor segmentation is vital for delivering effective radiation therapy, as it ensures an optimal therapeutic dose to the tumor while sparing surrounding healthy tissues [2]. Traditionally, segmentation relied on computed tomography (CT) and positron emission tomography (PET) to provide both anatomical and functional information, but these modalities often face limitations due to inter-observer variability and inconsistency in tumor delineation [3, 4]. To address this, MRI T2 has emerged as a powerful modality, particularly for head and neck cancers, as it offers superior soft tissue contrast, which is essential for more precise tumor visualization and improved treatment planning [5]. MRI T2 is particularly useful in the context of OPC, where accurate identification of tumor boundaries in pre-radiotherapy and mid-radiotherapy stages can significantly impact treatment outcomes [6, 7]. Leveraging MRI T2 in these critical phases can enhance the accuracy of tumor tracking and reduce variability in segmentation, thus improving radiation targeting, especially in the adaptive radiotherapy planning setting [8].

This work presents the results of our OPC auto-segmentation model utilizing MRI T2 sequences, focusing on the pre-radiotherapy and mid-radiotherapy setting, predicting both gross tumor and nodal volumes, as part of the 2024 HNTS-MRG challenge.

2 Methods

We utilized a deep learning architecture, Auto3DSeg, in conjunction with the MONAI framework to develop auto-segmentation models for oropharyngeal cancer (OPC) patients [9, 10]. For Task 1, the model was trained to predict tumor and node volumes using pre-radiotherapy MRI T2 images for segmentation. Task 2 extended this approach by incorporating both registered-pre-radiotherapy and mid-radiotherapy MRI T2 images with the registered-pre-radiotherapy ground truth GTVp and GTVn labels to predict GTVp and GTVn volumes at the mid-radiotherapy stage.

2.1 Imaging Data

The dataset used in this study was released through the HNTS-MRG 2024 challenge at MICCAI 2024, as available on Zenodo [11]. This dataset includes MRI scans and corresponding segmentation masks for patients with head and neck cancer, primarily oropharyngeal cancer, who underwent radiotherapy at The University of Texas MD Anderson Cancer Center. The dataset consists of pre-radiotherapy T2-weighted MRI scans taken 1–3 weeks before the start of radiotherapy and mid-radiotherapy T2-weighted MRI scans taken 2–4 weeks into treatment. For each patient, segmentation masks for the gross tumor volumes and involved lymph nodes are provided, derived from multi-observer STAPLE consensus.

The HNTS-MRG 2024 challenge is divided into two tasks: Task 1 involves segmenting tumor and nodal volumes using pre-RT MRI, while Task 2 requires segmentation on mid-RT MRI using the registered-pre-radiotherapy image, mid-radiotherapy image, and the registered-pre-radiotherapy ground GTVp and GTVn masks. A single training dataset containing 150 unique patients has been provided for both tasks. During the testing phase, different test data will be used for each task and this data is private to the organizers.

All imaging data, including segmentation masks, are provided in Neuroimaging Informatics Technology Initiative (NIfTI) format to facilitate ease of use. Pre-RT and mid-RT MRI scans are available in either fat-suppressed or non-fat-suppressed versions, with no exogenous contrast agents used. The dataset also includes registered pre-RT images and segmentation masks aligned to the mid-RT image space, generated using SimpleITK with both rigid and deformable transformations for ease of comparison in real-world adaptive radiotherapy settings [12–14].

The dataset is structured in a standardized format and contains anonymized patient identifiers. Pre-RT and mid-RT data are organized by patient ID and timepoint, with consistent labeling conventions across cases.

In order to develop effective pre- and post-processing techniques for challenge test set evaluation for both Task 1 and Task 2, analysis of the training sets GTVp and GTVn instance-lesion mask volumes was performed. Specifically, radial shrinkage was performed at varying integer amplitudes from 0–5 mm for the registered-pre-radiotherapy instance-lesion mask volumes for comparison to the respective mid-radiotherapy instance-lesion mask volumes. Analysis of the GTVp and GTVn instance-lesion mask volumes for the entire 150 cases training set was performed and the associated analysis code is available at (github.com/dlabella29/HNTSMRG_2024_DL_Public). The code to perform the inner margin radial shrinkage is available under the erode_mask method of the calcDiceMarginsTask2allPreRT.py script.

2.2 Image Pre-Processing

Task 1
There was no pre-processing performed for Task 1 of any image or mask data. All of the unmodified pre-radiotherapy, registered-pre-radiotherapy, and mid-radiotherapy image-mask pairs were used in an initial pre-trained model as shown in Fig. 1. Only the unmodified pre-radiotherapy image-mask pairs were included in the fine-tuned models as shown in Fig. 1A.

Task 2
Task 2 pre-processing included modification of both the registered-pre-radiotherapy and the mid-radiotherapy T2 MRI, which are shown in Figs. 2A and 2B. The modifications were performed by radially expanding the GTVp and GTVn labels by 1 cm. This expanded region was then used as a foreground for cropping of the original registered-pre-radiotherapy and the mid-radiotherapy T2 MRI. Then, an additional 0.5 cm x, y, and z directional expansion was made on the bounding box of the new cropped foreground to make the final modified registered-pre-radiotherapy and the mid-radiotherapy T2 MRI, which are shown in Figs. 2C and 2D. This pre-processing was performed to allow for a smaller focused area for SegResNet training on the most pertinent areas of GTVp and GTVn involvement. Since the pre-radiotherapy masks are provided as input data for the Task 2 challenge, and since training set anaylsis showed that there was no significant evidence of GTVp or GTVn growth greater than 1 cm between registered-pre-radiotherapy and mid-radiotherapy cases, there was little concern for potential exclusion of GTVp or GTVn disease on the mid-radiotherapy MRI.

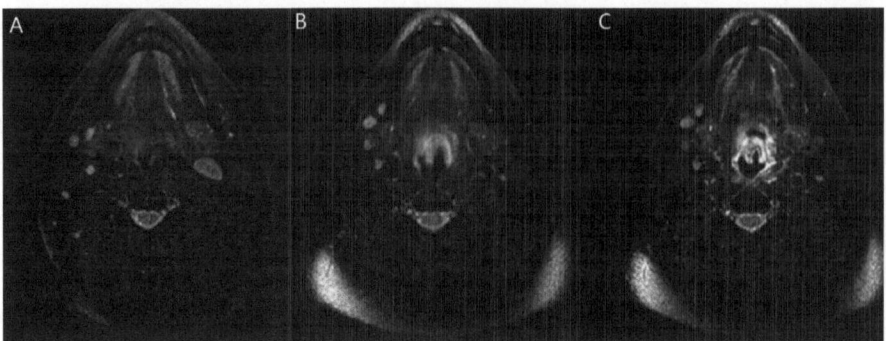

Fig. 1. An illustration of the unmodified axial T2 MRI of case number 155 with tumor label (red) and nodal labels (green) for the (A) pre-radiotherapy, (B) registered pre-radiotherapy, and (C) mid-radiotherapy image-mask pairs. (Color figure online)

2.3 Model Architecture and Implementation

We employed a deep learning convolutional neural network model based on the Seg-ResNet architecture, implemented using the MONAI and Auto3DSeg frameworks [9, 10]. The Auto3DSeg framework is an automated medical image segmentation platform designed to streamline the development, training, and inference of 3D segmentation models [9, 10]. It integrates state-of-the-art deep learning algorithms with a modular, pipeline-based architecture to efficiently handle the entire workflow. Auto3DSeg leverages advanced techniques such as automated hyperparameter tuning, architecture search, and multi-fold training strategies to optimize model performance for specific datasets [9, 10]. Its flexibility allows for customization while offering pre-defined configurations for various segmentation tasks [9, 10]. The framework supports diverse input formats, multi-class labeling, and domain-specific constraints, making it a robust solution for clinical and research applications [9, 10]. SegResNet is an encoder-decoder based semantic segmentation network, with the initial filters set to 32 [9, 10]. The encoder used 5 ResNet blocks with instance normalization. The downsampling included 5 stages with 1, 2, 2, 4, and 4 convolutional blocks, respectively. Data augmentation included random flipping on all axes, random rotation and scaling, random smoothing, noise, intensity scale and shifting. We trained the model using global (patient level or traditional volumetric) Dice scores and focal cross entropy loss. We used an AdamW optimizer with a learning rate of 0.0002 and a weight decay of 0.00001. We used a batch size of 1 and 1 image per batch was used. The MRI T2 images served as input, with each voxel classified as either background, GTVp, or GTVn in the output mask. In Task 1, only the pre-radiotherapy T2 MRI was used as input. In Task 2, the modified registered-pre-radiotherapy and modified mid-radiotherapy T2 MRI were used as input. No further data-augmentation was performed in order to limit the size of the training dataset to speed up training time. All training was conducted on a laptop computer with an RTX 2070 GPU and 16 GB of available RAM.

For Task 1, the model training involved two stages. First, a pre-trained model was developed over 802 epochs, with a planned total of 900 epochs. Pre-training was not

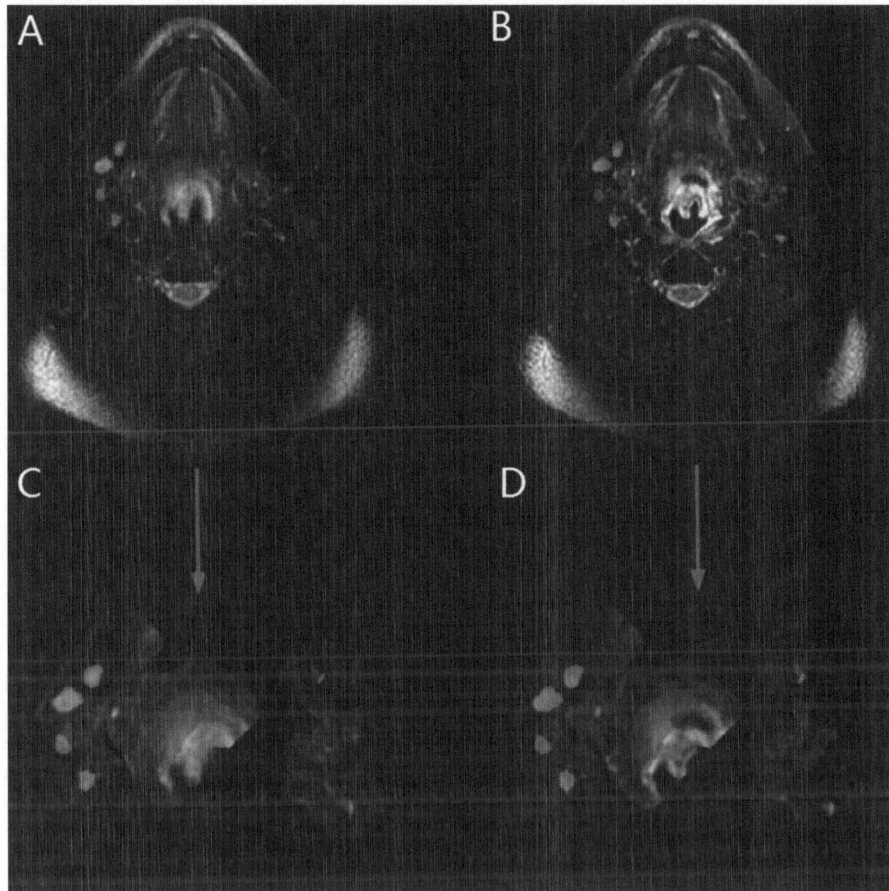

Fig. 2. An illustration of the axial T2 MRI of case number 155 demonstrating pre-processing completed for the Task 2 training dataset which included initial (A) registered-pre-radiotherapy T2 MRI and (B) mid-radiotherapy T2 MRI that were reduced in size and foreground cropping to 1 cm radially around all GTVp (red) and GTVn (green) labels as shown in (C) and (D). (Color figure online)

completed to all 900 planned epochs due to model crashing during training and limitation on time to allow for re-training. The pre-trained dataset included each of the pre-radiotherapy, mid-radiotherapy, and registered-pre-radiotherapy image-mask pairs, all as separate training cases as shown in Fig. 1. Fine-tuning of this pre-trained model was conducted. The fine-tuning dataset only included the pre-radiotherapy image-mask pairs shown in Fig. 1A, as the pre-radiotherapy MRI are the only images provided as input data in the Task 1 challenge. There was a plan for 556 epochs using cross-validation with a planned 8 folds. The folds were split to equally have 12.5% of the total training set data, with no data left out for independent testing. The Json files for training splits, Python scripts to generate the Json files, and Python scripts to conduct model training are available at (github.com/dlabella29/HNTSMRG_2024_DL_Public). The number of epochs of 556 was predicted by the Auto3DSeg architecture [9, 10]. Unfortunately, due to training time limitations, only 6 of the planned 8 folds underwent fine-tune training. The models for the 7th and 8th folds were not run. Fold 0 utilized the same validation set's pre-radiotherapy MRI as the pre-trained model's pre-radiotherapy pre-training dataset, and this fold successfully completed all 556 epochs. Folds 1–5 used new validation set's pre-radiotherapy images compared to the pre-trained model's training dataset, but each of these respective validation set's images were a part of the pre-training datasets training images. Therefore, artificially high validation scores were achieved early on during training during the fine-tuning for folds 1–5 as shown in Table 1. Additionally, not all of the folds 1–5 made it to completion as shown in Table 1. Re-training was not feasible due to time limitations. A weighted majority voting ensembling technique was employed with the model weights shown in Table 1. Fold 1 was assigned the highest weight, since this fine-tuned model used the same validation pre-radiotherapy image-mask pairs as those utilized in the pre-trained model. Folds 1 and 5 had lower relative weights due to the models crashing early during the training process. The best validation scores were achieved early on during each of the folds 0–5 fine-tuning. This is likely due to over-fitting on the features associated with the validation cases that were previously used in the pre-trained model's training sets. Therefore, each fold's respective final model was used for inference instead of the best epoch models.

Table 1. Task 1 fold-wise overall accuracy for the fine-tuned SegResNet models. The overall accuracy (Best Validation Dice or Final Validation Dice) was computed as the average of the GTVp and GTVn global (patient level or traditional volumetric) Dice scores. The models were trained over various epochs with cross-validation. Note that fold 0 used the exact same validation cases as the pre-trained model, whereas folds 1–5 each had 12.5% of their validation cases within the pre-trained model's training sets; therefore explaining the artificially high Best Validation Dice.

Fold	Epochs Completed	Best Validation Dice (GTVp, GTVn)	Best Epoch	Final Validation Dice (GTVp, GTVn)	Model Weight for Ensembling
0	556	0.7292 (0.694, 0.765)	59	0.7004 (0.664, 0.737)	0.2
1	335	0.8766 (0.881, 0.872)	9	0.8522 (0.847, 0.858)	0.1375
2	556	0.8814 (0.873, 0.889)	9	0.8748 (0.858, 0.892)	0.175
3	556	0.8651 (0.868, 0.862)	19	0.8593 (0.848, 0.871)	0.175
4	556	0.8785 (0.863, 0.894)	9	0.8612 (0.850, 0.872)	0.175
5	465	0.8719 (0.866, 0.878)	9	0.853 (0.836, 0.870)	0.1375

For Task 2, the model training involved a single stage without any pre-training or fine-tuning. A total of 534 epochs were successfully completed for each of the 5 planned folds. The folds were split to equally have 20% of the total training set data, with no data left out for independent testing. Training summaries are provided in Table 2.

For Task 1 and Task 2, all the Python code for pre-processing, training, inference, ensembling, and post-processing are publicly available (github.com /dlabella29/HNTSMRG_2024_DL_Public). The inference code was modified from the Auto3DSeg architecture inference methods (github.com/Project-MONAI /tutorials/tree/main/auto3dseg).

Table 2. Task 2 fold-wise overall accuracy for the SegResNet models. The overall accuracy (Best Validation Dice or Final Validation Dice) was computed as the average of the GTVp and GTVn global (patient level or traditional volumetric) Dice scores. The models were trained over 534 planned epochs with 5-fold cross-validation.

Fold	Epochs Completed	Best Validation Dice (GTVp, GTVn)	Best Epoch	Final Validation Dice (GTVp, GTVn)	Model Weight for Ensembling
0	534	0.5967 (0.530, 0.663)	456	0. 5859 (0.513, 0.659)	0.2
1	534	0.6223 (0.559, 0.686)	466	0.6160 (0.546, 0.686)	0.2
2	534	0.6578 (0.521, 0.795)	340	0.6541 (0.510, 0.798)	0.2
3	534	0.5879 (0.433, 0.743)	300	0.5620 (0.368, 0.756)	0.2
4	534	0.6341 (0.492, 0.776)	332	0.6287 (0.488, 0.769)	0.2

2.4 Model Post-Processing

After ensembling was conducted to generate GTVp and GTVn preliminary masks, post-processing for both Task 1 and Task 2 was performed. Post-processing involved filtering out predicted instance-lesions based on volume thresholds to minimize false positives. Specifically, for Task 1, GTVp instance-lesion predictions smaller than 200 mm^3 and GTVn instance-lesion predictions under 60 mm^3 were excluded. In Task 2, slightly lower thresholds were applied, with GTVp instance-lesion predictions under 175 mm^3 and GTVn instance-lesion predictions below 50 mm^3 being removed. These volume cutoffs were clinically determined after analyzing GTVp and GTVn volumes on a per-lesion basis across the entire 150-case dataset as described in Sect. 3.1. Additionally, for Task 2, a 1 cm expansion was applied to the registered pre-radiotherapy ground truth mask to accommodate expected GTVp and GTVn boundary extremes as shown in Fig. 3.

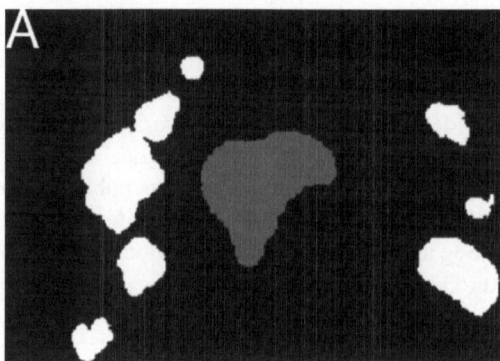

Fig. 3. An illustration of the axial registered-pre-radiotherapy mask of case number 155 which demonstrates the 1cm expansion around all ground truth GTVp and GTVn labels. This processed mask was used in the post-processing stage of inference for Task 2.

Predicted voxel labels for the mid-radiotherapy GTVp and GTVn outside of this expanded region were discarded, as they were presumed to be false positives. This post-processing step was crucial in enhancing prediction accuracy and ensuring that the GTVp and GTVn delineations remained clinically relevant for both tasks. Additionally, if there were any instances where a GTVp and a GTVn are directly touching, then post-processing determined whether a larger and separate GTVp lesion existed. If there was another GTVp instance-lesion, then the touching GTVp and GTVn was all converted to GTVn. If there was no other GTVp instance-lesion, then the touching GTVp and GTVn was all converted to GTVp.

2.5   Final Testing Evaluation

The HNTS-MRG 24 challenge final testing phase metrics included the DSCagg for GTVp and GTVn on the organizer's hidden testing set. DSCagg has historically been used for head and neck tumor and node automated segmentation challenges given its ability to account for multiple lesions [15, 16]. Additionally, the mean DSCagg was computed based on the average of the GTVp DSCagg and GTVn DSCagg. Equation 1 shows the formula for computation of the DSCagg, where N is the total number of test images, $y_{\{i,k\}}$ is the ground truth mask for either GTVp or GTVn for voxel k of image i, and $\hat{y}_{\{i,k\}}$ is the model's prediction mask [11, 15–17].

$$DSC_{agg} = \frac{2\sum_i^N \sum_k \left(\hat{y}_{\{i,k\}} \cdot y_{\{i,k\}}\right)}{\sum_i^N \sum_k \left(\hat{y}_{\{i,k\}} + y_{\{i,k\}}\right)} \tag{1}$$

All metrics were computed automatically on the Grand Challenge host website using the HNTS-MRG 24 organizers evaluation metric code (https://hntsmrg24.grand-challe nge.org/).

3 Results

3.1 Ground Truth Challenge Dataset Results

Analysis of the complete ground truth HNTS-MRG 24 dataset demonstrated that 12 GTVp instances in the registered pre-RT images had a volume under 16 mm^3 among cases 3, 60, 110, 125, 125, 125, 161, 164, 169, 169, 179, 193. The next smallest GTVp instance had a volume of 192 mm^3 for case 34. Analysis demonstrated that 13 GTVn instances in the registered-pre-radiotherapy images had a volume under 40 mm^3. The next smallest GTVn instance had a volume of 98 mm^3 for case 184. Analysis demonstrated that only 1 mid-radiotherapy GTVp instance had a volume of 49 mm^3 for case 149 and the next smallest mid-radiotherapy GTVp instance had a volume of 231.5 mm^3 for case 155. Analysis demonstrated that only 1 mid-radiotherapy GTVn instance had a volume of 0.66 mm^3 for case 191 and the next smallest mid-radiotherapy GTVn instance had a volume of 64.5 mm^3 for case 155. Note that all of the instance GTVp and GTVn volume calculations were performed on a lesion-wise level with 26-connected component analysis to determine distinct instance lesions. These findings suggest that post-processing thresholds should be used to remove potential false-positive instance-lesions.

Further training set analysis compared the ground truth registered-pre-radiotherapy tumor and node masks compared to the ground truth mid-radiotherapy GTVp and GTVn masks. Analysis demonstrated a lesion-wise aggregated DSCagg of 0.448 for GTVp and 0.725 for GTVn ground truth comparisons amongst the registered-pre-radiotherapy and the mid-radiotherapy masks. Further ground truth lesion-wise DSCagg comparisons were made of the registered-pre-radiotherapy ground truth masks when radially shrunken by 1–5 mm compared to the mid-radiotherapy ground truth masks. These values are shown in Table 3. These findings suggest that a simple radial reduction method from the registered-pre-radiotherapy mask could be utilized as an algorithm by itself, although this would be non-patient and non-adaptive-image specific.

Table 3. These data show the ground truth Dice comparisons of the registered-pre-radiotherapy ground truth masks when radially shrunken by 0–5 mm compared to the mid-radiotherapy ground truth masks.

Registered-pre-radiotherapy mask reduction amplitude	GTVp Dice	GTVn Dice
0 mm	0.448	0.725
1 mm	0.135	0.847
2 mm	0.490	0.651
3 mm	0.441	0.518
4 mm	0.370	0.412
5 mm	0.331	0.313

3.2 Final Testing Phase Results

The Task 1 average cross-validation global (patient level or traditional volumetric) Dice scores at the best epochs were 0.841, 0.860, and 0.851 for GTVp, GTVn, and overall accuracy, respectively. These values are the averages of the fold level values shown in Table 1. The submitted weighted majority voting ensembled model for Task 1 achieved testing phase DSCagg scores of 0.82, 0.72, and 0.77 for GTVp, GTVn, and overall accuracy, respectively.

The Task 2 average cross-validation global (patient level or traditional volumetric) Dice scores at the best epochs were 0.507, 0.727, and 0.617 for GTVp, GTVn, and overall accuracy, respectively. These values are the averages of the fold level values shown in Table 2. The submitted majority voting ensembled model for Task 2 achieved testing phase DSCagg scores of 0.49, 0.81, and 0.65 for GTVp, GTVn, and overall accuracy, respectively.

For both Task 1 and Task 2, given that there was no left out challenge cases for independent model testing evaluation of the DSCagg, it was not possible to perform comparison of local independent testing vs the provided testing phase performance for DSCagg since the loss used in the model's validation evaluation was the global (patient level or traditional volumetric) Dice score. However, it is hypothesized that the interim best epoch validation traditional DSC scores are higher than the testing phase DSCagg given that they report the best epoch validation DSC across the entire model training and there may not have been significant DSCagg penalties using the traditional DSC for smaller missed lesions.

The final testing phase metrics were provided by the challenge organizers. No other final test phase metrics are available at this time.

4 Discussion

This study utilized a SegResNet deep learning architecture as part of the MONAI and Auto3DSeg frameworks to evaluate and infer OPC primary tumor and nodal disease on T2 MRI in the pre-radiotherapy and mid-radiotherapy settings as part of the HNTS-MRG 24 challenge (hntsmrg24.grand-challenge.org/overview/) [9–11].

The achieved test set dice scores of 0.72 for GTVp and 0.82 for GTVn in Task 1, indicate robust performance in pre-treatment segmentation. These results align with recent advancements in deep learning-based segmentation, where ensemble approaches have consistently outperformed single-model predictions by mitigating individual model biases and enhancing overall accuracy [18, 19].

In Task 2, the DSCagg of 0.49 for GTVp and 0.81 for GTVn reveal a dichotomy in segmentation performance between GTVp and GTVn at the mid-radiotherapy stage. The superior performance in GTVn segmentation can be attributed to the distinct anatomical features and consistent response of lymph nodes to radiotherapy, which may present as more homogeneous changes compared to primary OPC tumors [20, 21]. Conversely, the lower dice score for GTVp in mid-radiotherapy highlights the inherent challenges in accurately capturing dynamic tumor responses, such as heterogeneity in tissue density and irregular shrinkage patterns [22]. This discrepancy underscores the necessity for further refinement of segmentation algorithms to better accommodate the complex morphological transformations of primary tumors during treatment. In future iterations of this study, implementing a larger radial expansion during the pre-processing phase should be considered to encompass a greater portion of the surrounding anatomical structures. This approach may enhance the model's contextual understanding and improve segmentation accuracy. However, in the current study, the application of bounding boxes to generate smaller image regions was essential to address the limitations imposed by training time constraints associated with processing larger images.

When comparing this study's DSCagg performance for GTVp and GTVn based on MRI compared to prior head and neck segmentation challenge performance based on PET/CT imaging, notable differences are appreciated. Salahuddin et al. reports an DSCagg of 0.774 and 0.760 on the test set for GTVp and GTVn, respectively [23]. Additionally, Chu et al. utilized a Swin-UNETR CNN architecture for head and neck tumor automated segmentation based on PET/CT and reported a DSCagg of 0.642, 0.670, and 0.656 for GTVp, GTVn, and overall accuracy, respectively [17]. This study's Task 1 performance DSCagg of 0.82 for GTVn was higher than the prior work on PET/CT automated segmentation, but the DSCagg of 0.72 was similar to the prior work on PET/CT automated segmentation [17, 22]. Additionally, this challenge's Task 2 performance DSCagg of 0.81 for GTVn was higher than the prior work on PET/CT automated segmentation, but the DSCagg of 0.49 was notably lower than the prior work on PET/CT automated segmentation [17, 22]. Further investigation should evaluate the reasoning for differing performance in GTVn and GTVp automated segmentation on PET/CT based imaging compared to pre-radiotherapy and mid-radiotherapy MR imaging.

The lesion-wise instance average Dice scores presented in Table 3 offer valuable insights into the prognostication of tumor and nodal size reductions during radiotherapy for head and neck cancers. Specifically, the GTVp Dice score peaks at a 2 mm radial reduction (0.490), suggesting that a 2 mm decrease in GTVp size from pre-radiotherapy to mid-radiotherapy aligns most closely with the ground truth segmentation. This indicates that, on average, GTVp may undergo a significant reduction of approximately 2 mm during the early phases of radiotherapy, providing a measurable benchmark for treatment efficacy. Conversely, the GTVn Dice score is highest at a 1 mm reduction (0.847), implying that GTVn structures exhibit a more modest decrease in size, with a 1 mm reduction being the most accurate reflection of their true anatomical changes during treatment. These differential reduction patterns between tumors and nodes underscore the heterogeneous nature of tissue responses to radiotherapy. Interestingly, this study's Task 2 model's test set mid-radiotherapy GTVn DSCagg of 0.81 was lower than the simple registered-pre-radiotherapy mask reduction amplitude of 1 mm, which had a Dice score of 0.847 averaged across the entire training dataset when compared to the respective case's mid-radiotherapy GTVn mask. In spirit of the challenge, this study did not utilize the simple 1 mm radial mask reduction, as this should not be a method used in the clinic for adaptive radiotherapy planning as it is not patient specific or on-treatment image specific. Similarly, the Task 2 model's test set mid-radiotherapy GTVp DSCagg of 0.490 was identical to the simple registered-pre-radiotherapy mask reduction amplitude of 2 mm across the whole training set. Further analysis of the testing set using these simple radial reductions algorithms would be interesting for comparison to this challenge's participant's deep learning models. Further studies should also evaluate the inter-observer variability between annotators for each case in the challenge training dataset to determine whether human-human Dice is similar, worse, or better, than the challenge's participant's automated segmentation models when compared to the STAPLE consensus ground truths.

Utilizing the Dice scores associated with radial shrinkage of registered-pre-radiotherapy masks as prognostic markers could allow clinicians to better predict treatment outcomes, tailor radiotherapy plans, and adjust therapeutic strategies in real-time to enhance precision and effectiveness using adaptive radiotherapy planning [8, 24, 25]. The adaptive radiotherapy automated segmentation models could use the radial reduction masks for GTVp and GTVn as additional data when inferring the real time GTVp and GTVn target volume delineation. However, it is essential to consider individual patient variability and the potential influence of factors such as tumor biology and treatment modalities, which may affect the generalizability of these findings. Future studies should aim to validate these prognostic thresholds in larger, diverse cohorts and explore the underlying mechanisms driving the differential responses of tumors and nodes to radiotherapy.

Notably, the DSCagg in the HNTS-MRG 24 challenge is used similarly to the modified lesion-wise DSC used in the 2024 BraTS meningioma radiotherapy automated segmentation challenge as shown in Eq. 2, and it is similar to the lesion-wise DSC used in the BraTS preoperative meningioma, glioma, metastasis, sub-saharan Africa glioma, and pediatric brain tumor automated segmentation challenges as shown in Eq. 3 [26–31].

$$Lesion\,wise\,DSC = \frac{\sum_i^L \text{Dice}(I_i)}{\text{TP} + \text{FN} + \text{FP}} \qquad (2)$$

$$Modified\,lesion\,wise\,DSC = \frac{\sum_i^L \text{Dice}(I_i)}{\text{TP} + \text{FN}} \qquad (3)$$

In Eqs. 2 and 3, L is the number of ground truth lesions. Predicted true positive (TP) + false negative (FN) lesions is equal to L. A predicted lesion is counted as a TP if at least 1 predicted voxel overlaps with the respective ground truth's respective region of interest mask. A lesion is counted as a FN if the model does not predict any voxels within the ground truth's respective region of interest mask. A predicted lesion is counted as a false positive (FP) if the model predicts a distinct lesion that does not overlap with any ground truth lesions' voxels [26].

The modified lesion-wise DSC used in the 2024 BraTS meningioma radiotherapy automated segmentation challenge does not penalize for false positive instance lesion predictions, as radiotherapy planning scans are typically not used for diagnostics, but rather for the contouring workflow [28]. Additional predicted instance lesions on radiotherapy images can easily be removed during the contouring workflow using a Boolean tool or other similar tools and would not significantly detract from the amount of time needed to contour, and therefore false positives can be considered to not need to be penalized in performance metrics. Additionally, localizing and contouring additional lesions that were not automatically contoured could theoretically take excess time and therefore false negatives should still be considered in performance metrics. Future automated segmentation studies and challenges using radiotherapy planning images, should consider not penalizing false positive instance lesions as these can be removed relatively easily during the radiotherapy planning process.

Unfortunately, due to participant time limitations, only the SegResNet network was used as part of the Auto3DSeg architecture, whereas previous studies have utilized each of the SegResNet, DiNTS, and Swin-UNETR CNN and transformer-based components with sliding window inferential ensembling [9, 10, 32]. If additional time was available, then additional models using the DiNTS and Swin-UNETR CNN architectures with an iSTAPLE ensembling technique would have been utilized and future studies should consider using this method.

Future studies should also consider using a larger GPU than this study's RTX 2070 to allow for more complex model structures, as well as more consistent and thorough training. By increasing the number of parameters, increasing the number of epochs, and utilizing additional data augmentation, a greater model accuracy will likely be achieved for both Task 1 and Task 2.

5 Conclusion

This study developed and validated SegResNet deep learning models for the automated segmentation of GTVp and GTVn in OPC using pre-treatment and mid-treatment T2 MRI using the Auto3DSeg and MONAI frameworks [9, 10]. By integrating model ensembling with specific pre- and post-processing techniques, our models testing DSCagg scores of 0.72 and 0.82 for GTVn and GTVp in pre-treatment MRI (Task 1), and 0.81 and 0.49 for GTVn and GTVp in mid-treatment MRI (Task 2). These results highlight the effectiveness of deep learning-based automated segmentation in enhancing radiation oncology workflows. Future studies will focus on improving mid-treatment GTVp and GTVn segmentation performance and exploring the clinical implications of automated tumor delineation in adaptive radiotherapy.

Acknowledgments. This study was made possible by HNTS-MRG 24 and its organizers. We would like to extend our sincere thanks to the organizers for hosting the challenge and for their continuous support, particularly with debugging code throughout the competition. Their efforts have been invaluable to the success of this work.

Disclosure of Interests. We have no disclosures.

References

1. Chaturvedi, A.K., et al.: Human papillomavirus and rising oropharyngeal cancer incidence in the United States. J. Clin. Oncol. **41**(17), 3081–3088 (2023). https://doi.org/10.1200/JCO.22.02625
2. Peters, L.J., Ang, K.K., Jr.Thames, H.D.: Accelerated fractionation in the radiation treatment of head and neck cancer: a critical comparison of different strategies. Acta Oncol. **27**(2), 185–194 (1988). https://doi.org/10.3109/02841868809090339

3. Avery, E.W., Joshi, K., Mehra, S., Mahajan, A.: Role of PET/CT in oropharyngeal cancers. Cancers (Basel) **15**(9), 2651 (2023). https://doi.org/10.3390/cancers15092651

4. Foster, B., Bagci, U., Mansoor, A., Xu, Z., Mollura, D.J.: A review on segmentation of positron emission tomography images. Comput. Biol. Med. **50**, 76–96 (2014)

5. Vishwanath, V., Jafarieh, S., Rembielak, A.: The role of imaging in head and neck cancer: an overview of different imaging modalities in primary diagnosis and staging of the disease. J. Contemp. Brachyther. **12**(5), 512–518 (2020). https://doi.org/10.5114/jcb.2020.100386

6. Thiagarajan, A., Caria, N., Schöder, H., et al.: Target volume delineation in oropharyngeal cancer: impact of PET, MRI, and physical examination. Int. J. Radiat. Oncol. Biol. Phys. **83**(1), 220–227 (2012). https://doi.org/10.1016/j.ijrobp.2011.05.060

7. Nishioka, T., Shiga, T., Shirato, H., et al.: Image fusion between 18FDG-PET and MRI/CT for radiotherapy planning of oropharyngeal and nasopharyngeal carcinomas. Int. J. Radiat. Oncol. Biol. Phys. **53**(4), 1051–1057 (2002). https://doi.org/10.1016/s0360-3016(02)02854-7

8. Pollard, J.M., Wen, Z., Sadagopan, R., Wang, J., Ibbott, G.S.: The future of image-guided radiotherapy will be MR guided. Br. J. Radiol. **90**(1073), 20160667 (2017). https://doi.org/10.1259/bjr.20160667

9. The MONAI Consortium: Project MONAI. Zenodo (2020). https://doi.org/10.5281/zenodo.4323059

10. Myronenko, A.: 3D MRI brain tumor segmentation using autoencoder regularization. In: International MICCAI Brainlesion Workshop, pp. 311–320. Springer, Cham (2018)

11. Wahid, K., Dede, C., Naser, M., Fuller, C. Training Dataset for HNTSMRG 2024 Challenge [Data set]. Zenodo (2024). https://doi.org/10.5281/zenodo.11199559

12. Beare, R., Lowekamp, B.C., Yaniv, Z.: Image segmentation, registration and characterization in R with SimpleITK. J. Stat. Softw. **86**(8) (2018). https://doi.org/10.18637/jss.v086.i08

13. Yaniv, Z., Lowekamp, B.C., Johnson, H.J., Beare, R.: SimpleITK image-analysis notebooks: a collaborative environment for education and reproducible research. J. Digit. Imaging **31**(3), 290–303 (2018). https://doi.org/10.1007/s10278-017-0037-8

14. Lowekamp, B.C., Chen, D.T., Ibáñez, L., Blezek, D.: The design of SimpleITK. Front. Neuroinf. **7**, 45 (2013). https://doi.org/10.3389/fninf.2013.00045

15. Andrearczyk, V., Oreiller, V., Jreige, M., Castelli, J., Prior, J.O., Depeursinge, A.: Segmentation and classification of head and neck nodal metastases and primary tumors in PET/CT. In: 2022 44th Annual International Conference of the IEEE Engineering in Medicine & Biology Society (EMBC), pp. 4731–4735 (2022).https://doi.org/10.1109/EMBC48229.2022.9871907

16. Andrearczyk, V., et al.: Overview of the HECKTOR challenge at MICCAI 2022: automatic head and neck tumor segmentation and outcome prediction in PET/CT. In: Head and Neck Tumor Segmentation and Outcome Prediction: Third Challenge, HECKTOR 2022, Held in conjunction with MICCAI 2022, Singapore, 22 September 2022, Proceedings, vol. 13626, pp. 1–30 (2023). https://doi.org/10.1007/978-3-031-27420-6_1

17. Chu, H., et al.: Swin UNETR for tumor and lymph node segmentation using 3D PET/CT imaging: a transfer learning approach. In: Andrearczyk, V., Oreiller, V., Hatt, M., Depeursinge, A. (eds.) Head and Neck Tumor Segmentation and Outcome Prediction. HECKTOR 2022 (Lecture Notes in Computer Science, vol. 13626, pp. 123–132. Springer, Cham (2022). https://doi.org/10.1007/978-3-031-27420-6_12

18. Wahid, K.A., Ahmed, S., He, R., et al.: Evaluation of deep learning-based multiparametric MRI oropharyngeal primary tumor auto-segmentation and investigation of input channel effects: Results from a prospective imaging registry. Clin. Transl. Radiat. Oncol. **32**, 6–14 (2022). https://doi.org/10.1016/j.ctro.2021.10.003

19. Naser, M.A., et al.: Head and neck cancer primary tumor auto segmentation using model ensembling of deep learning in PET/CT images. Head Neck Tumor Segm Chall. **13209**, 121–132 (2022). https://doi.org/10.1007/978-3-030-98253-9_11

20. Mastronikolis, N.S., et al.: Insights into metastatic roadmap of head and neck cancer squamous cell carcinoma based on clinical, histopathological and molecular profiles. Mol. Biol. Rep. **51**(1), 597 (2024). https://doi.org/10.1007/s11033-024-09476-8
21. Koroulakis, A., Jamal, Z., Agarwal, M.: Anatomy, Head and Neck, Lymph Nodes. StatPearls Publishing, Treasure Island (2022)
22. Vogel, D.W.T., Zbaeren, P., Thoeny, H.C.: Cancer of the oral cavity and oropharynx. Canc. Imaging **10**(1), 62–72 (2010). https://doi.org/10.1102/1470-7330.2010.0008
23. Salahuddin, Z., et al.: From head and neck tumour and lymph node segmentation to survival prediction on PET/CT: an end-to-end framework featuring uncertainty, fairness, and multi-region multi-modal radiomics. Cancers **15**(7), 1932 (2023). https://doi.org/10.3390/cancers15071932
24. Surucu, M., Shah, K.K., Roeske, J.C., Choi, M., Small, W., Jr., Emami, B.: Adaptive radiotherapy for head and neck cancer. Technol. Cancer Res. Treat. **16**(2), 218–223 (2017). https://doi.org/10.1177/1533034616662165
25. Morgan, H.E., Sher, D.J.: Adaptive radiotherapy for head and neck cancer. Cancers Head Neck. **5**, 1 (2020). https://doi.org/10.1186/s41199-019-0046-z
26. LaBella, D., et al.: Analysis of the BraTS 2023 intracranial meningioma segmentation challenge. arXiv preprint arXiv:2405.09787 (2024)
27. Moawad, A.W., et al.: The brain tumor segmentation (brats-mets) challenge 2023: brain metastasis segmentation on pre-treatment mri. arXiv preprint arXiv:2306.00838 (2023)
28. LaBella, D., et al.: Brain tumor segmentation (brats) challenge 2024: meningioma radiotherapy planning automated segmentation. arXiv preprint arXiv:2405.18383 (2024)
29. de Verdier, M.C., et al.: The 2024 brain tumor segmentation (BraTS) challenge: glioma segmentation on post-treatment MRI. arXiv preprint arXiv:2405.18368 (2024)
30. Kazerooni, A.F., et al.: The brain tumor segmentation in pediatrics (BraTS-PEDs) challenge: focus on pediatrics (CBTN-CONNECT-DIPGR-ASNR-MICCAI BraTS-PEDs). arXiv preprint arXiv:2404.15009 (2024)
31. Adewole, M., et al.: The brain tumor segmentation (brats) challenge 2023: glioma segmentation in sub-saharan Africa patient population (brats-africa).ArXiv (2023)
32. Myronenko, A., Yang, D., He, Y., Xu, D.: Auto3dseg for brain tumor segmentation from 3d mri in brats 2023 challenge. In: MICCAI, Vancouver, Canada (2023)

Correction to: Head and Neck Gross Tumor Volume Automatic Segmentation Using PocketNet

Awj Twam, Adrian Celaya, Evan Lim, Khaled Elsayes, David Fuentes, and Tucker Netherton

Correction to:
Chapter 19 in: K. A. Wahid et al. (Eds.): *Head and Neck Tumor Segmentation for MR-Guided Applications*, **LNCS 15273, https://doi.org/10.1007/978-3-031-83274-1_19**

The original version of the chapter was published without an Acknowledgement section. This has been corrected.

The updated version of this chapter can be found at
https://doi.org/10.1007/978-3-031-83274-1_19

© The Author(s) 2025
K. A. Wahid et al. (Eds.): HNTS-MRG 2024, LNCS 15273, p. C1, 2025.
https://doi.org/10.1007/978-3-031-83274-1_22

Author Index

K. A. Wahid et al. (Eds.): HNTS-MRG 2024, LNCS 15273, pp. 275–276, 2025.
https://doi.org/10.1007/978-3-031-83274-1